W0254318

LCP for Microwave Packages and Modules

This comprehensive overview of electrical design using liquid crystal polymer (LCP) gives you everything you need to know to get up to speed on the subject. It describes successful design and development techniques for high-performance microwave and millimeter-wave packages and modules in an organic platform. These were specifically developed to make the most of LCP's inert, hermetic, low-cost, high-frequency (DC to 110+ GHz) properties.

First-hand accounts show you how to avoid various pitfalls during design and development. Extensive electrical design details are given in the areas of broadband circuit design for low-loss interconnects, couplers, splitter–combiners, baluns, phase shifters, time-delay units, power amplifier modules, receiver modules, phased-array antennas, flexible electronics, surface mounted packages, microelectromechanical systems (MEMS), and reliability. Ideal for engineers in the fields of RF, microwave, signal integrity, advanced packaging, material science, optical, and biomedical engineering.

Anh-Vu H. Pham is a Professor at the University of California in Davis, where he leads the Microwave Microsystems Laboratory. He has published around 100 peer-reviewed papers, several book chapters, and one book, is currently the Vice Chair of the IEEE International Microwave Symposium Technical Committee on Power Amplifiers and Integrated Devices, and is IEEE Distinguished Microwave Lecturer on Microwave LCP Packaging for the term 2010–2012.

Morgan J. Chen is a Staff Engineer at Futurewei Technologies, an R&D US subsidiary of Huawei Technologies. He has worked in both academia and industry for over a decade, advancing high-frequency packaging from DC to past 60 GHz.

Kunia Aihara is a Signal Integrity Engineer at Hirose Electric USA.

The Cambridge RF and Microwave Engineering Series

Series Editor
Steve C. Cripps, Distinguished Research Professor, Cardiff University

Peter Aaen, Jaime Plá, and John Wood, *Modeling and Characterization of RF and Microwave Power FETs*
Dominique Schreurs, Máirtín O'Droma, Anthony A. Goacher, and Michael Gadringer, *RF Amplifier Behavioral Modeling*
Fan Yang and Yahya Rahmat-Samii, *Electromagnetic Band Gap Structures in Antenna Engineering*
Enrico Rubiola, *Phase Noise and Frequency Stability in Oscillators*
Earl McCune, *Practical Digital Wireless Signals*
Stepan Lucyszyn, *Advanced RF MEMS*
Patrick Roblin, *Nonlinear FR Circuits and the Large-Signal Network Analyzer*
Matthias Rudolph, Christian Fager, and David E. Root, *Nonlinear Transistor Model Parameter Extraction Techniques*
John L. B. Walker, *Handbook of RF and Microwave Solid-State Power Amplifiers*

Forthcoming titles:
Sorin Voinigescu, *High-Frequency Integrated Circuits*
David E. Root, Jason Horn, and Jan Verspecht, *X-Parameters*
Richard Carter, *Theory and Design of Microwave Tubes*
Nuno Borges Carvalho and Dominique Scheurs, *Microwave and Wireless Measurement Techniques*

LCP for Microwave Packages and Modules

ANH-VU H. PHAM
University of California, Davis

MORGAN J. CHEN
Huawei Technologies

KUNIA AIHARA
Hirose Electric USA

CAMBRIDGE UNIVERSITY PRESS
Cambridge, New York, Melbourne, Madrid, Cape Town
Singapore, São Paulo, Delhi, Mexico City

Cambridge University Press
The Edinburgh Building, Cambridge CB2 8RU, UK

Published in the United States of America by Cambridge University Press, New York

www.cambridge.org
Information on this title: www.cambridge.org/9781107003781

First published 2012

Printed in the United Kingdom at the University Press, Cambridge

A catalog record for this publication is available from the British Library

Library of Congress Cataloging in Publication data
LCP for microwave packages and modules / [edited by] Anh-Vu H. Pham, Morgan J. Chen, Kunia Aihara.
p. cm. – (The Cambridge RF and microwave engineering series)
Includes bibliographical references and index.
ISBN 978-1-107-00378-1 (hardback)
1. Microwave devices – Materials. 2. Microelectronic packaging – Materials.
3. Liquid crystal devices. 4. Polymer liquid crystals. I. Pham, Anh-Vu H.
II. Chen, Morgan J. III. Aihara, Kunia.
TK7876.L387 2012
621.381′3–dc23 2012007343

ISBN 978-1-107-00378-1 Hardback

Contents

Preface

Package design and fabrication techniques are critical to the high-frequency community. In building improved products, packaging developments are driven by economics, performance, and reliability. For example, the telecom industry as a whole is currently pushing to improve electrical performance and lower cost by replacing the current transceiver designs with new surface mount solutions. This book is intended for electrical engineers involved in designing microwave circuits. As operating frequencies rise with the emergence of high-speed products, engineers will increasingly need a good understanding of RF/microwave packaging.

This book presents engineering breakthroughs in liquid crystal polymer (LCP) applications to microwave-frequency electronics. It appears that LCP is a highly attractive platform to achieve low-cost hermetic devices that offer mechanical flexibility. These benefits are attractive for applications in gigabit wireless communication, radar and imaging systems. Liquid crystal polymer research is currently a very hot topic in microwave engineering, with contributions from several research groups and organizations on a global level.

As we will discuss, using LCP can be challenging at times. The inert chemistry of LCP, which provides its attractive electrical and mechanical properties, can also act to hinder actual circuit build. The book gives brief descriptions of the theory and provides deep insights into the practical issues of design and realization with LCP. Numerous real-world examples with expanded explanations of previously published works are included to create a comprehensive and cohesive volume. We hope to share tips and tricks that we have found for successfully processing LCP for microwave packages and circuit modules. We describe successful techniques in using LCP and how to avoid pitfalls.

In general, very few books on microwave packaging form an interdisciplinary bridge between electrical, mechanical, and chemical expertise. This will be the first book to discuss LCP packaging at the package, component, and system levels. The authors are perhaps uniquely positioned to describe and discuss novel LCP packaging techniques for microwave circuits, since they come from academia, government, and industrial research. All three authors have conducted research from 2003 to the present on advanced microwave packaging using liquid crystal polymer.

Book organization

The book is organized bottom-up, beginning with the basics of packaging and a description of LCP's fundamental material properties. Next, techniques for physically processing this new material and LCP packaging techniques for devices are discussed. The book gradually progresses into increasingly complex electrical circuits and designs. Numerous specific examples are provided to cover the wide range of microwave packages and circuits. Specific chapter focuses are described below.

Chapter 1 provides an overview of packaging. It includes a discussion of existing electronic packages, package requirements, and general package design process flow. This chapter is intended to provide a sufficiently broad overview of packaging and high-frequency topics.

In Chapter 2 we describe LCP material properties in terms of chemical, electrical, physical, and environmental properties. The chemistry of LCP is discussed in order to offer the reader a comprehensive material property overview. Then LCP materials are electrically characterized to show their excellent low-loss performance and stable operation under humidity. LCP packages are demonstrated to provide a fine-leak rate of less than 5×10^8 atm cm^3/s, which passes the hermetic requirements set by method 1014, Mil-Std-883 [3]. Hence LCP is a viable low-cost option to replace many traditional ceramic packages.

In Chapter 3 we discuss fabrication techniques for processing LCP, including in-depth detail on novel techniques. We hope to provide insight and to share techniques that we have used to process LCP. Material formats are introduced, and methods for metallizing laminates are presented. Standard PCB processes compatible with LCP are discussed. Detailed descriptions are also provided for flex laminate PCB processes, including metallization, etching, mechanical and laser via processes, and multilayer lamination. Further, novel processes for LCP are presented. These topics include special handling techniques, bulk LCP machining, selective sealing, and molding processes.

In Chapter 4 we show novel implementations for using LCP in MEMS chip-scale packages (CSPs). We explain how, using new packaging processes, we designed and implemented a two-bit phase shifter with RF MEMS switches in a multilayer organic module. This build employing LCP allows MEMS devices to be hermetically sealed with a Si base, LCP walls, and a Cu roof. Blind vias through LCP form first-level interconnects in this package. An application for these MEMS devices in a two-bit phase shifter circuit is shown as an extension of chip-on-flex (CoF) technology.

In Chapter 5 we describe a variety of feed-through designs for air-cavity surface mount technology (SMT) packages. A return loss greater than 20 dB is demonstrated at Ka-band frequencies for these feed-throughs. Air-cavity SMT packages are characterized using a low-noise amplifier (LNA). Further, a multilayer LCP implementation to create a multi-chip module (MCM-L) receiver front end in an air-cavity SMT package is shown. Novel bandpass filter feed-throughs are also developed. Bandpass filter designs are presented that give excellent high-frequency interconnection with minimal loss, which allows a DC block to be built directly into the packaging. These research efforts show that it is possible to use low-cost organic hermetic surface mount packages for millimeter-wave frequencies.

Chapter 6 presents LCP for passive components, including implementations for novel wide-band baluns, Wilkinson power combiner–dividers, and hybrid couplers. Many of these devices take advantage of multilayers and thin physical dimensions to achieve breakthrough performance. Our wideband multilayer balun structures make evident how LCP packaging provides an ideal high-density organic module technology with an embedded passive. The balun achieves less than 0.5 dB insertion loss, 0.5 dB amplitude imbalance, and 5° phase imbalance over 6–18 GHz. The broadband Wilkinson offers performance on LCP over 2–18 GHz with 1.6 dB excess insertion loss and 12 dB isolation. A novel, compact, hybrid coupler is demonstrated on multilayer LCP that offers a less than 7° phase imbalance and 15 dB isolation over 2–17 GHz.

In Chapter 7 we discuss LCP packaging for system integration and provide a design for package integration of a true-time delay (TTD) circuit, push–pull power amplifiers (PAs), and a full receiver module with phased-array antennas. Our DC-10 GHz broadband long-time-delay (LTD) circuit provides amplitude compensation over long time variation control from 0 to 600 ps delay in 200 ps increments (two-bit). The LTD is implemented with MEMS switches to provide less than ±0.5 dB amplitude imbalance over all frequencies. Using the broadband LCP baluns described in Chapter 6, a Ku-band push–pull amplifier is presented that achieves 20 dB second-order harmonics reduction over 6–18 GHz. Lastly, a design for a full receiver module that has an integrated phased-array antenna and active devices packaged into a novel LCP platform is given. In this antenna design, LCP is an ideal low-loss material with dimensions suited for microwave propagation.

In Chapter 8 we consider reliability aspects for LCP packaging. Qualification tests and results are provided to demonstrate a high level of robustness under proper process conditions. Typical reliability tests are derived from military and JEDEC standards. Since it passes these stringent life and stress tests, LCP packaging is clearly able to meet the required standards under varying heat, moisture, and temperature conditions.

Acknowledgements

This book has provided an opportunity for us to detail our work in RF/microwave LCP packaging and RF modules. We greatly appreciate the help given by Dr Julie Lancashire, Mia Balashova, and Sarah Finlay from Cambridge University Press, and by Dr Susan Parkinson, our copy-editor, in guiding us to the final text. We also wish to thank our collaborators and sponsors for enabling us to touch and push the boundaries of microwave research.

For additional technical contributions to this book, the authors would like to thank those associated with the University of California, Davis. Former students from the Microwave Microsystems Laboratory include Chris Aldritt, Dr Andy C. Chen, Cheng Chen, Eric Chen, Jia-Chi Samual Chieh, Dr Arvind Keerti, Chi Y. Law, Dr Chao Lu, Dr Mark P. McGrath, Mehmet Onsiper, Cuong Nguyen, Alexander Stameroff, Hai Ta, Yiren Wang, Mary Wu, Junyang Dring Xiang, John Yan, and Zhaonian (Evan) Zhang. We are grateful for their technical contributions, support, and camaraderie. We would like to thank Professor Rick Branner, Professor Jonathan Heritage, Professor David A. Horsley, Professor Saif Islam, Professor Andre Knoeson, and Professor Neville C. Luhmann for their advice and guidance.

Lastly, the authors would like to thank their parents and family members for their continued patience and support.

1 Introduction to electronic package engineering

Over the last few decades, major advancements in the semiconductor supply chain have occurred. These advancements have provided standard foundry processes, physical theories that explain device models, accurate and efficient software, and test equipment. Today, front-end components up to 95 GHz (E-band) and beyond are either commercially available or can be specifically commissioned. In all these advancements, packaging has been the one area where investments and technology have lagged.

In many cases, packaging presents the major bottleneck to overall performance. As trivial as the connection of components may sound, the unfortunate reality is that the signal integrity of interconnects quickly limits performance at high frequencies. Engineers in the digital world are now coming up against some of these limitations, and only recently has the field made the investments necessary for these signal integrity issues to be overcome. Even microwave engineers, whose whole world is high frequency, struggle to find acceptable packaging solutions. Many will be faced with designs riddled with crippling mismatch loss, coupling loss, and unacceptable resonances.

For many engineering projects, cost is also an important design criterion and differentiating feature. Without careful, directed analysis, engineers may find their projects behind schedule and over budget. Further, lower operating costs can be achieved with package designs that encourage simple fabrication, assembly, handling, and test. Packaging research helps to advance new design techniques and package processes. For these reasons, packaging is rapidly gaining attention as a necessary growth field.

Liquid crystal polymer (LCP) is a new material that has emerged as a "holy grail" which could hold the key to a packaging revolution. It is a plastic with barrier properties similar to ceramic crystal. Since LCP is a plastic, multilayer structures using it are easy to process at low cost. Because it has crystal properties, this material exhibits low electrical losses, is near-hermetic, and has low moisture absorption and mechanical flexibility properties. This chapter is intended to provide a basic overview of high-frequency packaging. In section 1.1 we briefly discuss various packages that were previously popular but may not be suitable for high-frequency operation. Section 1.2 lists common packaging requirements. Then, in section 1.3 we describe the general process flow for package design.

1.1 A brief discussion of standard packages

Countless package types exist, but most fall into a few different standardized package styles. At low frequencies, i.e. < 1 GHz, plastic packages such as a single/dual in-line package (SIP/DIP) or a small-outline integrated circuit (SOIC) are used. A DIP is mounted by inserting its two parallel pin rows through contact vias on a mother board. An SOIC package is similarly mounted, by soldering its leads to motherboard pads. As the application frequency increases, these packages become unsuitable. In a DIP a number of features, including the bond wire connections, package lead legs, and solder joints, present features that are detrimental to electromagnetic power transmission at high frequencies. Further, DIP and SOIC package leads have lengths on the order of several millimeters. At microwave frequencies these features are electrically large (i.e. greater than one-twentieth of a wavelength) and cause significant mismatch. Further, the above-mentioned packages are not precisely defined in regard to their dimensions on the motherboard when mounted. These large tolerances make high-frequency use impossible. To get an idea of the scale, one-twentieth of a wavelength at 6 GHz corresponds to 2.5 mm. At 60 GHz, one-twentieth of a wavelength corresponds to 250 μm, which is shorter than most advanced wire bonds. Hence, it is apparent that RF packaging requires greater consideration as operating frequencies increase.

Efficient signal power transfer requires an equivalent matched (typically 50 Ω) transmission line between source and destination. As operating frequencies increase, this becomes difficult owing to proportionally challenging distributed transmission line effects. Compensating these parasitic effects becomes very difficult with the limited space inside a package and/or on a motherboard. Other metal-can-type packages, such as TO-8, suffer similar frequency limitations. Quad flat packages (QFPs) and pin grid arrays (PGAs) are generally formed with shorter leads (or pins) to minimize parasitics and increase input/output density. These types of package may offer reasonable performance. It is possible to use such leaded packages at high frequencies in non-hermetic applications.

To remove lead parasitic effects completely, leadless packages have been developed that are attractive for higher-frequency work. Leadless packages in rough chronological order of their appearance include ball grid array (BGA), quad flat no-lead (QFN), land grid array (LGA), flip-chip, and wafer-scale packages. These packages have recently been adopted as a JEDEC standard, and they are at the forefront of advanced packaging research. A BGA package contains a rectangular surface on which the leads have been replaced with solder balls. This package is surface mounted (SMT). Lead elimination greatly reduces inductance from bumps and package pad connections as compared with that found with in-line and small-outline integrated circuit (SOIC) packages. Similarly, QFN and LGA technologies are packages with metal surfaces exposed for contact in SMT application. The QFN and LGA technologies differ, as one has contact pads along the periphery while the other has pads arrayed across the entire surface. The QFN package may be formed with an over-mold lead frame or printed as an organic substrate, while

LGAs are only available in the latter style. Flip-chip packaging involves attaching a die with bumps that directly attach the die pads to the board. Bumps are formed of either solder with a controlled collapse chip connection (C4) process or gold bumps formed using ball bonders. Additional mechanical support is optionally provided with under-fill material. When diced, the individual packaged chip is the same size as the die. Lastly, wafer-scale packaging consists of laminate film placed on a wafer to form interconnects and may be mounted by means of either the SMT or the flip-chip method. Today, a number of chips are packaged in these technologies because they offer lower costs, smaller form factors, and higher-frequency performance over their leaded predecessors; LCP can and will be shown to have unique advantages when one is building these types of package.

1.2 Package design requirements

One goal for package engineers is to make a package transition look like a 50 Ω through line. Electrical interconnects need a 50 Ω matched package transition to reduce power reflections. In this book, an electrical interconnect from chip to PCB will be referred to sometimes as a package *feed-through*. In any feed-through design, there are conductor and dielectric losses that contribute to the total loss. These losses are due in large part to the material used; LCP is one material that provides low-loss performance. In order to discuss losses for high frequencies, one needs to use S-parameters, the most common electrical measure of package performance. They will be reviewed insofar as they relate to electrical packaging in section 1.2.1, in an intuitive manner. Other common requirements in package design besides electrical performance and material involve weight (section 1.2.2), size (section 1.2.3), thermal effects (section 1.2.4), and reliability (section 1.2.5).

1.2.1 Electrical requirements

To characterize power transfer characteristics, S-parameters are used in lieu of Z-parameters since *scattering* becomes a more practical metric to characterize in a consistent 50 ohm termination system. The *insertion loss* (IL) is the decibel form of the ratio S21 of the output at port 2 and the input at port 1. Thus we have

$$S21 = \frac{\text{power out of port 2}}{\text{power into port 1}}, \tag{1.1}$$

$$\text{IL} = -10 \log S21 = -10 \log\left(\frac{\text{power out of port 2}}{\text{power into port 1}}\right) \text{ (dB)}, \tag{1.2}$$

where the powers are in absolute values, i.e. watts (W). Insertion losses are often tracked to within 0.1 dB accuracy. At any juncture, power entering port 1 either can be (1) transmitted through to port 2, (2) reflected back to port 1, or (3) dissipated in between port 1 and port 2. Note that in a microwave package there is no strict convention for port designation, but typically port 1 is used to indicate the input. The S-parameters should be accompanied with port-definition information when three or more port networks are characterized. Transmission lines have losses that are largely dominated by conduction loss and dielectric loss. Conduction loss generally dominates at low frequency and is proportional to the square root of the frequency. Dielectric loss is proportional to frequency and hence may become the dominant loss mechanism at sufficiently high frequencies. Other, usually negligible, loss effects include radiation and dielectric conduction loss. For packaging, one goal is to minimize insertion loss at desired frequencies.

Another important parameter to consider is the *return loss* (RL), which is represented by S11 where

$$\mathrm{S11} = \frac{\text{power out of port 1}}{\text{power into port 1}}; \tag{1.3}$$

the return loss is then given by

$$\mathrm{RL} = -10 \log \mathrm{S21} = -10 \log\left(\frac{\text{power out of port 1}}{\text{power into port 1}}\right) \text{ (dB)}. \tag{1.4}$$

The return loss thus measures the amount of power reflected from a port. Backward power reflections are unacceptable, and can cause component damage, in many systems. Generally, however, for packaging the goal is to maximize the return loss or, in other words, to have a large negative S11 value. Many commercial components specify a 10 dB return loss. However, junction mismatches can add up in a phenomenon known as voltage standing wave ratio (VSWR) stacking. For junctions with a reasonable return loss, VSWR stacking may cause a worst-case return-loss degradation of 6 dB. For instance, back-to-back junctions, each with a 20 dB individual return loss, may give only a 14 dB combined return loss at some frequencies. For this reason, one often needs to design packaging with a much higher return loss than that required by the total system specification.

In package interconnects, a significant portion of the insertion loss can be attributed to the mismatch loss (ML), given as

$$\mathrm{ML} = -10 \log[1 - (10^{-\mathrm{RL}/20})^2] \text{ (dB)} \tag{1.5}$$

where RL is in dB units. For example, a 10 dB return loss has a 0.46 dB associated mismatch loss.

One way to improve the return loss is to add attenuating elements. This improves return loss at the expense of insertion loss. For example, consider the case of adding a 3 dB attenuator in front of a 10 dB return loss reference plane. The return loss will improve by 6 dB to 16 dB. The reason is that the power is reduced by 3 dB in both the forward and return directions. In this case, the insertion loss is only degraded by

3 dB as a result of forward power attenuation. However, to lower costs and increase performance, more careful package design is required, as we shall discuss.

Cross-talk, also referred to as *isolation*, may be yet another package requirement. Cross-talk generally is given as a large negative decibel value, e.g. –40 dB, while the isolation is reported as the absolute value, here, +40 dB. Cross-talk may be considered a special case of insertion loss where now an undesired signal appears. In this case, the aggressor signal may leak into another part of the package and add unwanted noise.

Other parameters that are relevant to microwave operation are the ABCD-parameters and the T-parameters (which determine the transmission). A number of well-known relations between the S-, ABCD-, T-, and Z parameters can be obtained. Lastly, a variation of the normal S-parameters is presented by mixed-mode S-parameters, which apply to differential systems. In a set of mixed-mode S-parameters, differential and common-mode parameters are both present. For differential signaling, differential S-parameters indicate the useful signal being transmitted. Other parameters relating to common-mode performance can be used during the design process. Again, well-known relations enable one to convert between single-ended and mixed-mode S-parameters.

The noise figure (NF) is an electrical parameter critical to low-noise-amplifier (LNA) and receiver design. The noise figure is directly affected by the electrical packaging, where a package insertion loss in front of the LNA degrades the noise figure by the same amount. In other words, 1 dB of package loss before the LNA correspondingly degrades the NF by 1 dB.

The output power is also directly degraded by any back-end packaging losses. For instance, 1 dB of package loss after a power amplifier (PA) means that maximum output power is reduced by 1 dB. This would mean that a 1 W (30 dBm) PA chip would only offer 0.8 W (29 dBm) after packaging. Note that 0.2 W power is 23 dBm. Also, note that the output-side packaging losses similarly directly degrade the third-order intercept (IP3 or TOI) and 1 dB compression point (P_{1dB}), which are measures of linearity. Hence, packaging can significantly affect electrical performance and has become a major interest of electrical engineers.

1.2.2 Weight requirements

It is well known that lightweight packages are desirable to cut down operation cost in mobile systems such as handsets. Less known is the need to reduce weight in airplanes and ground vehicles. A phased-array radar may be constructed using ceramic packages and weighs about 250 kg in a military vehicle application [9]. Packaging materials such as LCP are attractive for weight-sensitive applications since its mass density is only around 1 gram per cubic centimeter. If ceramic is replaced with LCP then the total weight of a vehicle may be reduced by around 160 kg, and this improves the efficiency of the vehicle. For space applications, lightweight gains are even more significant, where the ratio of the rocket fuel weight and the load must be an overwhelming 100 :1 during launch. Hence, weight can be a significant criterion for performance, and reduction in weight can lead directly to reduced operation costs.

1.2.3 Package physical requirements

High-density packaging capability is highly desirable. Multilayer structures achieve this by allowing the same footprint to have more functionality. Additional adhesive layers are required to achieve multilayer structures using substrates such as FR4, high-frequency laminates [1], or organic films [2]. Adhesive layers are not guaranteed to have low moisture-absorption characteristics, and this may allow moisture to flow freely into the package cavity. Fortunately, LCP allows for multilayer laminated builds, has a low moisture uptake, and so provides an ideal match for high-density requirements. LCP is conveniently available in varying film thicknesses (0.5 mil to > 4 mil) with various Cu cladding thicknesses. A common metal thickness for LCP is 1/2 oz (18 μm). Incidentally, while PCB terminology uses "oz" for thickness, it should more appropriately be given as oz/ft^2, referring to the weight of the metal over one square foot. Lastly, other constraints may occur from the limitation in the X, Y, or Z dimensions dictated by the system requirements.

1.2.4 Package thermal requirements

Packaging high-power active components requires thermal management. Performance and reliability degrade when the recommended junction temperatures are exceeded. Further, packages such LGAs are sensitive to mismatch of the coefficient of thermal expansion (CTE) at the epoxy interface between a package and the semiconductor die, which could be around 6 ppm/°C. These packages are also sensitive to CTE mismatch at solder joints between the package and the motherboard (17 ppm/°C). Given that chips are usually quite small compared to the overall package dimensions, it is usually recommended that package materials have a CTE closer to that of the motherboard PCB than that of the die. Thus LCP is attractive as it has a CTE of 17 ppm/°C, which has been engineered to match Cu and PCBs. Further, LCP is a flexible material, so that it can yield easily and may accommodate CTE mismatches without the material cracking. This allows LCP packages to have low solder-joint stresses.

Depending on the particular chip technology, 135, 150, and 175 °C are commonly specified temperatures at which junctions must be maintained in order to provide 10^6 hours of mean time to failure (MTTF) with 85 °C ambient [10]. Emerging microwave chip technologies involving SiC and GaN may have junction temperatures above 200 °C. In order to remove heat from active components, thermal conduction and convection mechanisms may be explored. To reduce junction temperature, one can utilize a copper die-pad that externally conducts heat away from the package. Examples of multilayer thermal schemes are provided later in this book. In a scheme based on LCP, a cavity is drilled entirely through soft LCP dielectric so that a chip sits directly on a thin copper surface to allow efficient heat dissipation. Convection can also be used to cool packages with either passive circulation as heat rises, or active forced-air cooling to remove heat. Heat sinks are commonly used to provide a large stable material that can be conveniently temperature regulated and monitored. Heat sinks can also be machined to allow greater surface area for convection cooling and are commonly incorporated outside the package design. Thermoelectric coolers (TECs) based on the Peltier effect may also be used to cool packages. In this method, electric current facilitates the removal of heat through a semiconductor junction. Note that TECs are usually placed externally to RF package modules, for, placed internally, they would add significant cost and power consumption.

The thermal dissipation in a package may be calculated as the DC power plus the input RF power minus the output RF power. Power density is an important aspect to consider, as thermal concerns are too often limited to the output-stage transistor junctions in a power amplifier chip. This means that a significant amount of power is often dissipated in a very small area of a chip. Hence, for thermal considerations, power and area are important parameters to note during the thermal design of packages. Analysis may be performed to calculate the first-order junction temperature from a knowledge of the material's thermal resistance. For instance, with 15 °C/W thermal impedance from junction to case (θ_{JC}) and 2 W power dissipation, one would expect a 30 °C heat rise from case to junction. With an 85 °C case temperature, the junction temperature is expected to be 115 °C. This may be summarized as

$$T_J = T_C + \theta_{JC} P_{DISS} \quad (^\circ\mathrm{C}), \tag{1.6}$$

where T_J is the junction temperature in °C, T_C is the case temperature in °C, θ_{JC} is the package thermal impedance from junction to case in °C/W, and P_{DISS} is the power dissipation in W. The thermal impedance θ_{JC} may be approximated to first order by the equation

$$\theta = \frac{L}{KA} \quad (^\circ\mathrm{C/W}) \tag{1.7}$$

where L is the length (in m), K is the thermal conductivity in W/(m °C) or W/(m K), and A is the cross-sectional area in m^2. Additional first-order approximations for planar multilayer structures may incorporate heat-spread angles to estimate the thermal impedance [11]. For a more accurate analysis, full three-dimensional simulation tools are available. Additional details on these tools are discussed in section 1.3.

1.2.5 Package reliability requirements

An electronic package should provide long-term reliability and sufficient protection from external environment changes, such as temperature, humidity and pressure or force. Moisture is a significant aggravator degrading electrical performance. When moisture enters a package, it can corrode metal pads or traces on MMICs and the constituents of package and chip materials. These may change electrical properties, and this may in turn lead to system failures. Therefore, a package material should have very low moisture absorption and permeability in order to stop moisture from entering the package cavity. LCP is known especially for having low-moisture-uptake and hermetic properties. A package is considered to be hermetic if its cavity leak rate is less than 5×10^{-8} atm/(cm^3 s). LCP has been demonstrated by several groups to satisfy this low-leak-rate requirement for hermeticity. It has been demonstrated to satisfactorily package MEMS devices, which are one of the most environmentally sensitive technologies in existence.

Commonly used reliability tests have been standardized by various organizations [6–8]. (Interestingly, despite their imposing sound, military standards may not always be the strictest available for a given type of test.) Since it is impractical to perform every possible test, tests are selected on the basis of potential reliability concerns as determined by engineering experience. The tests selected are often based on those that historically have been performed for similar products. Common tests include thermal cycling, life testing, and highly accelerated stress testing (HAST). More specifics on actual tests are provided in Chapter 8.

Reliability testing is often expensive and time-consuming; some tests need to run for several months. A number of vendors offer services to perform these tests and are glad to offer their inputs. Ideally, every manufactured package would be put through the same gauntlet of tests; however, in order to save time, money, and allow faster time to market, a method known as "qualification by similarity" may be performed. This builds on prior successful qualification efforts on like packages. When conducting qualification by similarity it is important to confirm that the package being considered is actually similar in form and manufactured by the same suppliers. Qualification by similarity is only valid for less challenging package designs and is generally possible for packages slightly smaller than prior-qualified parts. This method should be used carefully and sparingly, as any error in method application may call into question packages previously qualified as well as engineering integrity.

1.2.6 Package material requirements

Ceramic, glass, and metal hybrid packages are frequently used to provide low loss and hermeticity at microwave frequencies. Processes such as low-temperature co-fired ceramic (LTCC) have been developed to lower processing temperature and enable multilayer packaging. Even so, LTCC is still processed at 1000 °C, and in a small size, that may not be cost effective. Furthermore, screen-printed conductor on green tape does not have as high a conductivity as copper, and high-frequency performance is limited owing to losses. Poor thermal dissipation is another disadvantage of ceramic packages. Exceptional ceramics known for good thermal dissipation are aluminum nitride (AlN) and beryllium oxide (BeO), but these materials are extremely costly and suffer from poor fabrication. In the case of BeO, the dust generated from processing is known to be poisonous and may require special disposal.

Low-cost lightweight hermetic packages are needed in order to save space, an important aspect of microwave circuits. Organic materials have a low mass density, around ~1.2 g/cm^3, compared with ceramics, which have mass density 4 g/cm^3. Many low-cost packages are constructed using over mold with plastic. However, such over mold causes electrical detuning of the characteristic impedance and phase and increases the chip junction capacitance. To some degree, such a capacitance increase can actually help offset the electrical package's inductance, but generally it complicates the chip design considerably and is only viable when extremely challenging electrical specifications are being met. Further, their high dissipation factors and affinity to moisture make most plastics unattractive for high-frequency packaging. However, LCP is unique as a thermoplastic for building new, sophisticated, high-frequency packages.

1.3 Package design process

In this section we provide a generic set of instructions that should serve the reader undertaking a new package design. It is our hope that the packaging industry can benefit from standardization, which has served to advance the electronics industry as a whole. Package design often begins with technology selection. Then, drawings are created to specify how the package will appear. Electromagnetic and circuit simulations are conducted. Lastly, physical packages are built and tested. Any experienced engineer can attest that it is necessary to iterate within and across the steps outlined above. Also, note that this section focuses on the package or module build. There are additional considerations such as the assembly of electronics into the package, and many vendors are ready to provide guidance.

1.3.1 Technology selection

The process to design a package often begins with technology selection. It is in this first step that the material properties and physical capabilities are clarified. Often a selection is made on the basis of institutional experience combined with ad hoc changes of certain features, depending on what the market offers. Once a process is selected, the engineer must identify package manufacturers and obtain a set of design rules. Non-recurring engineering (NRE), volume unit pricing, and lead times are important, and should be discussed, for any engineering project.

Unlike in the case of semiconductor foundries, package design rules are often not prepackaged for a customer. Often it is up to the engineer to discuss needs and see whether a particular concept is viable. Common parameters and dimensions of interest include the minimum signal width, the gap, the via diameter, pad, anti-pad, pitch, and the material thicknesses.

Ceramic packaging is considered highly reliable in regard to thermal and hermetic requirements. Common ceramic package materials include alumina (Al_2O_x), beryllium oxide (BeO), and aluminum nitride (AlN). The drawbacks of ceramic packages are that they are generally bulky, expensive, have high dielectric constants, and suffer from shrinkage, which limits the precision of the final package dimensions. The shrinkage can be as much as 20% in length, depending on the specific process conditions. An absorber may also be required to be inserted if package resonances are an issue. Ceramic packages are created as air cavities and may either be dominantly metal or ceramic along the edge surface. Once a package is available, the electronics are assembled and lid sealing is performed. This is often done using seam sealing, which involves rollers that conduct a high electrical current to heat and melt the metal lids along their edges.

Plastic packaging is another technology highly regarded for being low cost and easy to process. Packages may either be air-cavity formed or fully over molded. A large number of plastic varieties are readily available. The metal for signals can be provided through printing or lead frame techniques. The drawbacks of plastic packages include high dielectric loss and non-hermetic characteristics, which can lead to moisture ingress and popcorning. Owing to the process limitations of plastics, it can be difficult to obtain precise physical features down to 100 μm. Further, over-mold parts may be sensitive to wire sweep owing to the possibility of plastic flow.

Direct chip attach (DCA) is yet another package technology that has the benefit that it can be made with low-cost printed circuit board (PCB) [5]. In DCA, electrical chips can be directly mounted onto PCB. Direct chip attach techniques encompass chip-on-board (CoB), chip-on-flex (CoF), chip-on-glass (CoG), and flip-chip and have been used to realize a number of package form factors including land grid array (LGA) and QFN. Depending on the application, DCA can present assembly challenges and can give rise to CTE chip-to-substrate or substrate-to-motherboard mismatches.

For ceramic, plastic, and DCA packages we would invite the reader to consider LCP as a viable alternative to any of the processes considered in this subsection; the quest for hermetic organic surface mount packages has led to the investigation of LCP for packaging. It has a permeability close to that of glass and can be used to construct hermetic cavities. Further, it retains most of plastic's benefits as a low-cost packaging solution. One concept using LCP for a highly functional system-in-package (SiP) is illustrated in Fig. 1.1. This book will go into the details of why LCP is an appropriate choice that offers most of the benefits presented by ceramic and plastic technologies.

1.3.2 Computer-aided design (CAD)

Once the general package concept and associated design rules have been determined, a model package must be drawn and analyzed. Manufacturers should specify the electronic formats that they can accept. Common formats include AutoCAD(.dxf) and Gerber. In these cases it may be necessary to do somewhat redundant work, namely to draw package models for both manufacturing and electromagnetic design purposes. Standardization and sophistication are increasing, and there is now a push to align these tasks better. As package houses become more sophisticated,

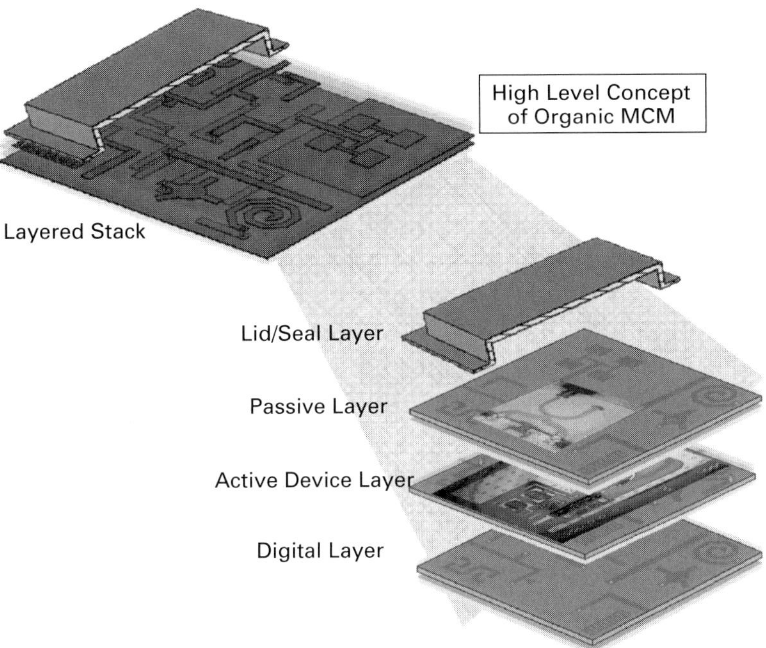

Fig. 1.1 A conceptual organic multichip module (MCM) package.

more are accepting CAD drawings directly exported from electromagnetic design software.

Electronic analysis is required for packages operating at microwave frequencies. For electromagnetic design, a number of tools are commercially available [12–18]. Electrical design may consist of frequency- and time-domain analysis, and it may employ solving techniques such as analytic, method-of-moments (MoM), finite element method (FEM), or time-domain techniques (which include FDTD and FIT). Varying degrees of accuracy may also be involved, including mesh size, bond wire features, and metal boundary conditions.

It is important to keep the simulation structures as small as possible. Generally, key portions of the design are analyzed in a piecewise fashion in order to keep the design tractable in view of the inevitable limitations in computer processing capabilities. The simulation time grows somewhat exponentially with increasing complexity, and any means of keeping simulations small should be implemented. One common technique applied to symmetric transmission-line structures is to create a perfect-H boundary condition along the split plane. This allows that only half the physical structure is analyzed, and hence, this method requires half the meshing. Another benefit of keeping the simulation size small is the reduction in potential box resonances that may result. Such resonances may or may not exist in the physical structure; they occur when the simulation assumes conductive-wall (perfect-E) boundary conditions, as required by a number of simulation tools. If the simulation results appear to show unknown resonances, box resonances are a likely culprit. If box resonances are ruled out then it becomes important to look for any features where metal regions are separated by half the resonance wavelength. Less common are quarter-wave resonances, which may be formed between metal and open, i.e. dielectric surfaces.

A grid or mesh is used to create a simpler substructure for analysis. Such a mesh may be represented either by regular rectangular gridding or more advanced tetrahedral structures for three-dimensional simulation. Meshing has largely been simplified by modern tools that allow an adaptive approach. In these algorithms, the mesh is iteratively increased until the results are stable, to within some tolerance. In HFSS, this tolerance is referred to as ΔS and is the largest difference, as surveyed across all frequency points and S-parameters, between successive iterations. When manual meshing is required, the meshing should include at least three segments across the feeding port. It is critical to define ports correctly, as different results can be obtained from different port definitions. In general, it is important to ensure that the correct mode is excited. This is done in a number of ways specific to the particular simulator being used. For three-dimensional simulators, waveports are commonly used to insert RF energy and capture the fields being generated. Waveports should be large enough to incorporate the relevant EM fields and should exhibit a negligible field along the port edges, which are commonly defined as ground. Additional meshing and port definition information is readily available in software manuals and online user forums.

As designs are specified with tolerances, generally it is useful to conduct electromagnetic simulation on the key dimensions at extreme tolerance conditions. This can be helpful in identifying risks and also can provide manufacturer feedback about areas with special tolerance requirements. Many manufacturers are receptive to RF-sensitive dimension requests and may offer solutions to increase yield and minimize costs.

Circuit models are often developed that largely involve a few stages of resistors, inductors, and capacitors. These models are beneficial for performing transient analysis. Further, in order to map physical features to electrical effects, it is useful to develop a circuit model; thus, one may understand a package design more clearly. It is common to integrate the microwave circuitry into the package technology, owing to the larger structure sizes of the latter. These circuits are often modeled with a combination of analytic, EM-based, and empirical approaches.

Thermal simulations may be performed using commercially available tools [12, 16, 18–21]. As with electrical simulation, it is important to be cognizant of the simulation size, as large aspect ratios between fine features and large structures will result in long simulation times. Heat dissipation in packaging is largely dominated by *thermal conduction*. In thermal conduction, heat is transferred across high-thermal-conductivity materials such as copper. A second mechanism to remove heat, usually to a lesser degree, occurs through *thermal convection*. Convection involves the movement of heat carried by either air or another fluid. Convection can be forced with cooling fans or can occur passively as heat rises. For this reason, convection effects can only be considered if the final orientation of the package with respect to gravity is known. Thermal electric coolers (TECs) can be modeled with ideal model equations. Unlike electrical simulation, it is generally accepted among mechanical engineers that, generally, thermal simulations provide only rough accuracy owing to the difficulties of modeling all effects. For instance, material interfaces are challenging to simulate since air is an excellent thermal insulator and is generally present at most interfaces if even only at a microscopic scale. Further, it may be simply impractical to simulate surface oxidations and finishes, owing to their small feature sizes. For this reason, thermal simulations require feedback from significant experience and testing on actual packages. Thermal cameras can be used to provide actual measurements, and care should be taken to ensure that the infrared image resolution is acceptable for the device junctions being examined. In some unique package designs, thermal–mechanical simulations are also conducted to understand the package's structural integrity as well as package shape changes that could affect solder assembly.

1.3.3 Test and measurement

Once the package has been built, its performance needs to be characterized. This may be simply a matter of assembling the electronics and testing the overall packaged performance. More often than not, a careful analysis of the package itself is conducted. This usually begins with verification of the physical dimensions to see whether they fall within the specified tolerances. A package may require destructive analysis in order to view features such as buried vias, interlayer signals, and plating quality. Tools to aid analysis include, but are not limited to, high-power microscopes with dimension gauges, scanning electron microscopes (SEMs), acoustic imaging, and X-rays.

Electrical testing at microwave frequencies is often conducted with probe cards in an on-wafer probe station. This technique requires that reference planes be available where the ground and signal pads are available in a coplanar format. When such a reference plane is not available, it may be possible to provide package ports with additional substrates to allow microstrip-to-coplanar conversion [4]. During probing, care should be taken to ensure that the probes land evenly on package surfaces, since in practice package parts may be mounted slightly unevenly. This will prevent the probes from premature wear or compromising results.

Package S-parameters may be captured with a network analyzer. In order to perform calibration, probes are generally calibrated to an impedance standard substrate (ISS). An ISS is a substrate that contains known circuit elements such as shorts, opens, through lines, or loads. Some calibration techniques include SOLT (short–open–load–through), TRL (through–reflect–line), or LRM (line–reflect–match), and each are solved using different algorithms. With calibration, equipment effects can be removed up through the probe tip. Additional de-embedding techniques can be performed if the S-parameters of the additional fixtures are known.

Another common technique involves the implementation of a back-to-back structure. In many cases, one end of a feed-through is difficult to access or would cause a physical change that would affect package performance. In these instances, it may be possible instead to build a back-to-back structure consisting of two feed-through transitions. Measurements in this case may double the insertion loss. The return loss may decrease by 6 dB in some parts of the structure depending on how well the transition is designed.

Time-domain characterization is another type of measurement that is proving useful for identifying discontinuities. When a signal is fed forward, reflections appear at feed-through discontinuities. In time-domain reflectometry (TDR) these signals are measured at twice the propagation time from input to discontinuity. The reason is that the signal must make a round-trip journey. Gating techniques can be employed to isolate known reflections, and thus to minimize their effects. With gating, a Fourier transform can be performed to obtain S-parameters relating to a specific time-domain segment. Note that S-parameters obtained in this way should be used only as a part of a qualitative design and analysis technique, as significant amounts of relevant data may be removed through the gate-windowing process. These techniques are generally limited to larger package structures and are limited by time resolution.

1.4 Concluding remarks

This chapter has provided an introduction to all the packaging aspects discussed in this book. A brief history of packages allowed the reader a reasonable overview of the major package types used in electronics. The considerations involved in package design were discussed and the relevant theory given, to supplement the reader's knowledge. Different package technologies were presented, and it was suggested that LCP is in a unique position to replace incumbent materials. Practical computer-aided design and test considerations were also presented.

References

[1] http://www.rogerscorp.com
[2] http://www.kapton-dupont.com
[3] "Test Method Standard Microcircuits: Mil-Std-883," June 18, 2004.
[4] http://www.jmicrotechnology.com
[5] http://www.siliconfareast.com/cob.htm
[6] http://www.jedec.org/about-jedec/jedec-history
[7] http://www.ipc.org/ContentPage.aspx?pageid=IPCs-Name
[8] http://www.dscc.dla.mil/programs/milspec/
[9] H. Wakabayashi, Y. Osawa, K. Toda, T. Hamazaki, R. Kuramasu, "A SAR system on the ALOS," in *Proc. Conf. on Int Archives of Photogrammetry and Remote Sensing*, vol. XXXI, part B1, Vienna 1996, pp. 193–196.
[10] H. Fukui, "Thermal resistance of GaAs field-effect transistors," in *Proc. IEEE IEDM*, 1980, p. 118.

[11] D. Meeks, "Fundamentals of heat transfer in a multilayer system," *Microwave Journal*, vol. **1**, no. 1, pp. 195–172, January 1992.
[12] http://www.ansys.com
[13] http://www.agilent.com
[14] http://www.cst.com
[15] http://www.sonnetsoftware.com
[16] http://www.mentor.com
[17] http://web.awrcorp.com
[18] http://www.comsol.com
[19] http://www.solidworks.com
[20] http://www.cfdesign.com
[21] http://www.thermalsoftware.com

2 Characteristics of liquid crystal polymer (LCP)

Morgan J. Chen, Kunia Aihara, Cheng Chen, Anh-Vu H. Pham

Liquid crystal polymer (LCP) has interesting characteristics that have garnered the attention of RF and microwave circuit designers. One major difference from typical high-frequency materials is that LCP is a thermoplastic. With LCP's advent, engineers have discovered how to integrate high-performance electronics directly into the thermoplastic package. The reason why LCP is attractive for RF/microwave electrical design is its unique combination of excellent electrical and mechanical properties. LCP is extremely stable in the presence of moisture and exhibits near-hermetic properties. Given its relatively low cost, LCP is rapidly becoming the material of choice for new generations of electronics requiring increasing integration and performance. This makes LCP a serious candidate for multi-chip module (MCM), system-in-package (SiP), and advanced packaging technology.

In section 2.1, we will first discuss LCP's chemical properties. The basic chemistry will be introduced, and LCP's composition will be described and discussed. This section is presented from an application stance and is intended to provide a brief working background to why LCP behaves as it does.

In section 2.2 we describe LCP's electrical properties. LCP has been characterized as having a very low dielectric constant and loss factor over the frequency range from below 1 GHz up to past 110 GHz [2]. Its low dielectric constant allows reasonable line impedances to be formed on thin-film material and, further, minimizes the impact on the nearby electronics as well as the capacitive detuning effects of packaging. Methods for electrical characterization and study results are presented. The test results of LCP in moisture are also provided, to demonstrate its stable electrical characteristics over humidity changes.

In section 2.3 we discuss LCP's physical characteristics. It is a low-cost dielectric material that has low moisture absorption (equivalent to that of glass) and is commercially available as single sheets, laminated substrates, or small pellets for molding. The advantages of LCP compared with other organic substrate materials are a low moisture absorption, low coefficient of hydroscopic expansion (CHE), low permeation, and the adjustability of its coefficient of thermal expansion (CTE) through thermal treatment processes. Aspects including its dimensional stability and its adhesion strengths to different materials will be described. Test results of LCP's electrical properties in moist environment are also provided to demonstrate the stability of LCP's performance over a range of humidities.

In section 2.4 we describe the environmental characteristics of LCP. One of the most unique properties of LCP is its ability to be used for hermetic packaging. Topics will include outgassing tests and near-hermetic properties. Special attention has been given to characterizing LCP in the context of electronic assemblies. Several methods to evaluate LCP's near-hermetic capability will be described and discussed.

2.1 LCP chemistry

The field of LCP for electronics is now almost a decade old. However, the study of LCP by material scientists and chemists dates back to as early as 1923, when Vorländer synthesized LCP using benzene rings linked through ester groups [1]. While this book focuses on applications of LCP for novel RF and microwave circuitry, it is also worthwhile briefly to study LCP's chemistry. Useful insight can be gleaned by connecting LCP's unique properties with its fundamental composition. In this section we attempt to cover the basics of LCP chemistry that may be relevant to practising electrical engineers. The interested reader is referred to the book [1] for more details on LCP chemistry than can be presented in this text.

2.1.1 Liquid crystalline characteristics

The field of liquid crystals has existed in chemistry for over 100 years. The term "liquid crystal" refers to a phase of matter that exhibits partly solid and partly liquid properties. Liquid crystals can flow like traditional liquids, but the molecules retain a preferred orientation. In comparison with that of liquid crystals, the field of liquid crystal polymers (LCP) has been studied by chemists only for the last few decades, and the science on this subject continues to evolve. LCPs are generally solid at room temperature. If liquid crystals are observed to depend on temperature, they are referred to as *thermotropic*. This means that the liquid-crystal material can be heated up from the solid phase to the liquid-crystal mesophase. Alternatively, if liquid crystals are observed when it is in solution then the material is referred to as *lyotropic*. Thus far, the LCPs that have been used for advanced RF/microwave integration have been exclusively thermotropic. This ability to reshape LCP with heat lends itself well to the existing processes using thermocompression equipment found in today's printed circuit board industry. Figure 2.1 illustrates the organization of liquid crystal polymers in the solid phase and in the melted, fluid, phase.

The chemical term for a liquid-crystal state is a *mesophase*. Liquid crystals can be distinguished as *main-chains* and *side-chains*. In LCP, a *mesogen* or *mesogenic* unit refers to the fundamental repeating unit that forms the liquid crystal. Main-chains are liquid crystals with mesogens along a chain forming a sort of backbone structure; side-chains are those with mesogens that branch off the main line. All the LCPs marketed for RF or microwave and described in this book are of the main-chain variety. As a consequence of this preferred orientation, LCP has inherent

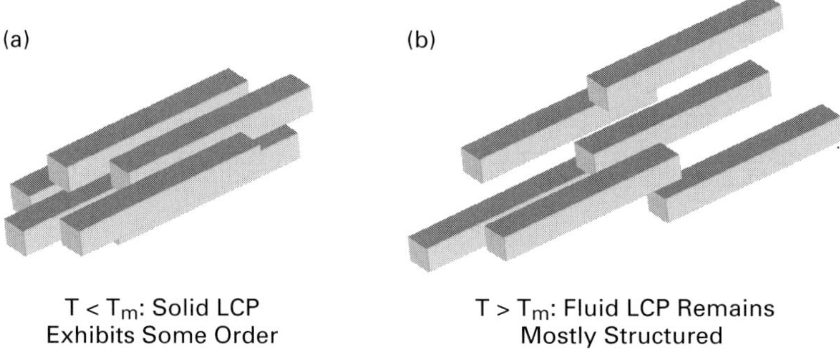

Fig. 2.1 (a) A liquid crystal solid and (b) a liquid crystal fluid.

anisotropy. The most prominent anisotropic characteristic appears in its coefficient of thermal expansion (CTE). Material providers have developed ingenious methods to mitigate these effects, and today, laminate LCP is commonly engineered to have a CTE of 17 ppm/°C in the *X* and *Y* dimensions, which is well matched to other PCB materials, while it is larger (150 ppm/°C) along the thin Z dimension. More on how these laminates are engineered to provide uniform *X* and *Y* properties is presented in chapter 3.

The way in which liquid crystals are arranged on the molecular level may be described as either *nematic* or *smectic*. In the nematic arrangement, snake-like molecules are aligned along one dimension. Smectic arrangements show a two-dimensional alignment and form layers. Smectic liquid crystals are often characterized as having a soapy feeling. If a nematic arrangement shows *chirality*, which refers to rotational asymmetry, then the liquid crystal is referred to as being *cholesteric*. The cholesteric molecular arrangement is sometimes classified as a third molecular arrangement in liquid crystals. Liquid crystals exhibit anisotropic properties because these crystal arrangements have preferred orientations. The LCPs used for electronics are nematic in nature.

2.1.2 Polymeric characteristics

Polymers are made up of repeating molecular chain structures. The word "polymer" translates literally from the Greek to "many parts." LCPs are characterized by their high strength and low weight. One of the most well-known LCP structures is Kevlar, which is used in bullet-proof vests. There is no single chemical structure that can characterize all LCPs. The most common LCP marketed to electrical engineers for high-frequency usage is Vectra, which is a nematic, main-chain, thermotropic, co-polymer blend of p-hydroxybenzoic acid (HBA) and 6-hydroxy-2-napthoic acid (HNA) monomers. This LCP contains only the atoms carbon, hydrogen, and oxygen and hence is environmentally benign. Owing to its thermoplastic nature, LCP may be used as an adhesive material and can be recycled. Chemical diagrams for LCP's repeating units are provided in Fig. 2.2. The HBA

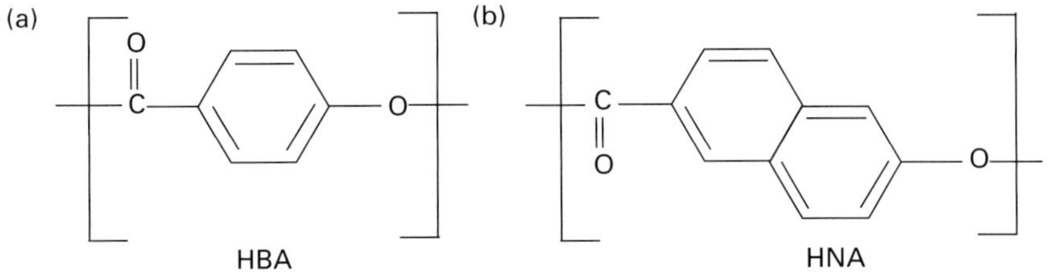

Fig. 2.2 (a) p-hydroxybenzoic acid and (b) 6-hydroxy-2-napthoic acid.

and HNA groups are each linked by ester bonds (-O-CO-), which allow long-range rotational flexibility, like a mechanical joint. In chemistry, this motion is referred to as *crankshaft* rotation. The *aromatic* rings, also known as benzene rings, in HBA and HNA are shown as hexagons in the chemical diagrams presented in Fig. 2.2. The term "aromatic" came into usage from the sweet scents observed during early studies on this structure. LCP laminates are neutral as far as scent is concerned. Each benzene ring is made up of six carbon atoms and exhibits a resonance structure where the double and single bonds constantly exchange places. This resonance structure effectively gives each bond the same length, and so the structure has a bonding distance that is intermediate between the bonding distances for single and double bonds.

An important offshoot of this aromatic feature is that it may be a major contributing factor to LCP's great stability. Aromatic rings also have the property that they can be efficiently stacked on top of each other, much like a tower of pancakes. This may play a major role in LCP's low diffusivity and permeability by allowing for high-density polymer packing. The HNA monomer, in particular, is known for its high degree of chain packing and its low solubility. The presence of crystalline domains in LCPs gives them unique properties such as low permeation and chemical resistance. The crystalline regions act as impermeable barriers for moisture and other chemicals. Thus the key to LCP's low permeation rate is its high molecular order. It is believed that each "crystal" formed is as close-packed as a pure crystal [3]. Hence, diffusion effects are dominated by transmission along grain boundaries, where the separate LCP crystals meet. LCP permeation is so low that it is on the same order of magnitude as glasses, which form a subclass of ceramics that are amorphous and are also used in near-hermetic packaging.

LCP has been chemically engineered and manufactured to have varying properties. Its CTE, which refers to changes in physical dimension resulting from changes in temperature, is in the range 8–17 ppm/K. LCP has near-hermetic characteristics, with a water absorption less than 0.004% [4, 5]. Its melting temperature, T_m, is known to vary from 280 °C to well over 350 °C, depending on the make and model. Since LCP is a thermoplastic polymer it has a melting temperature, which is different from the glass transition temperature for epoxy-based boards. The melting temperature is the threshold temperature where a material switches between the solid and liquid states. Alternatively, a related specification is the heat distortion temperature (HDT), which refers to the temperature at which a material begins to deform under a specified load. For LCP, the HDT may be around 260 °C, which is about 20 °C lower than T_m. In order to make LCP flow, laminate processes often slightly exceed T_m when high pressure processing is required. This allows the surfaces to melt while the bulk is intact, so that the LCP will not lose its intended shape. Also, at temperatures above T_m, LCPs adhere to themselves under lamination pressure.

In general, more aromatic rings in the composition result in higher LCP melting temperatures. The compositional ratios of HBA and HNA may be varied as another way to modify the melting temperature. For instance, HNA is known to exhibit a lower melting temperature than HBA owing to HNA's kinked structure, which causes extra crankshaft motion. Other techniques for affecting LCP's melting temperature are of particular interest to chemists, and more information can be readily found in [1]. Through these engineering methods, the material suppliers are able to offer *core* and *ply* LCP laminates that provide convenient usability and maintain the excellent dimensional stability needed for RF/microwave circuit design.

2.2 Electrical characteristics

LCP has a characteristically low dielectric constant and loss tangent. A low dielectric constant is desirable because it allows the practical fabrication of controlled impedance transmission lines on ultra-thin substrates (i.e. 20 μm) with low attenuation loss. In contrast, 50 Ω lines built on ceramic have a high dielectric constant, narrow line widths, and correspondingly higher attenuation. A low loss tangent is similarly desirable because then less energy is absorbed into the dielectric. The loss tangent, tan δ, of LCP is equal to 0.002–0.004 up to 100 GHz [6–9]. Hence, both the dielectric constant and the loss tangent play a direct role in RF losses. In this section we describe LCP's dielectric constant and loss tangent under nominal conditions from DC to 110 GHz using various measurement techniques. We will also discuss measurements showing LCP's stable electrical properties in the presence of moisture.

2.2.1 Dielectric and loss tangent characterization from DC to 60 GHz

Cavity resonators have been used to characterize the LCP dielectric constant and loss tangent. Cavity resonators are a benchmark standard for loss tangent (dissipation factor, DF) measurement. Resonances are determined by the inner cavity dimensions, as larger cavities have lower frequency resonances. The resonant frequencies of cavity resonators with and without dielectric loading are measured in order to characterize the dielectrics. A frequency shift defines the dielectric constant, which can be calculated through an iterative process. The 3 dB resonance bandwidth divided by the resonant frequency, i.e. the inverse quality factor 1/Q, approximates the loss tangent. With this assumption, the unloaded loss tangent

(a)

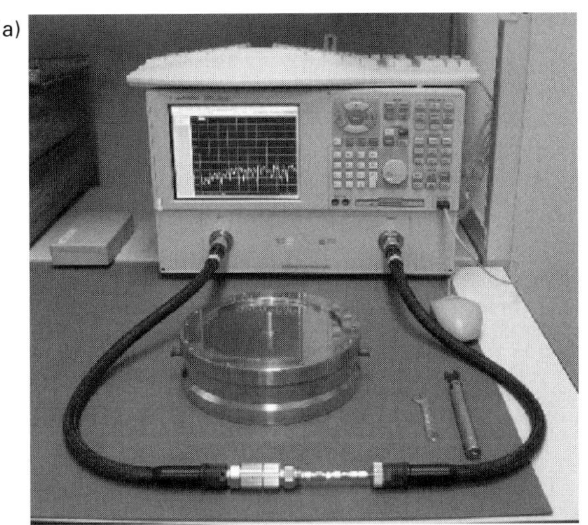

(b)

Fig. 2.3 (a) Measurement setup for dielectric characterization, (b) cavity resonators [11] (© 2007 IEEE).

contribution is designed to be negligible. Example equipment to perform cavity resonator measurements is shown in Figs. 2.3–2.5. Figure 2.3a shows a measurement setup using a two-port network analyzer to obtain transmission resonances from S-parameter measurements [11]. Figure 2.3b shows various cavity resonators used in the measurements. The materials are sandwiched within a cavity resonator to perform measurement. Characterization on LCP from DC to 40 GHz shows the dielectric constant to vary between 3.0 and 3.2. Figures 2.4 and 2.5 show the measured dielectric constant and loss tangent, respectively. The results indicate that consistent characteristic impedances can be designed over a wide frequency range, which, in turn, allows for broadband microwave circuit design. The loss tangent is characterized as around 0.0025 and remains fairly constant over DC to 40 GHz.

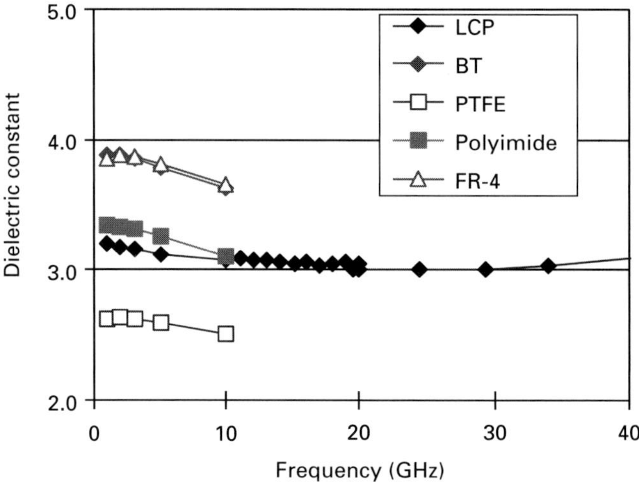

Fig. 2.4 Dielectric constant measurement of LCP up through 40 GHz.

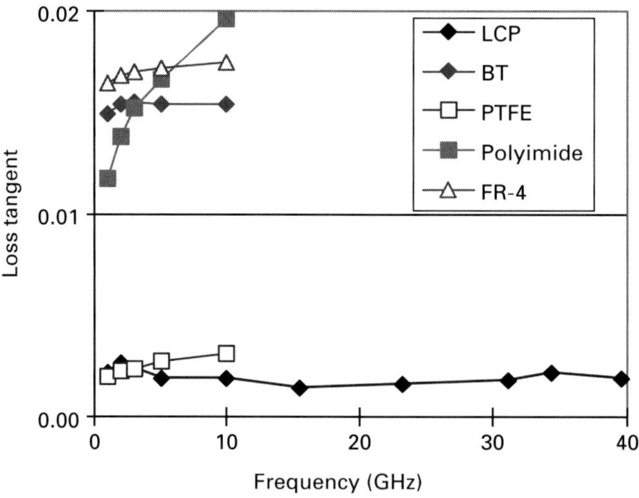

Fig. 2.5 Loss tangent measurement of LCP up through 40 GHz.

2.2.2 Dielectric and loss tangent characterization from 30 to 110 GHz

Higher-frequency electrical characterization on LCP has been conducted from 30 to 110 GHz [6]. In this work, various structures were designed in order to characterize LCP. Three methods were employed, including respectively the use of ring resonators, cavity resonators, or transmission line characterization.

Ring resonators are used for the careful characterization of a dielectric on the basis of its resonant peaks. Careful analysis is conducted, with consideration of the substrate thickness, ring diameter, dielectric constant, and skin depth as potential

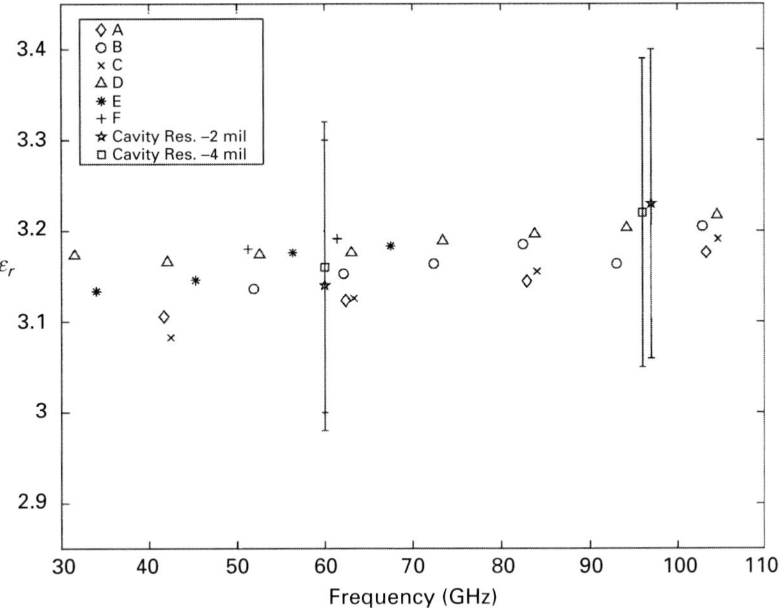

Fig. 2.6 Extracted LCP dielectric constant [6] (© 2004 IEEE).

causes for peaks in the ring resonator characterization. The relative dielectric constant is found to vary between 3.1 and 3.25 over 30 to 110 GHz. Unfortunately, ring resonators are impractical for calculating the loss tangent since this method is based on loss, and one requires more accurate conduction and radiation loss-accounting than can reasonably be performed.

Instead, cavity resonators are implemented to extract the loss tangent. These measurements showed that the loss tangent for LCP is around 0.005 from 30 to 110 GHz. Measurements on the dielectric constant using cavity resonators were also performed, and the results agree well with the ring resonator measurements discussed above.

Further, transmission lines can also be implemented for LCP characterization. Figures 2.6 and 2.7 show the extracted LCP dielectric constants and loss tangents versus frequency. In Fig. 2.6, plots labeled A through F correspond to various transmission lines with impedances varying from 60 to 90 Ω. The plots are overlaid with the cavity resonator measurements. As stated above, for the lower-frequency measurements these plots indicate that LCP maintains a constant performance over a wide range of frequencies. This constant performance allows microwave circuits to be designed over very broad bandwidths on LCP. Measurements show less than 2.6% variation in dielectric constant at 10 GHz. Over 80 GHz bandwidth there is less than 4.3% variation, with dielectric values 3.6 ± 0.05 and slope ~0.05/(50 GHz) or 0.001 GHz^{-1}. Measurements performed across a range of structures, including transmission lines, ring resonators, and cavity resonators, showed loss tangents ranging from 0.003 to 0.0085. The most accurate of these is believed to

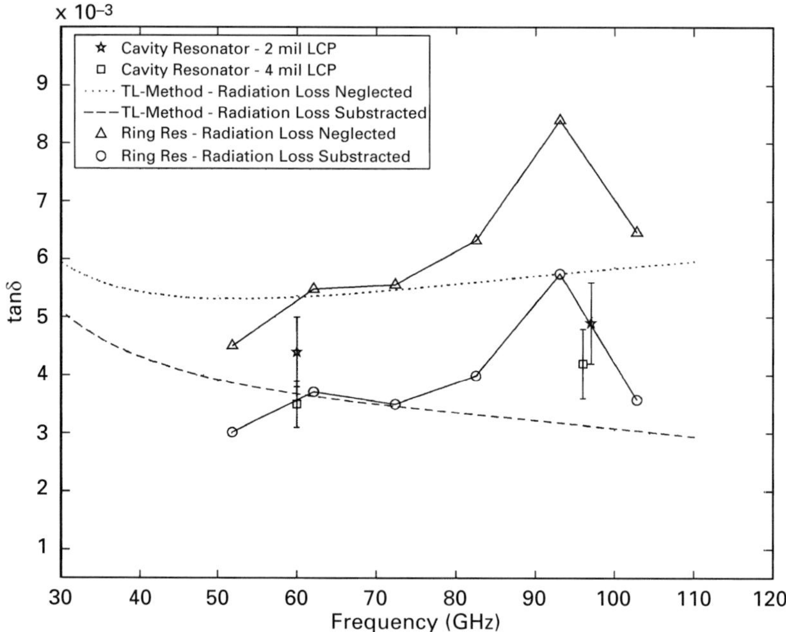

Fig. 2.7 Extracted LCP loss tangent versus frequency [6] (© 2004 IEEE).

be that from the cavity resonator. Two resonators were used, and measurements show a loss tangent in the range 0.0035–0.0045 at 60 GHz and 0.0042–0.0049 at 97 GHz. Figure 2.8 shows the results of attenuation measurements conducted on LCP microstrip lines with various dielectric thicknesses. Wider line widths result in lower conduction loss; for this reason, a low-dielectric-constant substrate is advantageous because then any particular characteristic impedance is realized for a wider line width and, in turn, a lower loss. Moreover, thicker substrates require wider line widths to realize a particular impedance. As shown in Fig. 2.8, 50 Ω microstrips built on 2, 3, 4, and 5 mil substrates respectively exhibit 2.5, 2, 1.5, and 1.35 dB/cm at 110 GHz. The plot shows linear attenuation versus frequency and indicates that the microstrip mode is maintained across the band. Other examples of microstrip lines are provided in Chapter 6 of this book, where we discuss a novel LCP-based long time-delay circuit that implements a tuned attenuation as low as 0.01 dB/mm at 10 GHz.

2.2.3 Electrical characteristics over humidity

LCP dielectric characterization over a range of humidities was also performed, to demonstrate stable operation in the presence of moisture [11]. Measurements were performed at 1 GHz. The measurement setup is shown in Fig. 2.9, which illustrates the material analyzer and, below it, the specimen chamber. The relative dielectric constant was measured to be 3.0. Graphs are plotted for various materials

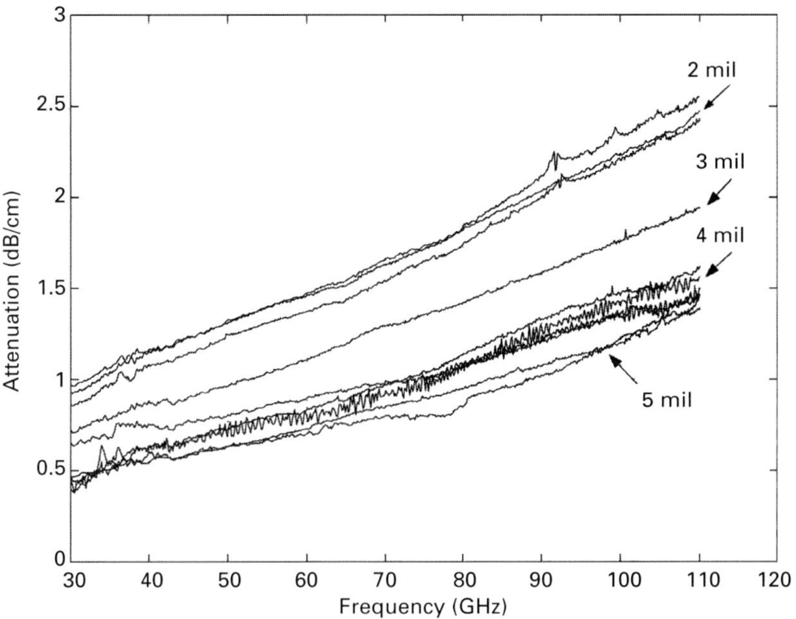

Fig. 2.8 Microstrip line losses on LCP substrates of 2–5 mil [6] (© 2004 IEEE).

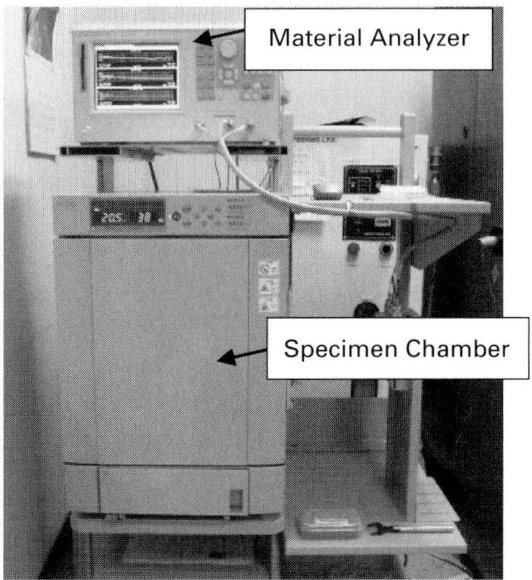

Fig. 2.9 Setup for LCP dielectric characterization over a range of humidities.

for comparison: bismaleimide-triazine (BT), polyimide, and the FR4 laminate all demonstrate a higher dielectric constant and loss tangent than LCP. Polyimide, often used for microwave flex circuitry but known to perform poorly in humidity, demonstrates the greatest variation with humidity of all the materials shown. The measurements on LCP show that the dielectric constant remains virtually the same

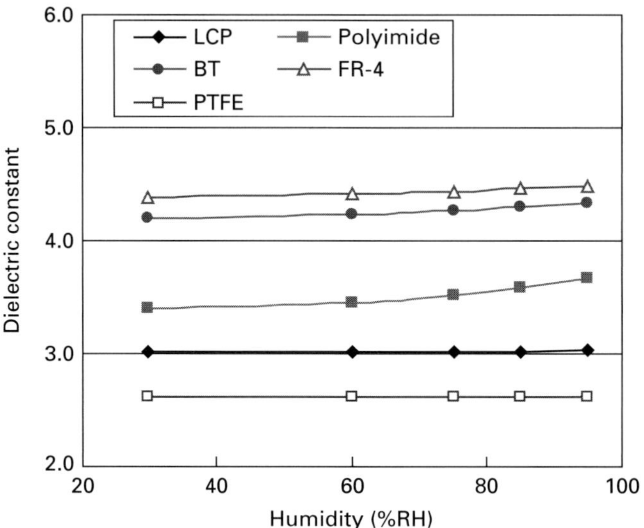

Fig. 2.10 Dielectric constant of LCP versus humidity [11] (© 2007 IEEE).

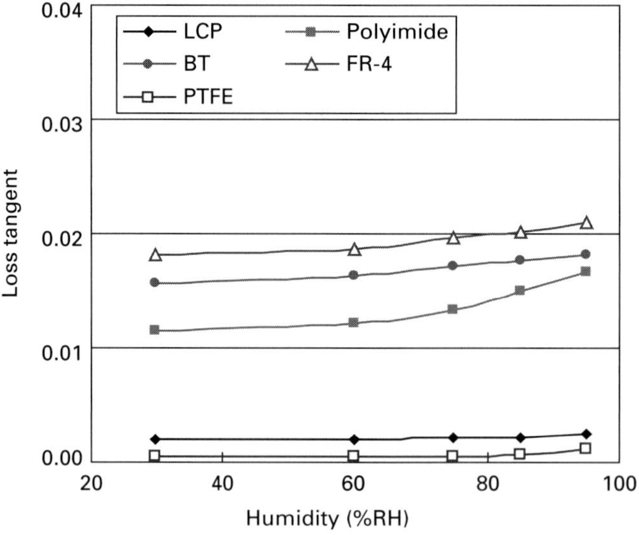

Fig. 2.11 Loss tangent of LCP versus humidity [11] (© 2007 IEEE).

over 30% to 95% relative humidity (RH), as shown in Fig. 2.10. The loss tangent is similarly stable at 0.0025, as shown in Fig. 2.11. For other materials such as BT, polyimide, and FR-4, the dielectric constant and loss tangent change for high relative humidity; in actual operation, the impedance and loss start to change. At high frequencies, owing to the smaller wavelengths, the electrical changes become more severe and make these materials inappropriate. However, LCP's electrical characteristics remain virtually constant with respect to humidity, as mentioned above. This demonstrates that LCPs have similar properties to ceramic and glass, which are widely used for millimeter-wave packaging.

2.3 Physical characteristics

LCP is commonly found in its natural state as a tan or light-yellow opaque material. It is also commercially available in color-modified versions, including black, blue, brown, green, and red. LCP laminate sheets may appear glossy or shiny, and so provide certain mirror-like qualities common to other smooth plastics. In this section we discuss LCP's dimensional stability, adhesion strengths to other materials, and flexibility.

2.3.1 Dimensional stability

The dimensional stability and thickness uniformity of LCP thin films are important parameters, which may need careful quantification and control for successful circuit design. LCP has low viscosity above its heat distortion temperature; under such a condition it becomes malleable. This generally means that it can transform during thermal processes and even more so under mechanical pressure. Thus, a concern with LCP is providing controlled processes so that the physical dimensions of a microstrip line are predictable after lamination under pressure. For instance, the use of a thinner or thicker dielectric than was intended will respectively decrease or increase the transmission line's characteristic impedance. In the design of a balun, the dielectric thickness is the single most critical parameter to be controlled as in turn it directly controls the impedance matching at the unbalanced port. Insufficient matching at this port will cause degradation of the balun's return loss. In Fig. 2.12, scanning electron microscope (SEM) photographs of laminated samples demonstrate the uniform thickness of the LCP layer. These samples have undergone reactive ion etching (RIE) to etch the LCP layer in various portions of the circuit, as well as subsequent thermocompression lamination. Figure 2.12a illustrates 2 mil high-temperature LCP jacketed with M1 (bottom layer) and M2 (top layer) Cu. Below layer M1 are multiple layers of LCP that have been laminated in order to achieve a particular desired thickness. Figure 2.12b is similar except that 1 mil thick high-temperature LCP was used instead of 2 mil; note the difference in scale between Figs. 2.12a and 2.12b. Since RIE was used to remove LCP in sections within the circuit shown in Fig. 2.12 (for a balun design), there is a potential concern that subsequent lamination could cause the LCP to reflow and the thickness to decrease under RF traces. If reflow were to occur, this would cause the line impedance to change and detune the circuit. A profile meter was used to measure LCP's thin-film thickness uniformity. The dielectric thickness of one such LCP sample after a

(a)

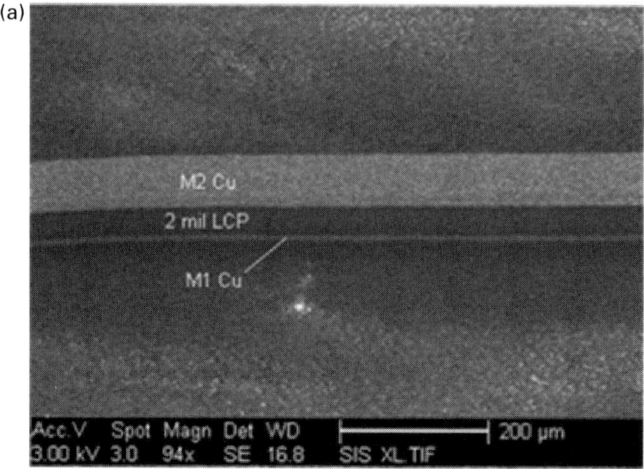

(b)

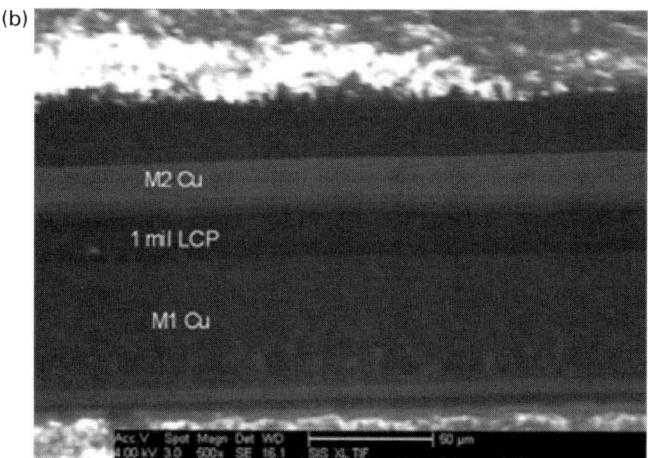

Fig. 2.12 Scanning electron microscope photographs of microstrip lines having (a) 2 mil LCP thickness and (b) 1 mil LCP thickness after RIE. The photographs show uniform thicknesses for these high-melting-temperature LCPs. The structures shown can also be used to measure dimensional distortion after lamination [15] (© 2007 IEEE).

lamination process was measured to be 52.1 μm ± 8%. This result agrees well with the manufacturer's specification of 50 μm ± 5%, indicating an unchanged dielectric thickness after lamination.

The same technique was repeated to measure other structures fabricated on the same substrate. The results show an average thickness of 36 μm for an intermediate layer of low-melting-temperature LCP layer after lamination. This low-temperature layer originally had a thickness of 50 μm before lamination. This loss in LCP thickness is expected owing to its reflow during the lamination process. A designer should take this thickness change into account during the design phase and utilize high-melting-temperature LCP layers having higher dimensional stability for microstrip lines or structures requiring tight dimensional control.

2.3.2 Adhesion strengths for package integrity

One advantage of LCP films compared with other laminates is the ability to adhere to materials without external adhesives in a lamination process. This feature not only simplifies packaging processes but also reduces the electrical losses associated with adhesive materials, which are normally plagued by high loss tangents. Furthermore, a multilayered LCP board will be homogeneous and can serve as a barrier against moisture. If epoxy-based pre-preg is used in a multilayered LCP board, it will create a path by which moisture can get into the package or board. Currently, LCP adhesion mechanisms are not fully understood by chemists but are believed to be due to a combination of mechanical locking, Van der Waals' forces, and chemical bonding. In mechanical locking, LCP physically molds and locks around niches and surface micro-defects; varying degrees of surface roughness will always exist along a surface even if only at a molecular level. Under Van der Waals' forces, molecular-level electrostatic effects create attractive forces. These micro-forces may produce partial adhesion on a macro-scale. Covalent C=O bonds have been observed, at interfaces where LCP is bonded to copper, that appear to correlate with adhesion strengths [12]. Regarding reliability concerns, adequate adhesion strengths are required because weak surface adhesion can mean that vias are at risk of splitting open and exposing delicate electronics.

Peel strength-pulling tests are conducted to evaluate the adhesion strengths between LCP and semiconductor materials. Peel tests are performed using a Chatillon pull tester. Photographs of a peel-test setup are presented in Fig. 2.13. A cross-sectional diagram showing how peel-testing experiments are conducted is provided in Fig. 2.14. In this figure, the peel width is the dimension orthogonal to the page. In this peel test, the flexible layer to be separated is gripped and pulled upwards at 90° at a constant rate while the rigid adhered base (i.e. silicon) is kept fixed. Testing is conducted on uniform-width strips in order to provide manageable force readings.

This test is used to ensure that the processes being developed provide the required adhesion strengths on all layers. A strain gauge attached to samples measures the resistant tensile force. Small test samples were peeled at 0.1 in/min, a slow rate, to provide sufficient measurement time on our relatively small samples. Slower rates are known to give higher peel strengths. In general, it follows that wider peel strips measure higher forces because the greater width provides proportionally more adhesive strength. Therefore, peel strengths are reported normalized to the peel-layer width. The units for peel strength are N/in. Several standards exist, from various groups, that specify commonly used peel-test conditions [20, 21].

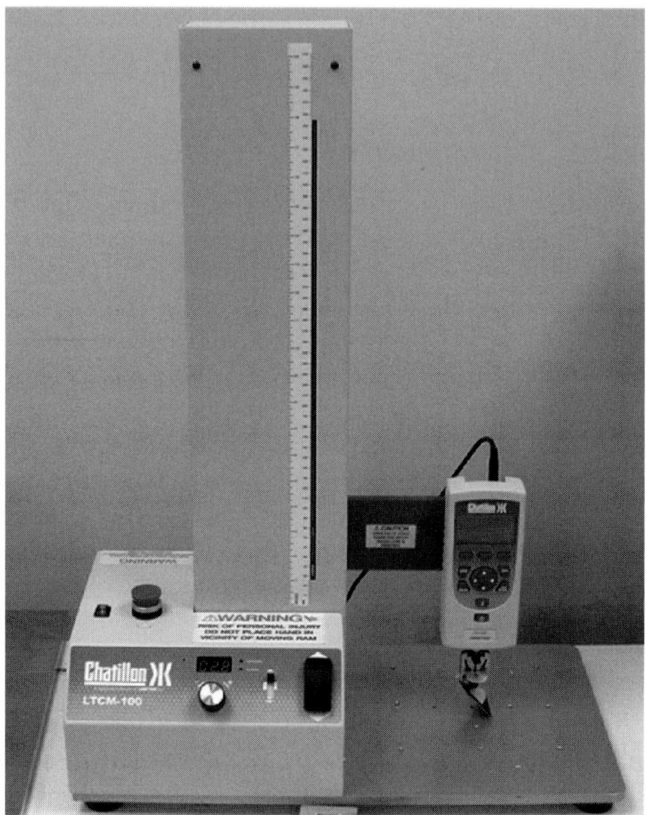

Fig. 2.13 Peel-strength test stand.

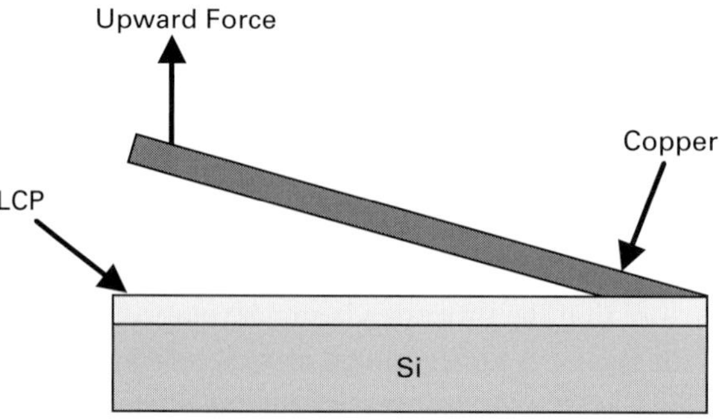

Fig. 2.14 Peel strength test, showing copper being pulled from LCP/silicon substrate [16] (© 2006 IEEE).

Gallium arsenide (GaAs) and more recently, silicon (Si) are common substrates used for RF/microwave chips. LCP provides excellent adhesion to both materials after lamination and may be ideal for use in wafer-level packaging. Samples of polished GaAs and Si laminated to LCP are shown in Fig. 2.15. The silicon

(a)

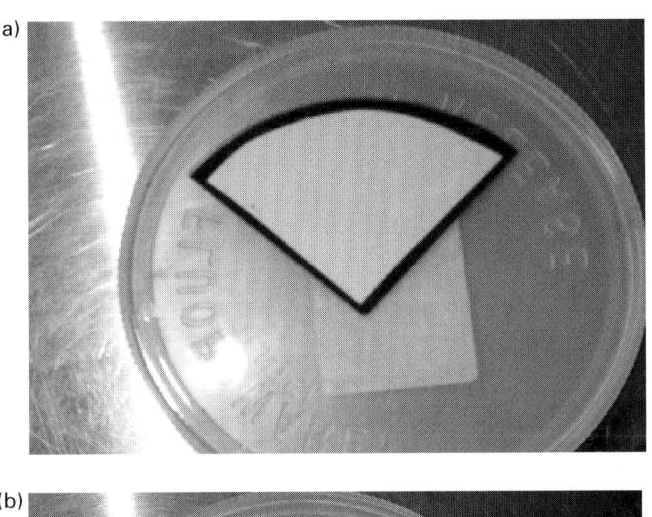

(b)

Fig. 2.15 LCP laminated to (a) GaAs and (b) Si.

sample with 675 μm original thickness was thinned to 100 ± 10 μm, whereas the 550 μm thick GaAs sample was thinned to 150 ± 10 μm. Peel-strength histograms are provided in Figs. 2.16 and 2.17 for Si and GaAs, respectively, under different conditions. This method of measuring peel strength versus lamination pressure and substrate thickness provides information needed to develop optimal manufacturing processes. On both substrates, peel-strength measurements are greater than 3 lb/in, which represents a lower limit commonly used by industry. Results are reported normalized to peel strip width. These tests indicate that LCP may provide sufficient adhesion for use in near-hermetic packaging applications.

2.3.3 Flexibility

IPC flexibility measurements on LCP were performed using the setup in Fig. 2.18a. Figure 2.18b shows how the stroke cycle is performed. Measurements

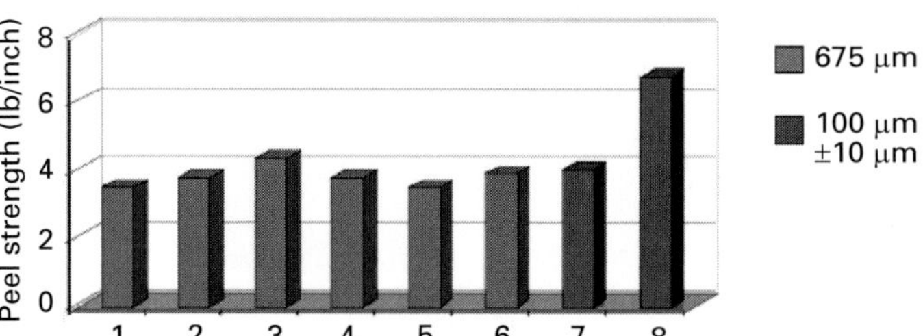

Fig. 2.16 Peel strength histogram for LCP samples laminated to Si under various lamination pressures and two different Si wafer thicknesses. The lamination pressures were as follows: sample 1–3, 200 PSI; sample 4, 150 PSI; samples 5 and 6, 0 PSI; samples 7 and 8, 100 PSI.

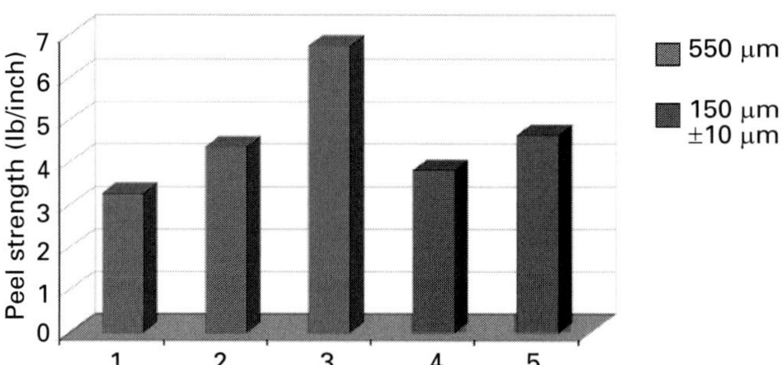

Fig. 2.17 Peel strength histogram for LCP samples laminated to GaAs for various lamination pressures and GaAs wafer thicknesses. The lamination pressures were as follows: sample 1, 150 PSI; samples 2–5, 100 PSI.

were made with a 1.25 mm flex radius and 30 mm stroke. The results show that LCP withstands greater than 50 000 flex cycles, and its performance compares reasonably well with polyimide. To put this number into perspective, 50 000 flex cycles amount to 6.8 flex cycles per day over a 20 year lifespan. For consumer applications with lifespans closer to five years, such as flexible laptop display connectors, this corresponds to over 27 flex cycles per day. Flex cycle results for 1 mil thick LCP are plotted in Fig. 2.19, together with results for two types of 1 mil thick polyimide. The results indicate approximately 55 000 cycles for LCP compared with 59 000 for the left-hand polyimide. This mechanical flexibility indicates that LCP can be implemented in unique multidimensional applications that may allow for higher-density or novel form factors for electronics. A related specification is Young's modulus, the tensile modulus, which is specified as 2255 MPa for LCP under IPC 2.4.16 [18]. The tensile strength for LCP is specified to be 200 MPa under IPC 2.4.19.

(a)

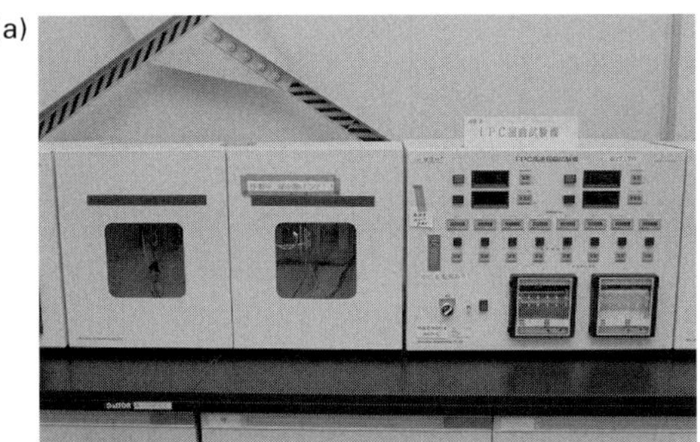

(b)

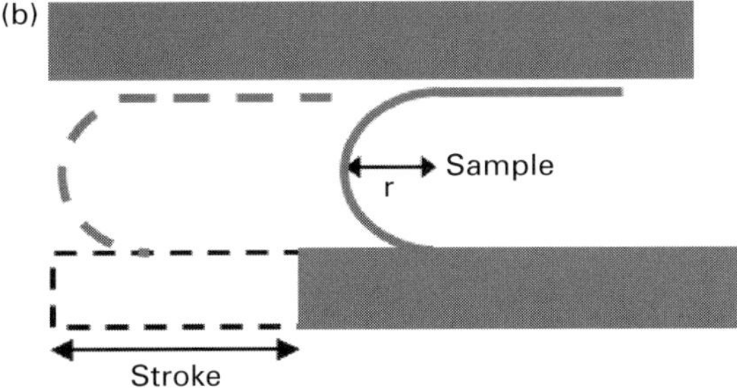

Fig. 2.18 (a) IPC flexibility setup, (b) flexibility stroke cycle.

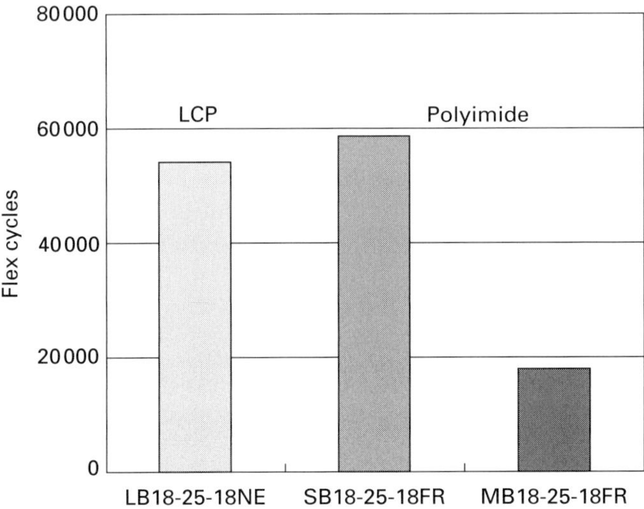

Fig. 2.19 IPC flexibility measurements for LCP and polyimide. The cover material was CVK0525KA (Arisawa Mfg).

2.4 Environmental characterization

LCP is known to be chemically inert and to have low moisture absorption and near-hermetic properties. This means LCP is not prone to interact with other materials and offers attractive barrier properties. In this section, we discuss the quantitative characterization of LCP's barrier properties. This is performed using outgassing tests, and permeability and leak-rate measurements and analysis. These will be discussed in more detail below.

LCP has been reported to have a permeability of 1.617×10^{-9} cm^2/s for helium in LCP [14]. This value is comparable with that for helium permeation through SiO_2 (quartz) at 1×10^{-9} cm^2/s. Package hermeticity may be quantitatively analyzed using a diffusion leak rate closed-form approximation equation [19],

$$R = \frac{KA\Delta P}{d}, \tag{2.1}$$

where R is the leak rate (in atm cm^3/s), K is the permeability (in cm^2/s), A is the exposed package area (in cm^2), ΔP is the pressure difference (in atm), and d is the package wall thickness (in cm).

Figure 2.20 shows the permeabilities of LCP and other packaging materials such as metal, glass, epoxies, and other organic polymers. The vertical axis gives the package wall thickness in cm. The upper horizontal axis gives the material permeability in mol/(m s MPa), and the lower horizontal axis gives the time in the package cavity environment. Altogether, this chart illustrates the time required to reach 50% of the external-environment humidity for a given package-wall thickness. LCP in this chart lies close to glass, which is often used in near-hermetic packaging. From the chart, LCP has a permeability of better than 10^{-13} mol/(m s MPa); a smaller permeability equals better hermetic properties. Note that, in the chart, the permeability is normalized to the pressure; this means that more pressure will cause more moles of gas to diffuse through LCP. The permeability is also normalized to time in seconds. Regarding the normalized physical dimension, m, this is not the thickness as one might think at first glance. Rather, the normalization is such that m(thickness)/m^2(exposed area) = 1/m. Hence, more area exposed will allow more diffusion to occur.

In conclusion, since the permeability of LCP is less than 10^{-13} mol/(m s MPa), as shown in the chart, a near-hermetic packaged environment using LCP may be realized.

2.4.1 Outgassing tests

Outgassing is a phenomenon whereby gas trapped in a material is released. Outgassing significantly concerns polymer materials used for packaging RF microelectromechanical systems (MEMS) because organic materials readily uptake moisture. During polymer processing in RF MEMS packaging, polymer materials

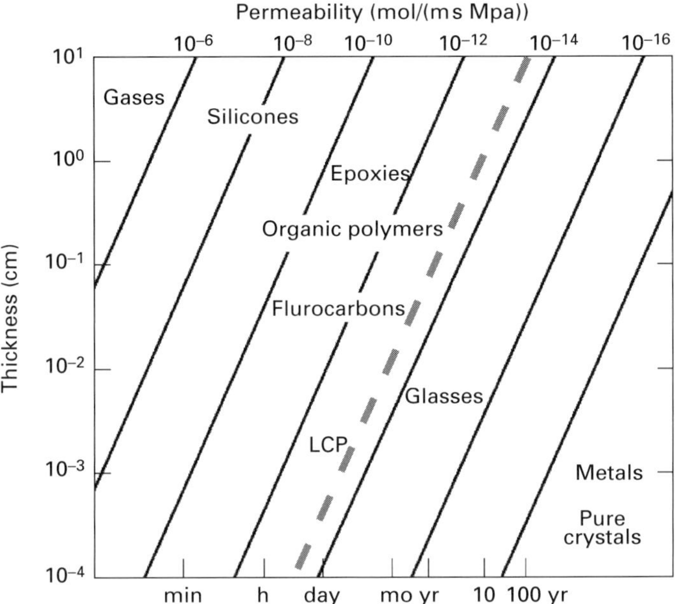

Fig. 2.20 Package wall thickness versus time or permeability. Each type of material is located in the general area of its permeability value.

may release gas particles that would degrade the RF MEMS switch's reliability. Polymers are especially prone to absorbing moisture, but this can be driven out by heating samples to above boiling point. Moisture in plastics has the effect of degrading electrical performance by increasing the high-frequency losses. This is a major reason why most plastics are not suitable for microwave electronics. LCP is a unique plastic as it has an extremely low moisture uptake, which allows reliable performance over time in the field without requiring additional expensive hermetic packaging.

Standard testing is followed to evaluate the outgassing characteristics of LCP materials [13]. This test measures LCP mass changes at 125 °C under vacuum for 24 hours. Under these conditions, moisture absorbed into the material will have converted from liquid to gas and have evaporated away onto a condenser plate. Outgassing test results are given in terms of total mass loss (TML), collected volatile condensable materials (CVCM), and water vapor regain (WVR). The TML is the measured mass percentage difference before and after testing. The CVCM is the condensed mass measured on a collector plate as a percentage of the initial specimen mass. The WVR is obtained by placing the measured specimens in an environment where there is an additional 50% relative humidity at 23 °C for 24 hours; the WVR value is calculated as the percent increase in the specimen mass after humidity conditioning. Historically, a TML of 1% and a CVCM of 0.1% define the maximum levels for spacecraft materials.

Table 2.1. Outgassing specifications and results for LCP [16]

ASTM E595 (1999) Outgassing tests	Specification	LCP
Total mass loss (TML)	< 1%	0.038%
Collected volatile condensable materials (CVCMs)	< 0.1%	0.004%
Water vapor regain (WVR)	n/a	0.031%

As seen from Table 2.1, the experimental results demonstrate that LCP has 0.038% TML and 0.004% CVCM, and thus satisfies spacecraft requirements. More importantly, even though LCP is a polymer material it has negligible outgassing levels. This means LCP is suitable for stable microwave performance and RF MEMS switch packaging.

2.4.2 Permeability and leak-rate package testing

Large-cavity test packages built fully with LCP were subjected to gross- and fine-leak testing in order to show directly that LCP can be used for near-hermetic packaging. There was a high surface exposure of LCP, so that the reported leak rates represent a worst-case scenario, whereas in practice electronic packaging has a significant metal covering in order to limit radiation effects as well as to offer good ground connection. Further, devices with significant hermetic concerns may be fabricated with integrated getters and dissicant for good measure. Representative test LCP packages have been fabricated as shown in Fig. 2.21. In order to fabricate the test part, a 1 mm thick LCP piece having an area of 1 in × 1 in is first formed by laminating 10 layers of 4 mil thick LCP. This is done four times to create four identical 40 mil pieces. Secondly, a cutout with area 0.5 in × 0.5 in is punched out and discarded from two LCP pieces. Thus a cavity is formed where LCP has been removed. Thirdly, four LCP pieces are arranged so that the pieces with cutouts are placed between two uncut pieces. In this manner, the cavity will be fully enclosed by LCP when the seams are sealed. Lastly, the stack is laminated through thermo-compression, so that the LCP material melts across the seams. The final package outer dimensions measure 1 in × 1 in × 4 mm, and the package contains a 0.5 in × 0.5 in × 2 mm cavity. A package is shown in Fig. 2.22. The packages are intentionally surface marked with various indentations, for traceability. The perspective view clearly shows the different layers that were individually and manually cut to create the test package stack-up.

Leak testing is performed following Mil-Std-883, Method 1014. In order to check the leak rates, two tests, known as the gross- and fine-leak tests, need to be performed. Fine-leak testing is performed under a 75 PSIA (pounds per square inch including atmospheric pressure) helium bomb for 125 minutes. Results are taken from a mass spectrometer, which is calibrated daily to be accurate to 2×10^{-8} atm cm^3/s helium. Since LCP is a thermoplastic material, its solubility is nonzero. In

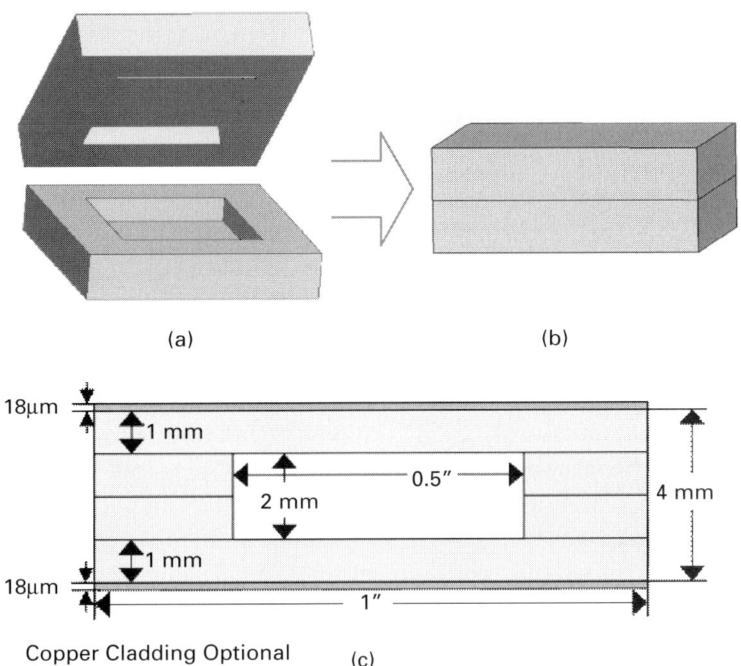

Fig. 2.21 (a) Pre-stack perspective package diagram, (b) perspective package view, and (c) cross-sectional diagram of LCP package.

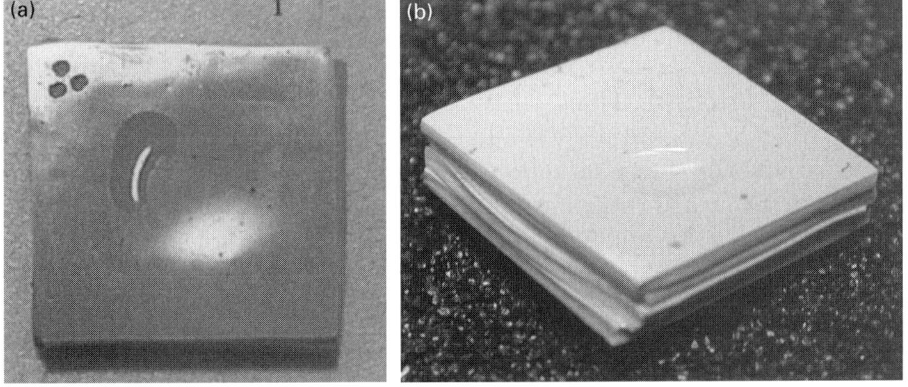

Fig. 2.22 (a) Top-down view of LCP package, and (b) perspective photograph of package.

other words, during bombing helium will adsorb onto the LCP package surfaces. Test packages have a ~0.3 cm^3 cavity volume, which is much larger than would typically be required in component-level packaging. In this case, we designed the cavity to be large enough to be compatible with traditional leak testing. The volume requirements for leak testing are set by the minimum space required to store a substantial amount of helium. A problem with testing smaller cavity devices is

Table 2.2. Leak-rate test results after immediate removal from bombing and after a three-hour dwell

	Measured immediate leak rate (atm cm^3/s He)	Measured leak rate after 3 hr dwell (atm cm^3/s He)	Gross leak (P or F)
Solid block	3.2×10^{-7}	5.0×10^{-8}	P
Package 1	3×10^{-7}	3.6×10^{-8}	P
Package 2	3×10^{-7}	3.6×10^{-8}	P
Package 3	3.2×10^{-7}	4.0×10^{-8}	P

that, for larger leaks, all the indicant helium would disappear before a measurement could be performed. To check that large leaks do not exist, gross-leak testing is performed. Gross-leak testing is a pass or fail test performed after fine-leak testing, as it is the more aggressive of the two tests. Gross-leak testing is performed under a 75 PSIA perfluorocarbon fluid bomb for 125 minutes for cavities under 0.05 cm^3. Samples are then placed into a bubble chamber containing a second, higher-boiling-temperature, perfluorocarbon fluid. The temperature of the bubble chamber is set to above the boiling temperature of the indicant fluid in order to convert all that fluid to gas. Unlike fine-leak testing, gross-leak results are pass or fail. Observations for bubbles from the sample are performed; the presence of bubbles indicates that a gross leak exists.

Leak-rate results are shown below in Table 2.2 with, for comparison, results for a solid LCP block which does not have a cavity that can be affected by leaks. The packages were found to pass the gross-leak tests, indicating that they were free from large leaks. The LCP cavity packages were found to have an average fine-leak rate of $(3.07 \pm 0.2) \times 10^{-7}$ atm cm^3/s helium. By comparison, the solid LCP block was found to have $(3.2 \pm 0.2) \times 10^{-7}$ atm cm^3/s helium. The measured difference between the solid and cavity block leak-rates is $(1.3 \pm 0.4) \times 10^{-8}$ atm cm^3/s. This difference reflects the portion of the actual leak-rate originating from the cavity and is significantly below the minimum required by military standards. Moreover, the differences are even lower than the equipment's 2×10^{-8} atm cm^3/s helium accuracy limit.

A further comparison was made between leak rates conducted immediately after testing and those conducted three hours later. Interestingly, the solid-block virtual leak rates only reach acceptable levels after a three-hour dwell time, as shown in Table 2.2. Using an appropriate three-hour dwell to remove virtual leak effects, all the cavity packages showed leak rates less than 5×10^{-8} atm cm^3/s He, which meets military standards.

2.5 Conclusions

In this chapter, we have shown that the chemistry of LCP directly provides the attractive properties of flexibility and near-hermetic structure. Electrically, LCP has a low dielectric constant and loss tangent, which facilitates microwave packaging, and transmission lines have been demonstrated on LCP to more than 100 GHz. Physically, LCP laminates provide a very convenient form factor, allowing for sturdy planar construction. Environmentally, LCP shows excellent moisture resistance and low leak rates. The outgassing rates for LCP are found to meet historical requirement levels for space application. Further, we have demonstrated that LCP cavities can achieve a leak rate of about 5.0×10^{-8} atm cm^3/s and can be used as reliable packages for microwave and millimeter-wave applications.

References

[1] A. Donald, A. Windle, S. Hanna, *Liquid Crystalline Polymers*, 2nd edition, Cambridge University Press, 2006.

[2] M. M. Tentzeris, J. Laskar, J. Papapolymerou, *et al.* "3-D integrated RF and millimeter-wave functions and modules using liquid crystal polymer (LCP) system-on-package technology," *Transactions on Advanced Packaging*, vol. **27**, no. 2, pp. 332–340, May 2004.

[3] Stefanie Romhild, "Transport properties and durability of LCP and FRP materials for process equipment," Dissertation, University of Stockholm, 2010.

[4] M. Chen, A. Pham, C. Kapusta *et al.* "Development of multilayer organic modules for hermetic packaging of RF MEMS circuits," in *Proc. IEEE MTT-S Int. Microwave Symp. Dig.,* San Francisco, June 2006, pp. 271–274.

[5] B. Farrell, M. Lawrence, "The processing of liquid crystalline polymer printed circuits," in *Proc. IEEE Electronic Components and Technology Conf.*, May 2002, pp. 667–671.

[6] D. Thompson, O. Tantot, H. Jallageas, G. Ponchak, M. Tentzeris, J. Papapolymerou, "Characterization of liquid crystal polymer (LCP) material and transmission line on LCP substrates from 30 to 110 GHz," *IEEE Transactions on Microwave Theory and Techniques*, vol. **52**, no. 4, pp. 1343–1352, April 2004.

[7] K. Jayaraj, T. Noll, D. Singh, "RF characterization of a low cost multichip packaging technology for monolithic microwave and millimeter wave integrated circuits," in *Proc. URSI Int. Signals, Systems and Electronics Symp.*, October 1992, pp. 443–446.

[8] G. Zou, H. Gronqvist, P. Starski, J. Liu, "High frequency characteristics of liquid crystal polymer for system in a package application," in *Proc. IEEE 8th Int. Advanced Packaging Materials Symp.*, March 2002, pp. 337–341.

[9] G. Zou, H. Gronqvist, J. Starski, J. Liu, "Characterization of liquid crystal polymer for high frequency system-in-a-package applications," *IEEE Transactions on Advanced Packaging*, vol. **25**, pp. 503–508, November 2002.

[10] Z. Wei, A.-V. Pham "Liquid crystal polymer (LCP) for microwave/millimeter wave multilayer packaging," in *Proc. IEEE MTT-S Int. Microwave Symp. Dig.*, Philadelphia, June 2003, pp. 2273–2276.

[11] K. Takata, A.-V. Pham, "Electrical properties and practical applications of liquid crystal polymer flex", in *Proc. IEEE Polytronic 2007 Conf., Tokyo*, January 2007, pp. 67–72.

[12] K. Nanbu, S. Ozawa, K. Yoshida, K. Saijo, and T. Suga, "Low temperature bonded Cu/LCP materials for FPCs and their characteristics," *IEEE Transactions on Components and Packaging Technologies*, vol. **28**, no. 4, December 2005.

[13] "Standard test method for total mass loss and collected volatile condensable materials from outgassing in a vacuum environment," ASTM-E 595–93, 1999.
[14] D. H. Weinkauf, "Gas transport properties of copolyesters II," *Journal of Polymer Science: Part B: Polymer Physics*, vol. **30**, no. 8, pp. 837–849, 1992. The values given therein assume that leakage is entirely due to permeation effects.
[15] A. C. Chen, M. J. Chen, A.-V. Pham, "Design and fabrication of ultra-wideband baluns embedded in multilayer liquid crystal polymer flex," *IEEE Transactions on Advanced Packaging*, vol. **30**, no. 3, August 2007.
[16] M. J. Chen, A.-V. H. Pham, N. A. Evers *et al.* "Design and development of a package using LCP for RF/microwave MEMS switches," *IEEE Transactions on Microwave Theory and Techniques*, vol. **54**, no. 11, November 2006.
[17] M. Chen, N. Evers, C. Kapusta *et al.* "Development of a hermetically sealed enclosure for MEMS in chip-on-flex modules using liquid crystal polymer (LCP)," in *Proc. ASME Interpack '05, Part C*, San Francisco, July 2005, pp. 2057–2060.
[18] http://www.rogerscorp.com/acm/products/17/ULTRALAM-3000-Series-Liquid-Crystalline-Polymer-Circuit-Materials.aspx
[19] Greenhouse, Hal, *Hermeticity of Electronic Packages*, New York, Willam Andrew Publishing, 2000.
[20] IPC-TM-650 Test Methods Manual.
[21] ASTM D6862 Standard test method for 90 degree peel resistance of adhesives.

3 Fabrication techniques for processing LCP

In this chapter we describe and discuss the processes used to fabricate liquid crystal polymer (LCP) for microwave packaging. The normal fabrication techniques have had to be in some cases modified or even reinvented for LCP owing to the unique challenges in working with this inert material. While LCP's chemical inertness makes it attractive for maintaining electrical performance over wide-ranging environmental conditions, it also means that LCP can be difficult to use in wet chemistries. These issues are complicated further by the fact that LCP is a plastic that can be easily flexed in thin-film form. An inherently anisotropic chemical structure adds an additional complication.

This chapter will progress through the different techniques available to form multilayer circuits and packages using LCP. In section 3.1 we discuss the available LCP formats, which include pellets and laminates. In section 3.2 we explain how copper cladding is added to LCP laminates and made ready for printed circuit board (PCB) processing. Section 3.3 covers the whole process, from laminate material to complete printed circuit board, using standard PCB technology. In section 3.4 we describe advanced process techniques that allow for complex specialized constructions. Lastly, section 3.5 gives a chapter summary.

3.1 Introduction to LCP material availability

LCP materials are available in resin, pellet, and laminate form. The resin and pellet forms are attractive for injection molding. Laminate LCP circuits are planar sheets that are ready for board manufacturers to process; they offer a convenient flat format for most microwave circuits.

3.1.1 Resin and pellet form

LCP resins and pellets are available from a number of commercial vendors [7, 20,21]. The pellets are essentially small cylinders (about 1 to 3 mm^3 in volume) of LCP that can be used to form LCP on different platforms. LCP materials may be heated and cooled in order to melt and reshape their physical form. This is useful since laminates can be formed out of pellets. In one simple laminate construction, pellets may be placed onto a baking tray and baked in an oven to form an LCP sheet.

Plaques are also available from LCP manufacturers for the purposes of characterization, e.g. for cavity resonators for dielectric and loss tangent characterization as discussed in the previous chapter.

The polymer strands within LCP are well known to align themselves along any direction of flow. When forming solid LCP pieces from the liquid state, special consideration may be required to prevent single-direction flow, which can cause parallel strands to form. A structure formed with parallel strands may easily tear along these lines and may exhibit weakness under transverse loads. Resin and pellet materials are attractive in applications where it is acceptable to melt LCP and reform the material within a mold. This can be performed using casting, transfer molding, or injection molding, the differences between which are based on how material is poured into their molds. Applications involving molding with resins and pellets will be discussed further in section 3.4.6.

3.1.2 Unclad LCP laminate form

The bulk of this chapter focuses on process techniques for LCP in laminate form. These are attractive for the flexible electronics industry because of their standard format, which is similar to rigid PCBs. They may be processed using techniques already in existence for flexible circuits. Numerous vendors now understand, and are comfortable with, printing multilayer LCP circuits at costs comparable with those for rigid PCBs. LCP laminates are available in both roll and sheet form.

Bare, unclad, LCP laminates are also commercially available [22]. Laminates may be formed in a process using a counter-rotating die [2, 23]. In this process, LCP is extruded through a rotating can having regular holes that allow LCP to enter. Immediately behind this first can is a second can rotating in the opposite direction. As LCP is forced out along the edges of these cans, the LCP is *biaxially* blown out and formed. The resulting film has polymer strands aligned along two directions, which gives LCP similar x- and y- dimension properties. The process is illustrated in Fig. 3.1 [1]. In this figure, LCP material is fed in sideways through a counter-rotating die seen at lower left. Material is then blown upwards, by means of the feedstock air, to form an LCP bubble. The LCP is cooled in bubble form and fed through pinch rolls that flatten the bubble. The flattened bubble is then cut on both sides in order to form two separate sheets (the figure shows one sheet), which are wound up into rolls.

3.1.3 Copper-clad LCP laminate form

In order for board manufacturers to build LCP circuits, they generally require laminate materials to be supplied with copper cladding. LCP laminates are commercially available in varying LCP thicknesses. Copper-clad LCP laminates are commercially available from a number of vendors [24, 25]. These laminates are

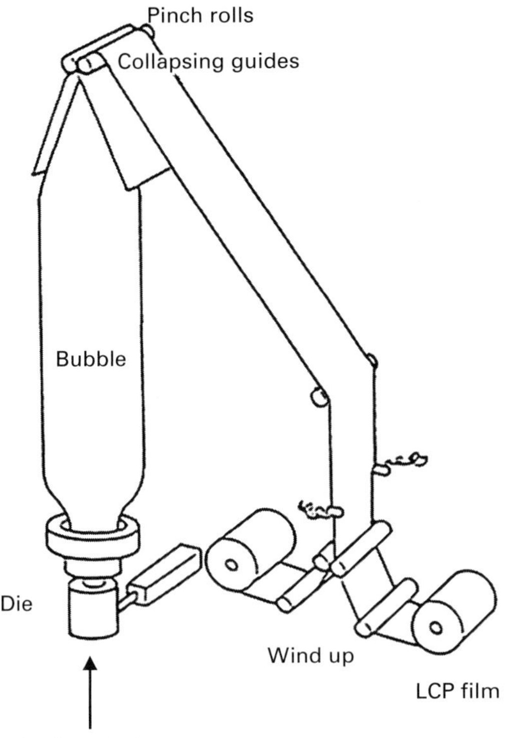

Fig. 3.1 LCP film formation [2] (courtesy of Nippon Steel Chemical Corporation).

designed to minimize the copper surface roughness and to provide excellent peel strengths between the copper and LCP. LCP may also be modified through additional thermal and chemical treatments to provide desirable qualities. Currently, vendors offer low- and high-melting-temperature materials with a variety of copper cladding configurations to accommodate most multilayer module structures. More information on how these manufacturers provide copper cladding on inert LCP is described in section 3.2 below.

3.2 Metallization on LCP

It has already been mentioned that LCP laminates with single- and double-clad Cu are commercially available for ease of use in a variety of thicknesses. In addition, as we will discuss in this section, an engineer may apply custom metallization, which offers the benefit of increased design flexibility. LCP laminate manufacturers may apply Cu cladding using a combination of chemistry, temperature, and rolled mechanical pressure. Casting, sputter-plating, and lamination are three methods that have been successfully used to create foil-copper-clad laminates (FCCLs).

3.2.1 Metal adhesion

Since LCP is chemically inert, metals do not readily adhere to its surface. Fortunately, researchers have developed methods to promote metal adhesion to LCP. Copper and gold are common metals for microwave circuits; other metallization is also available when needed for specialized applications. Without an adhesion layer, however, Cu and Au adhere poorly when deposited directly onto LCP.

In a common method of "adhesion promote," a titanium (Ti) seed layer 200 Å thick may be first deposited, using electron-beam (e-beam) physical vapor deposition (PVD). This initial layer adheres well to LCP and acts as an adhesion promoter for additional layers of thicker Au or Cu. Many processes described below either implicitly or explicitly have an adhesion-promote step.

3.2.2 Casting

In casting, a Cu foil roll is fed across two spindles and molten LCP is coated onto the foil. In this context, the Cu thickness and roughness are fairly precisely specified. The LCP thickness, however, is dependent on the roll-feed speed and the rate of material deposition. Furthermore, this technique has particular implications for material anisotropy, as LCP has a tendency to align itself in one direction at the molecular level.

3.2.3 Sputter-plating

In sputter-plating, a roll of LCP film is fed across two spindles within a vacuum chamber. Figure 3.2 illustrates a sputter-plating process that deposits copper onto LCP. The LCP film is fed from a roll and first subjected to plasma cleaning in an etching unit. This leaves the material slightly roughened and chemically activated. Sputtering then follows this cleaning. In this case, the sputtering material serves as the adhesion promote. As the spindles turn, the entire LCP roll is metallized with a seed layer along one side. This seed layer is very thin and serves as an adhesion layer between LCP and Cu. If desired, the LCP film may be removed at this point and electroplated with Cu to a desired thickness. Alternatively, as illustrated in Fig. 3.2, the LCP can continue on to a lamination step provided by the rolling unit.

3.2.4 Metal lamination

Lamination is a readily available technique to foil clad an LCP laminate. In lamination, LCP and Cu foil are pressed together to form the FCCL. Copper foil is readily available from a number of vendors in a variety of thickness and qualities. At the start of the process, the LCP is wound onto a carrier roll. Next, this LCP roll is simultaneously fed with a Cu foil roll through compression-lamination press rollers.

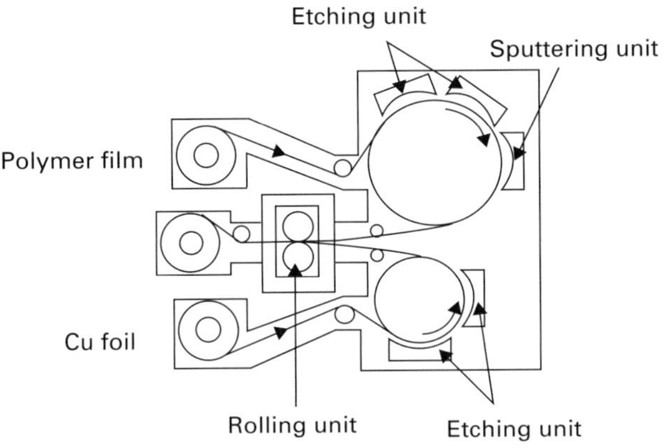

Fig. 3.2 LCP sputter plating [4] (© 2005 IEEE).

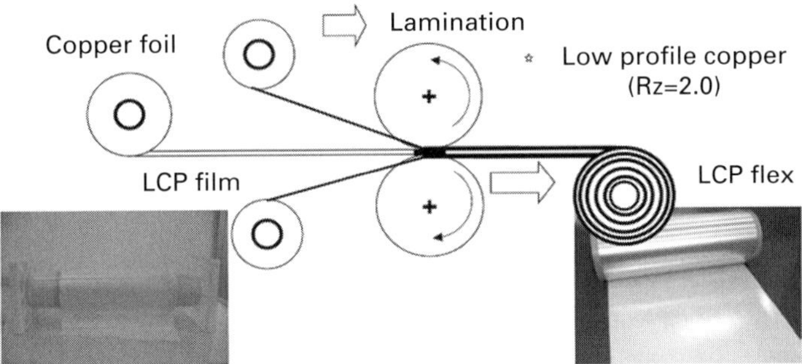

Fig. 3.3 LCP lamination copper cladding [2] (courtesy of Nippon Steel Chemical Corporation).

Heat is generally applied during lamination, but processing may also be performed at room temperature, as shown in [4]. Lastly, the carrier roll may be stripped away as necessary. This process is fairly common and provides excellent control of LCP's material properties. Further, lamination may be performed with two Cu foil rolls to form double-clad laminates. Figure 3.3 illustrates the double-sided copper cladding of LCP, with lamination using pre-formed LCP.

In a variation of this process, press-rolling is performed with liquid or molten LCP laminates. For LCP, this process may have strong implications for anisotropy if the raw LCP material is known to align strongly along the flow direction. This can mean that LCP laminated in such a process, which simultaneously forms a laminate and its Cu cladding, may suffer from a poor tear strength along certain directions. Further, these parts will exhibit an anisotropic CTE along various directions, which may or may not be an issue depending on the desired application.

Yet another variation of this process involves completely foregoing roll lamination equipment and instead using a lamination press. High-quality copper foils with a smooth surface are available for LCP but at a higher cost. These high-grade materials can be specifically requested from vendors. Other small-scale laboratory results have been shown to achieve a measured root-mean-square (RMS) roughness for Cu on the order of 0.1–0.35 μm. A 0.35 μm roughness is on the same order as the skin depth at 36 GHz [5].

3.3 Standard PCB processing

One of LCP's major advantages is its low-cost fabrication, which is possibly in part due to its being compatible with processes supported by many PCB fabrication houses. When identifying a new PCB fabrication partner, it may save time to deal with houses possessing prior LCP processing experience. Most PCB houses that support flex circuits with high-density interconnect (HDI) already own laser drill equipment and are either already familiar with LCP processing or can bring it online quickly.

This section supplies an overview of the necessary LCP processing steps that are performed by PCB fabrication houses. In section 3.3.1 we describe the steps in standard PCB processing, which first involves board cleaning. In section 3.3.2 we explain how, using photolithography, metal patterns are formed on each board surface, and in section 3.3.3 we explain how the top and bottom metal layers are electrically connected through laser via-drilling. Section 3.3.4 presents a description of how multiple boards are sequentially laminated to create a multilayer board. Lastly, in section 3.3.5, we consider how plating and finishing are performed.

3.3.1 PCB standard cleaning

LCP may readily attract dust particles, which can be difficult to see prior to processing. However, after lamination, significant amounts of dust can embed themselves into the LCP circuit and are readily visible unless proper precautions are taken. Contamination by particulates is of further concern because their interference with the subsequent PCB processes could cause suboptimal RF performance or even open or short circuits.

In order to aggressively remove organic contaminants, LCP may be wiped with acetone using the dust-free wipes common to cleanroom electronic processing. This process involves applying solvent to dust-free wipes and then wiping both sides of the LCP material. If one is working in a non-cleanroom environment, it may also be necessary similarly to wipe the working surfaces prior to setup. One may recognize this generic cleaning solvent as having odor fumes similar to nail polish remover, which is primarily acetone. In order to remove any residue that the acetone may leave, an additional isopropyl alcohol wipe is performed. Rubbing-alcohol is basically diluted isopropyl alcohol. Isopropyl alcohol evaporates cleanly leaving the LCP board pristine. Safety precautions similar to those above should also be applied when isopropyl alcohol is used.

3.3.2 PCB standard photolithography

Microwave applications commonly use Cu and Au because of their favorable conductivities, 5.8×10^7 S/m and 4.1×10^7 S/m, respectively. Metal shapes are formed using lithography. Lithography may be performed by first creating a mask pattern that controls the areas where light may pass. In a process known as *photolithography*, a sacrificial layer consisting of a light-sensitive material known as *photoresist* forms a pattern on metal. The photoresist is spun onto LCP and pre-baked to create a uniformly thick coating. Next, it is aligned to the mask and exposed to light. The areas of photoresist exposed to the light undergo chemical cross-linking. Both light and mask are removed, and a full bake is performed. Next, a *developer* solution is applied to create the photoresist pattern. At this point, the metal-clad LCP is etched with protective photoresist areas where the metal is desired. LCP materials are commonly etched involving *wet chemistry*. For example, ferric chloride is commonly used to etch copper, and a number of different chemistries may be used to etch gold. Once etching is complete, the photoresist is stripped away to give the finished pattern on the LCP.

If desired, a liquid photoimageable (LPI) solder mask may be screened-on in areas of non-microwave use to serve as wear-protection and to prevent accidental shorting during the later soldering processes. Solder mask makes typical PCBs appear green, although other colors are available. Further, there are solder masks specifically designed to be compatible with flex. Another term for this layer is the *covercoat*. Alternatively, a *coverlay* material is used with adhesive to encapsulate the traces.

When one is developing applications around LCP, achievable line width and space definition is a concern. Isotropic etch characteristics are inherent to wet etching. This generally means that line widths cannot be defined any more narrowly than the metal thickness. Fortunately, the thickness of the metal after plating is typically many orders greater than the skin depth: at high frequencies, current travels only along the metal surfaces. The conductor loss is mainly unaffected by the copper's thickness even when this gets down to 9 μm. This means that the metal on LCP may often be thinned down to micrometers in order to achieve a tight line width and spacing without any concerns arising about electrical losses at high frequency. Line-and-space width features down to 1 mil have been demonstrated. Figure 3.4 shows the patterning of 9 μm thick copper on LCP with 10 μm wide lines at a 40 μm pitch.

3.3.3 PCB standard drilling and milling

Vias serve as convenient structures to pass electrical signals from one layer to another. They consist of drilled holes that are metal plated. Vias built with LCP laminates may be categorized as blind, buried, or plated through hole (PTH). Blind vias are formed starting on an outer layer and may be defined as not piercing through the backside Cu. This provides a convenient technique to maintain a

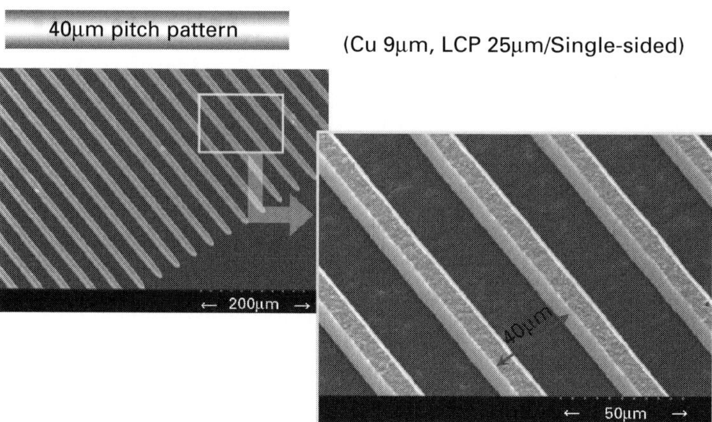

Fig. 3.4 Metal pattern of lines and spaces on LCP [16] (© 2007 IEEE).

near-hermetic structure. Buried vias are those located within a multilayer structure. They are formed in advance on core layers prior to lamination. Plated through hole vias are drilled through the entire stack, and the entire length is plated. Wet chemistry has not been found to be effective in etching LCP, owing to its inert chemical properties. Drilling may be performed with either a laser or a mechanical drill. Further discussion on via techniques for LCP and screening for excessive drilling on blind vias in manufacturing will be discussed below.

Laser ablation

Laser ablation is a convenient method for drilling through LCP. LCP readily absorbs laser energy, and ablation has been demonstrated on LCP using both YAG and CO_2 lasers. Lasing is also convenient when LCP is backed with a metal, for the metal acts as an etch stop by reflecting and scattering the laser light. A drawback with laser techniques is that ash residue and discoloration may be left behind. The discoloration is limited and has not been reported to present any problems. Ash may be removed by mechanical scrubbing with isopropyl alcohol. Laser ablation represents the most accurate method currently available for carving intricate volumes. However, when large volumes of LCP need to be removed through thick layers, laser ablation may become cost and time prohibitive. For such applications, mechanical drilling is an attractive alternative for removing the bulk of the material at reduced cost. Figure 3.5 demonstrates laser ablation on LCP to form both blind vias and through hole vias. Figure 3.6 demonstrates successful via plating on both blind and through hole vias.

Mechanical drilling and milling

Mechanical drilling with controlled *Z*-axis milling machines has also been demonstrated to work well in forming blind vias in LCP. Milling machines are able to cut laterally in order to route shapes. The mill bits spin and appear similar to drill

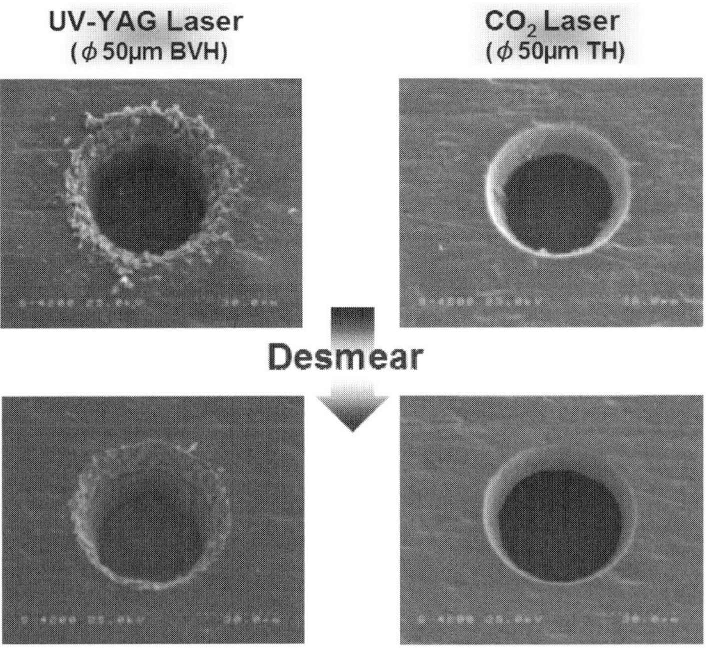

Fig. 3.5 Laser ablation of 50 μm vias in LCP [16] (© 2007 IEEE). The holes were smoothed using a desmear process.

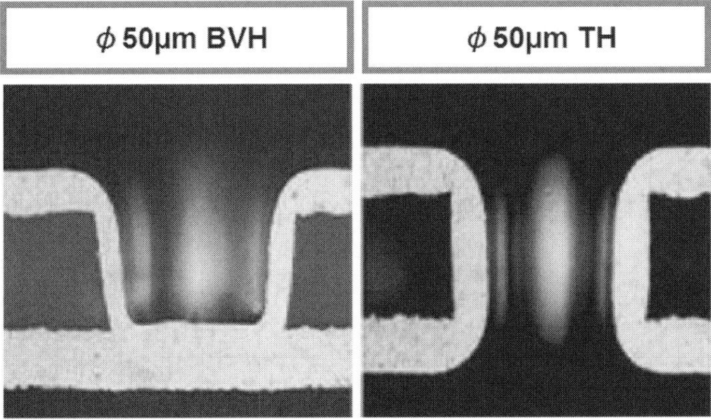

Fig. 3.6 Formation of a 50 μm blind via hole (BVH) and a plated through hole (TH) in LCP.

bits but are designed to wear appropriately given their lateral cutting action. This is attractive in reducing costs but typically requires a thicker copper backing owing to the lower tolerances compared with those in laser drilling. We have found consistent mechanical depth-controlled drilling when the drill is landing on 4 mil thick Cu. Thinner metal-jacket builds using 2 mil thick Cu have shown both open connection problems and insufficient drilling when forming vias, as well as telegraphed bumps on the opposing Cu side if the drilling is excessive. When removing large volumes of material, it is important to control drill speeds. This will prevent the LCP from gumming up the mill bit or forming "stringers," which are unintended strands of material stripped from the LCP by the drill.

Drill aspect ratios around 1 : 1 are offered on thin, i.e. 4 mil thick, LCP layers since the laser and minimum mechanical drill-bit diameters are on the same order. For thick multilayer substrates, 10 : 1 aspect ratios may be achieved. The aspect ratios on thick structures are limited by the requirement for sufficient circulation of the plating solution. If necessary, refined etch laser techniques can be used for trimming edges that have been mechanically drilled.

Evaluating LCP drilling

In manufacturing, parts may need to be screened for pinhole features resulting from excessive blind via drilling. In one sampling this was a low-occurrence event, which occurred in three of 144 inspected parts, or ~2% of parts. Material from a supplier needs to be explicitly screened for this defect. Figure 3.7 demonstrates a visual screening method, performed on optical comparator equipment, for quickly identifying over-drilling failures.

3.3.4 PCB standard multilayer lamination

A *thermoplastic* allows repeatable melting under repeated exposure to heat. By contrast, a *thermoset* material does not have any specified melting temperature owing to its manufacture by irreversible cross-linking curing; rather, thermosets will carbonize under applied heat before melting occurs. LCP is a thermoplastic material, which allows it to melt and be used as an adhesive layer. This means that lamination-process temperatures must exceed LCP's melting temperature so that the material is allowed to flow, and this needs careful control. There are additional challenges for alignment registration when using a opaque material. In this subsection we will describe processes for both uniformly planar lamination, which is useful for multilayer sheets, and selective lamination to form seals.

One advantage of LCP films compared with other laminates is their inability to adhere to materials without external adhesives in a lamination process. Hence, multilayer structures can be formed with a homogeneous LCP dielectric. This homogenous LCP build-scheme reduces the electrical losses associated with adhesive materials that are normally characterized by high loss tangents. Further, this feature simplifies the packaging and electrical design for passive structures. Significant progress has been made on multilayer LCP lamination. For example, 16

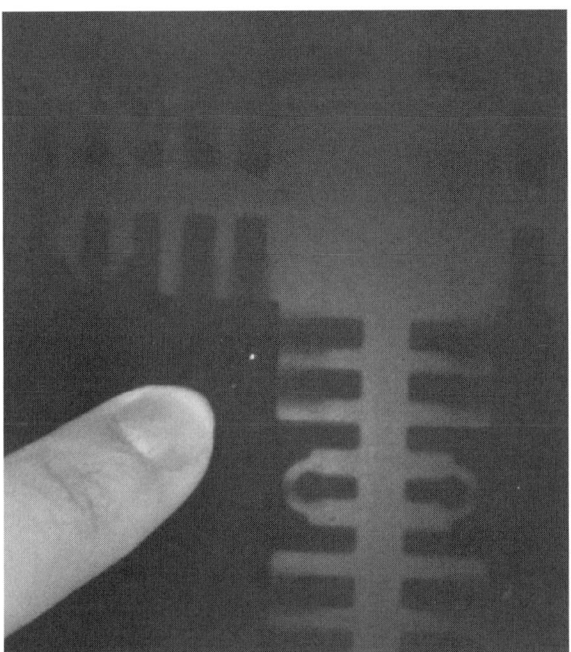

Fig. 3.7 Pin-hole detection (near finger tip) on an optical comparator (courtesy of Endwave Corporation).

LCP dielectric layers have been demonstrated containing various power, ground, RF, and resistive layers [11]. This multilayer build also showed novel via technologies, which allowed 60 μm diameter stubless vias to be formed using both conductive paste and plating techniques. These stack-ups have been shown to withstand pressure cooker tests and solder shock.

Bonding mechanism

Currently LCP adhesion mechanisms are not fully understood by chemists, but they are believed to be due to a combination of mechanical locking, Van der Waal's forces, and chemical bonding. In mechanical locking, LCP physically molds and locks around niches. Varying sizes of niche will always exist along a surface even if only at a molecular level. With regard to Van der Waal's forces, electrostatic effects at a molecular level create attractions that may produce partial adhesion at a macro scale. With regard to chemical bonding, covalent C=O bonds correlating with adhesion strengths have been observed at LCP–Cu interfaces [4].

Heterogeneous material lamination

LCP may be laminated to copper, gold, glass, silicon, gallium arsenide, and printed circuit boards and also to itself. A general all-purpose laminate adhesive has been found to successfully laminate different materials to LCP. Figures 3.8 and 3.9 show a sample and cross-sectional diagram, respectively, for copper-clad LCP laminated

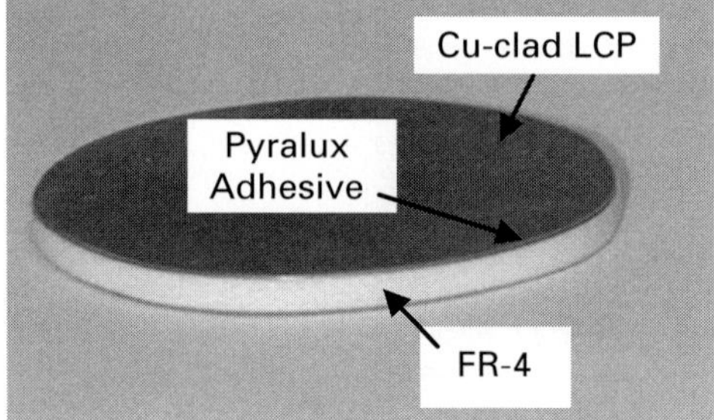

Fig. 3.8 Laminated one-inch disk consisting of LCP, adhesive laminate, and FR-4.

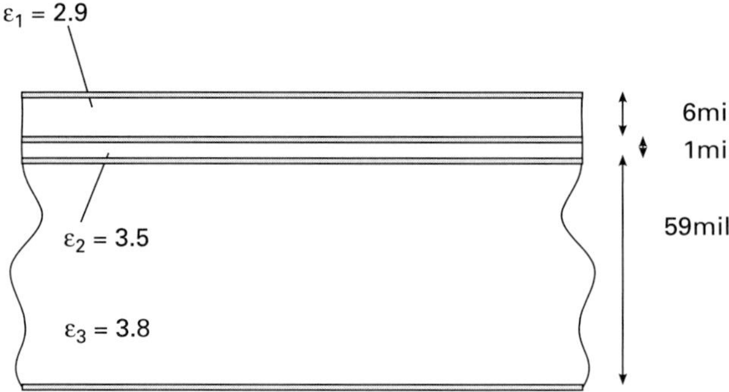

Fig. 3.9 Cross-section of the laminated one-inch disk of LCP, adhesive laminate, and FR-4.

to FR-4 using adhesive laminate. This technique allows an easy way to integrate microwave functionality on LCP with core PCBs carrying DC or low-frequency signals.

Homogeneous LCP lamination

In order to perform lamination with LCP as an adhesion layer, processing is required to exceed LCP's melting temperature. This will cause LCP to gradually change state from solid to liquid. Material datasheets on laminates will specify a *melting temperature* T_m. A common T_m value for LCP is around 240 °C, but this will vary depending on the particular blend. The T_m value for LCP is available as low as 180 °C. In actuality, the melting temperature is not well defined for LCP. For over a few tens of degrees around T_m, the viscosity will decrease as temperature increases, and LCP's morphology alters throughout the thermal process. For this reason, LCP lamination can be possible below T_m "melting" temperatures as long as increased pressure is applied. Consequently, the time, temperature, and applied pressure are critical parameters for controlling lamination.

A heated press may be employed in order to perform thermocompression lamination. Figure 3.10 shows a press developed for LCP lamination [12]. This press was

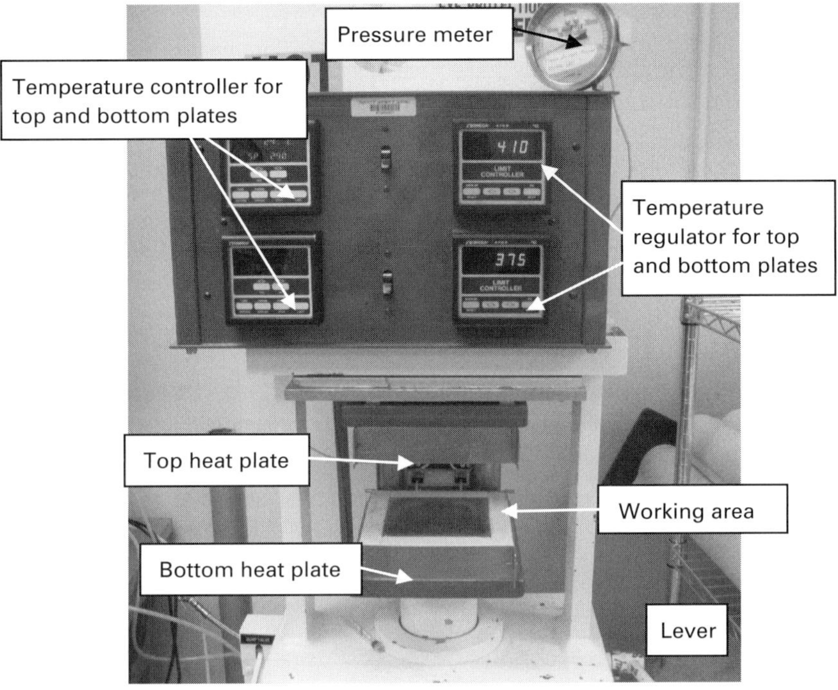

Fig. 3.10 Lamination press.

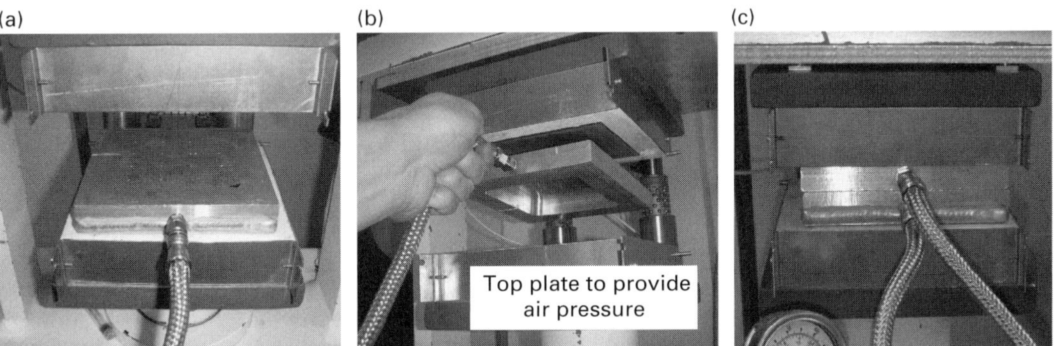

Fig. 3.11 (a) The bottom vacuum plate, (b) the top plate to provide air pressure, and (c) lamination in progress.

refitted from a wafer lamination press originally used for small outline integrated (SOI) wafer formation. Two heating plates are located above and below the working area. The press can provide up to 10 000 PSI. In the working area, a vacuum plate measuring 6 inches by 6 inches is inserted above the bottom heating plate (Fig. 3.11a), and another plate, also measuring 6 inches by 6 inches, with a drilled cavity in the middle and air pressure capability, is placed below the top heating

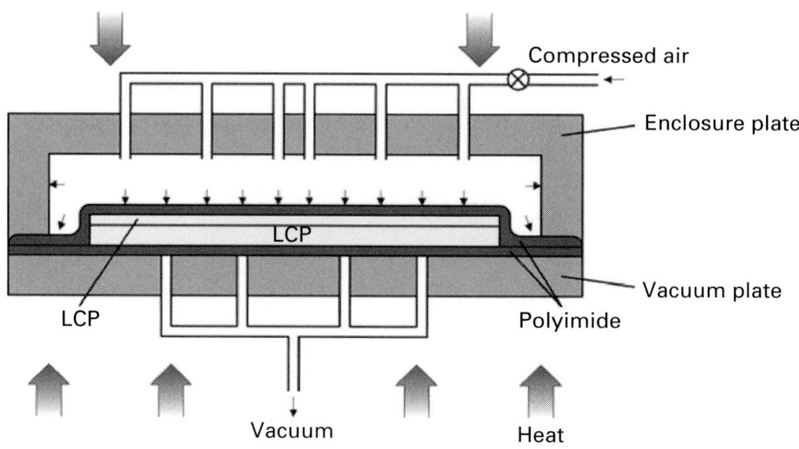

Fig. 3.12 Cross-section of the press during lamination [18] (© 2007 IEEE).

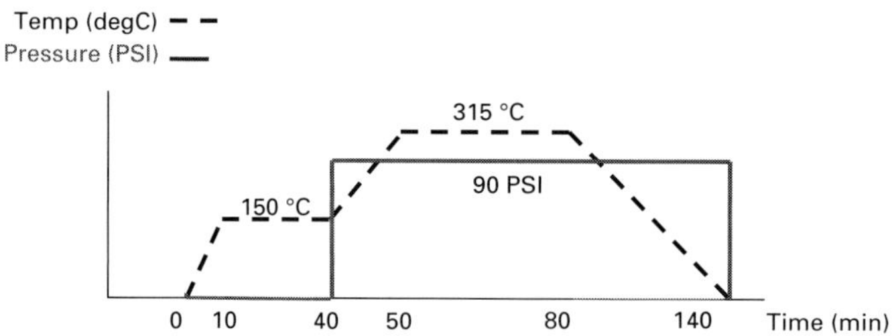

Fig. 3.13 Recommended lamination temperature and pressure profiles [19] (© 2011 IEEE)

plate (Fig 3.11b). Figure 3.11c shows the lamination press in use. Figure 3.12 shows a cross-section of the press during lamination. Polyimide films are placed above and below the LCP samples that are to be laminated, to prevent adhesion to the plates. During processing, a vacuum plate below the sample applies a negative air pressure to the polyimide films while a positive air pressure is applied from the top plate to constrain the sample. This application of pressure causes the polyimide films to form a mold around the sample that minimizes LCP reflow during the period when heat is applied. Figure 3.13 illustrates the recommended temperature profile in laminating high-melting-temperature LCP films. This specially controlled environment is critical because the window for quality LCP processing is relatively small; a few degrees of variation in temperature can make a large difference in the laminated stack quality. Appropriate heating element placement with less than 3 °C variation across the plates allows for uniform lamination.

Temperatures must be controlled to within 3 °C in order to obtain reliable, repeatable results. Thermal profiles are accurate to 1 °C variation across plate widths, as determined by direct temperature measurement, and this allows a repeatable process to be developed. Generally, samples are placed centrally on the lamination plates to minimize variations in temperature over their area. The lamination strength may be poor if the laminating temperatures are insufficient. However, it is found that if the laminating temperatures are excessively high then widely varying non-uniform lamination strengths occurr along the LCP–substrate interfaces. At extremely high temperatures, turbulent or boiling effects cause non-uniform lamination at adhesion interfaces and give an appearance of good bonds that are speckled in regions of generally poor lamination.

In multilayer stacks, materials should be interleaved with high- and low-melting-temperature LCPs. High-melting-temperature LCP layers should be used as the core, with double-sided metal printing; a low-temperature material should be used as the adhesive. If only PTHs are used then the entire stack may be pressed in one lamination step. However, it is more common in RF/microwave applications to require blind vias for performance reasons. This may mean that sequential laminations are needed to progressively build up the stack. When this is done, it is best to add material above and below the stack as symmetrically as possible for proper thermal control.

3.3.5 Surface finish

Solder masks are often unacceptable in areas where high-frequency signals pass. Instead, a plating finish is commonly applied in order to prevent surface corrosion and wear. Copper, in particular, is susceptible to oxidation corrosion. Gold is especially useful as a plated finish since it does not readily oxidize, can be directly electroplated onto copper, and provides a bondable surface. Common plating schemes for microwave circuits include, but are not limited to (1) electroless nickel and immersion gold (ENIG), and (2) electroless nickel, immersion palladium, and immersion gold (ENIPIG). The palladium in ENIPIG offers a diffusion barrier between Ni and Au to limit "black pad" effects that indicate nickel corrosion. The use of palladium means that less gold is required. Other variants of ENIPIG, such as ENEPEG etc., exist that differ in the actual mechanism for depositing palladium and/or gold. Both ENIG and ENIPIG plating offer a wire-bondable surface with excellent corrosion resistance. Hot-air solder leveling (HASL) is another process that may be encountered when one is developing flex cables; this involves dipping the PCB in solder in order to pre-tin each lead.

A phenomena known as gold embrittlement may occur where thick gold pads contact solder joints in component mounting. Embrittlement occurs where Au–Sn intermetallic compounds (IMCs) form. For this reason, the gold plating thickness should be minimized at any point where the solder is intended to contact the gold pads. A maximum of 3% Au to 97% SnPb solder, by weight, may be used before the onset of embrittlement. The density of gold is 19.3 g/cm^3 and the density of solder is 8.46 g/cm^3. This implies that the gold layer must be less than 1.3% of the solder thickness, assuming equal coverage of the gold and solder areas. For a solder joint 2 mil (50 μm) thick, the gold plating must be less than

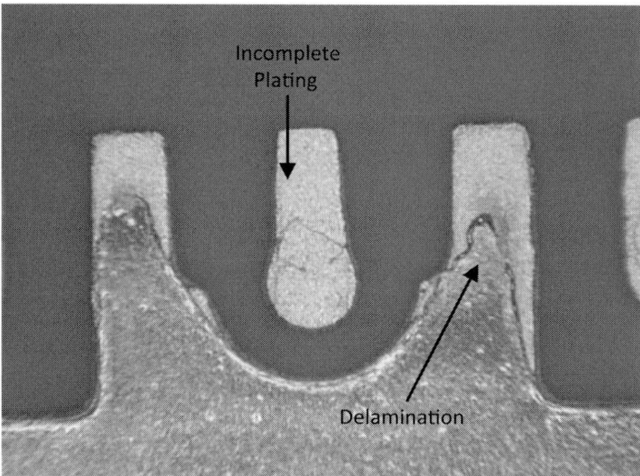

Fig. 3.14 Example of incomplete plating on QFN pads (courtesy of Endwave Corporation).

0.7 μm thick. The use of the new Pb-free solders is expected to change this ratio, but it is generally desirable to minimize the gold thickness for this reason and because of the higher material costs for gold. A scanning electron microscope and energy dispersive X-ray spectrometry (EDS) may be used to analyze the degree of embrittlement [6].

Another form of undesirable intermetallic compound (IMC) may form when aluminum contacts gold. This results in "purple plaque." This IMC is formed through Kirkendall voiding when growth rates differ between materials. Purple plaque is commonly introduced accidentally at wire bond interfaces where aluminum and gold contact. For simplicity, gold wires should be used with gold plating for microwave devices.

Figure 3.14 shows a photograph of an LCP QFN prototype package having a particularly unacceptable plating quality during initial prototyping. In this example, a Cu-clad LCP laminate was patterned and further electroplated to a desirable thickness. Besides cosmetic concerns, there may also be unknown long-term reliability issues related to surface treatment, cleaning, and plating delamination. The plating issues have since been resolved on the supplier side.

Pad flatness is an additional concern related to plating. This may have a direct negative impact on a package's electrical performance at microwave frequencies and above. Figure 3.15a shows an XYZ scatter plot of the pad surface positions on a reject 7 mm × 7 mm 28-lead air-cavity package, as measured with a microscope equipped with micrometer gauges. The Z-position was determined by adjusting for optical focus. Figure 3.15b shows a plot of the data in Fig. 3.15a along each of the four sides of the QFN. The data are shown normalized against the values for the outermost pads, to account for imperfect QFN mounting. The pad heights on the initial prototypes are given only to within ±0.5 mil, owing to variations in flatness due to inconsistent plating. However, this level of pad flatness is found

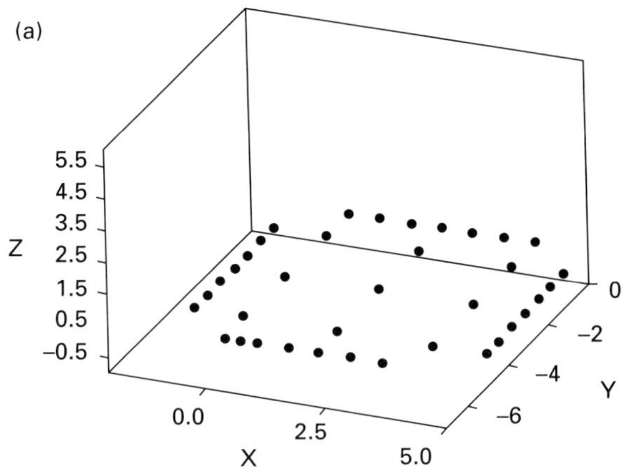

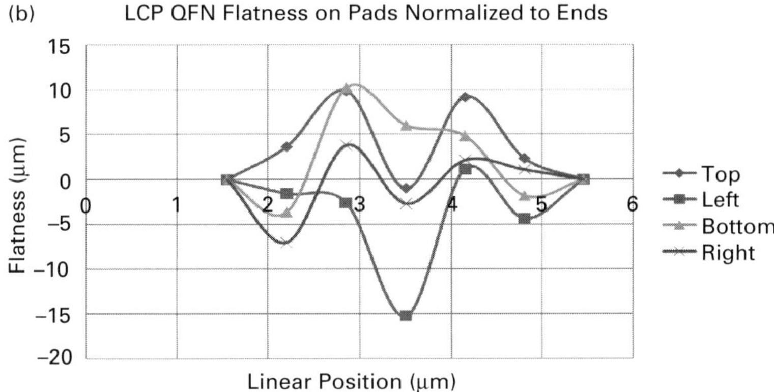

Fig. 3.15 (a) Three-dimensional scatter plot of QFN physical measurements and (b) two-dimensional plot showing the pad flatness (courtesy Endwave Corporation).

to maintain electrical performance, as the height differences can be absorbed into the solder joint.

3.3.6 Roll-to-roll processing

For volume or large-structure production, roll-to-roll technology may be used to handle flexible substrates [3]. Roll equipment can hold flexible material taut to provide accurate registration. Roll-to-roll processing allows for continuous metal deposition and patterning with the benefit of greater part throughput. LCP manufacturers accommodate this technology by providing customers with the option of receiving material in roll form. Essentially any process implemented to fabricate prototypes can be performed with roll-to-roll equipment to achieve attractive volume pricing. Figure 3.16 illustrates an example of roll-process lamination for LCP multilayer circuitry [2].

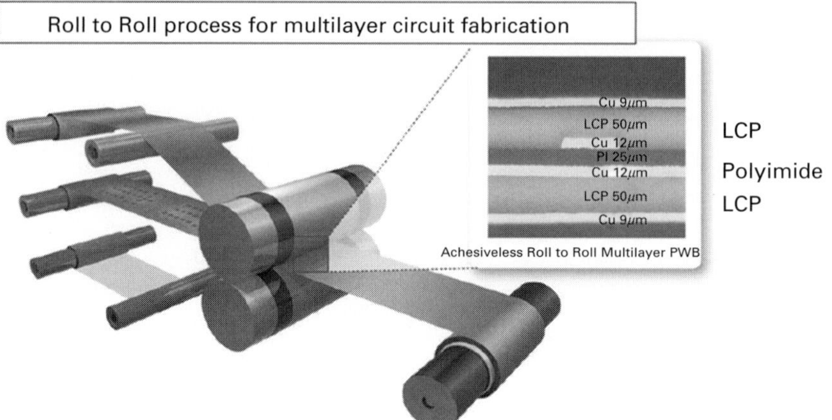

Fig. 3.16 LCP roll process demonstrating multilayer circuit fabrication [2] (courtesy of Nippon Steel Chemical Corporation).

3.4 Advanced LCP processes

LCP offers unique fabrication capabilities beyond what can be offered in standard PCB processes. In this section we describe various techniques developed for advanced RF/microwave circuits. These topics cover material handling, LCP machining, LCP air-cavity lid formation, selective LCP sealing, techniques to evaluate LCP lid sealing quality, and LCP molding.

3.4.1 Handling

Flexible materials require special handling because many processes and equipment rely on techniques based on rigid, planar, material. Special consideration needs to be given to prevent LCP from being wrinkled, bent, curled, or otherwise damaged when placed into equipment. We have observed that LCP laminates may roll up under their own CTE imposed stresses during metal deposition using either *electron beam physical vapor deposition* (EBPVD) or sputtering processes, when LCP is insufficiently mounted.

Photoresist-based mounting technique

One handling technique involves the use of photoresist to serve as an adhesion layer between LCP laminate and a rigid backing wafer. In essence, the photoresist is used to glue the LCP to a rigid substrate. This method has the benefit of being readily available and provides a convenient method to process LCP using wafer equipment. For this procedure, it is necessary to spin thick resist onto a rigid mechanical substrate using more solution and lower RPM spin rates than for semiconductor processes using photoresist. The LCP material may then be pressed onto the photoresist using a mask exposer in direct contact mode. The photoresist may be cured under standard oven conditions to allow the LCP to become fixed. Additional processing may be performed with LCP as a planar substrate. Once LCP processing is complete, the processed LCP laminates are removed using photoresist stripper. A drawback of this LCP mounting technique is that certain laboratory processes, including wet chemistry, require a stronger, more robust, mounting method than is offered by this photoresist adhesion method.

Frame mounting technique

A second, more robust, method for handling flexible laminates without large upfront tooling involves mounting LCP laminate panels into a rigid frame. The frames may be constructed using a rigid material such as stainless steel. An LCP panel may be fixed in its frame using either adhesive material or tooling holes. Fiducial features may be included in the frame in order to provide convenient alignment across multiple process steps.

The sawing steps present a situation where proper mounting is crucial. During the sawing steps, the LCP panels may be mounted using UV-curable tape in a similar manner to that in which wafers are mounted in a die saw. Sawing may be performed with either a mechanical saw or with laser ablation. With either drill technique, UV-curable tape provides a convenient format for maintaining the parts as an array.

3.4.2 Advanced LCP machining techniques

This section describes advanced LCP machining techniques that allow complex shapes and amounts of LCP to be removed in order to achieve a desired structure.

Mechanical stamping punch

A mechanical stamping punch tool can be used to cut specific LCP shapes. These tools consist essentially of two fitted shears, which allow for a particular shape to be cut away from sheets when the shears are clamped together. For research and development purposes, where the design is manually constructed and flexible, stamping tools can be picked up inexpensively at hobby stores that offer scrapbook supplies. This method provides a very convenient way to inexpensively create repeatable cut patterns. Very precise punch tools that allow for automation ability may also be designed and custom built for specific shapes.

Reactive ion etching

Reactive ion etching (RIE) may be used to remove the top thin-film LCP layers. Etch rates are very slow, and so etching can provide a very accurately controlled thinning technique on LCP to maintain specific thicknesses. It is possible to formulate the RIE recipe so as not to etch copper. Thus copper traces may be used as a mask to prevent the etching of the LCP material beneath it.

The RIE method under several hours of processing has been used to perform 25 μm of anisotropic LCP etching. Oxygen plasma etching under 350 W at 500 mT may be used to perform LCP etching at 0.22–0.27 μm per minute [9, 10].

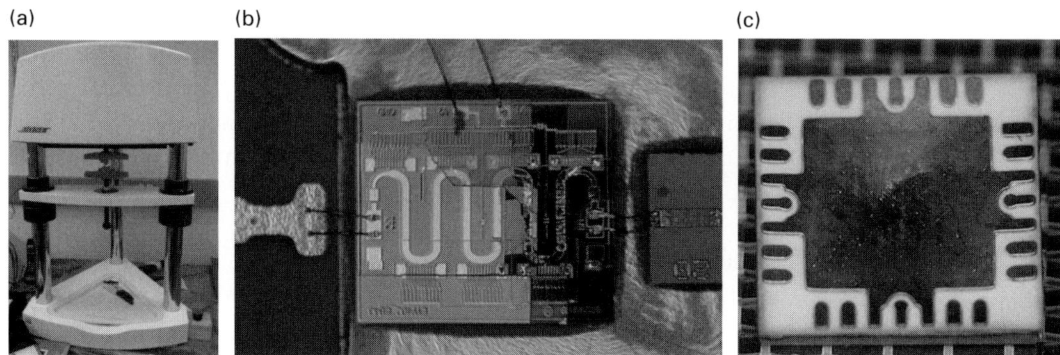

Fig. 3.17 (a) Mechanical load frame test setup, (b) cracked GaAs chip after test, (c) indented paddle surface after test (courtesy of Endwave Corporation).

Evaluation of machining for LCP packages

A common reason for milling LCP laminates is to minimize semiconductor bond wire lengths. One concern is whether these milled cavity packages offer sufficient robustness against mechanical damage. In an example discussed below, an LCP laminate was laser ablated for chip insertion, and the package was sealed with an LCP lid. In this case, the plastic domed lid protects the QFN on five of its six sides, while the Cu backing provides mechanical protection along the backside of the chip. A mechanical load frame, shown in Fig. 3.17a, is used to test whether the lid can withstand more than 250 N force without any damage. The remaining laminate surface is similarly measured to withstand 3.6 N force before chip crack (Fig. 3.17b) is observed. Note that after testing, which ran to 4.5 N, an indentation is observed, as shown in Fig. 3.17c. The measured force versus the displacement data is plotted in Fig. 3.18; the GaAs chip crack occurs at the discontinuity. The discontinuity represents the sudden relief as the chip cracks; this causes the force to drop as the displacement increases.

3.4.3 LCP air-cavity lid formation

In this section, case studies are presented on air-cavity lids formed with LCP. These are structures that physically provide a three-dimensional air cavity that may be attached to substrates containing circuitry. This provides the circuitry with the native-air dielectric environment required in many microwave applications. Firstly, we present a mechanically drilled air-cavity lid formed with LCP. Secondly, a version constructed using mixed laser and mechanical drilling techniques is discussed. In actual production the lid can be injection molded in high volume [26].

Mechanically drilled air-cavity lid

Mechanical machining has been demonstrated to form a low-cost LCP lid successfully. A lid is formed from a molded 3.2 mm (1/8 inch) thick LCP [7] board with a 355 °C melting point. Figure 3.19 shows an LCP sample. The LCP bulk is machined

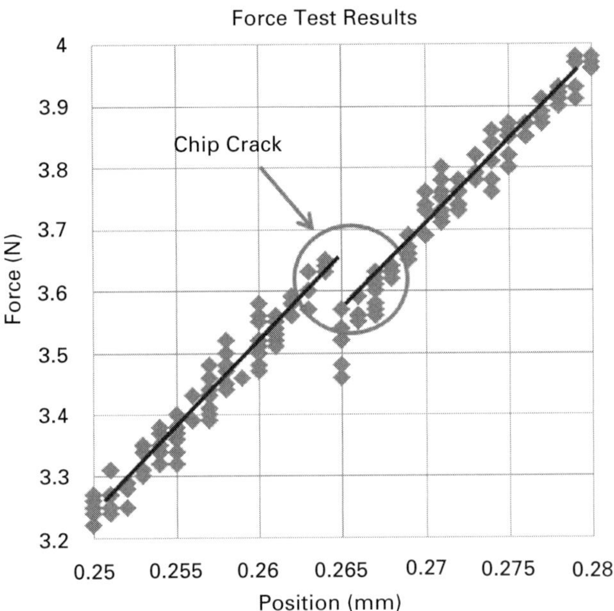

Fig. 3.18 Force versus displacement showing chip crack location (courtesy of Endwave Corporation).

Fig. 3.19 Sample image of an LCP board that has been mechanically machined.

to the individual lid size using a milling machine (Fig. 3.20). A 40 mil depth cavity, with dimensions on each side 0.16 inch smaller than the package base size, is drilled into the LCP bulk at its center. The sample is flipped, and the perimeter of the lid piece is machined down to form a lip with a 12 mil thickness. The lid lip should extend at least 15% of the outer wall-to-wall lid dimension to achieve reliable sealing. Figures 3.21a and 3.21b illustrate prototypes of a machined package lid.

Lid formed with combination laser and mechanical drilling

To demonstrate the process cost reduction through a combination of mechanical drilling and laser ablation, lid structures for use in an air-cavity package are

Fig. 3.20 Milling machine.

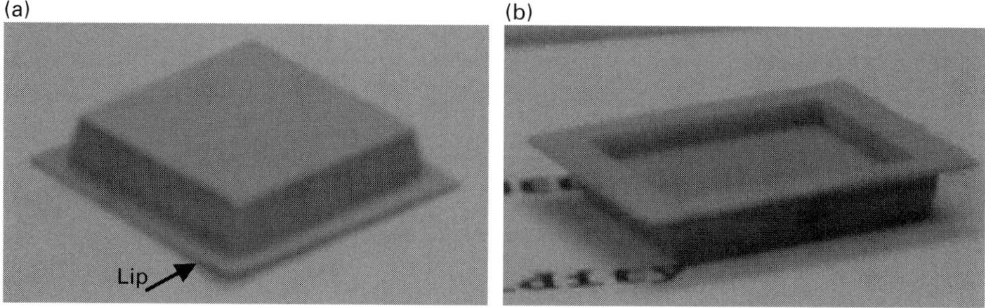

Fig. 3.21 (a) Top view and (b) bottom view of the all LCP lid [17] (© 2008 IEEE)

now presented. A dome-shaped lid is formed from multiple laminates consisting of 20 layers of 101.6 μm (4 mil) thick LCP. Initially, 20 high-melting-temperature LCP films 4 mil thick are taken and cut into three-inch squares. A single-clad 2 mil thick LCP film is inserted into the middle of the stack, sandwiched by 10 layers of 4 mil LCP films above and below. A single copper-clad LCP layer is positioned within the stack to serve as a visual layer stop for mechanical machining. Figure 3.22 illustrates the stack-up. The 21 layers are laminated at 315 °C and 90 psi pressure for 30 minutes in this configuration. Then, the package lid cavity is laser drilled on 80 mil thick bulk. Ninety-five percent of the cavity depth, i.e. 38 mil, is drilled mechanically to save cost and time, and the remaining five percent, 2 mil, is drilled with a laser for precision [8]. Once laser machining reaches the copper on the single-clad LCP in the middle of the bulk, machining is stopped. The thick lid cover provides

Fig. 3.22 Stack-up of multilayer LCP lid.

Fig. 3.23 Prototype of the multilayer LCP package lid in an array form.

rigidity to the package, and the cavity allows room for wire bonds. In laminating a lid to a package substrate, heat is applied locally at the package perimeter to avoid heating and damaging the internal active components. Figure 3.23 shows a multilayer LCP lid prototype.

3.4.4 Selective sealing

For cases where lamination along a particular seam is required, specific tools have been developed to allow localized heating. The ability to apply heat to specific locations is useful when forming large air cavities with "roofing" materials that can melt at LCP process temperatures. In particular, techniques have been developed to allow LCP-based air-cavity packages to be selectively sealed and formed.

One technique for laminating LCP material is laser welding [13]. In this method, laser energy is shone through a transparent material such as glass. This allows the LCP to absorb the energy, melt, and bond directly to the glass. A drawback of this technique is that a transparent bonding surface must be used, as the energy

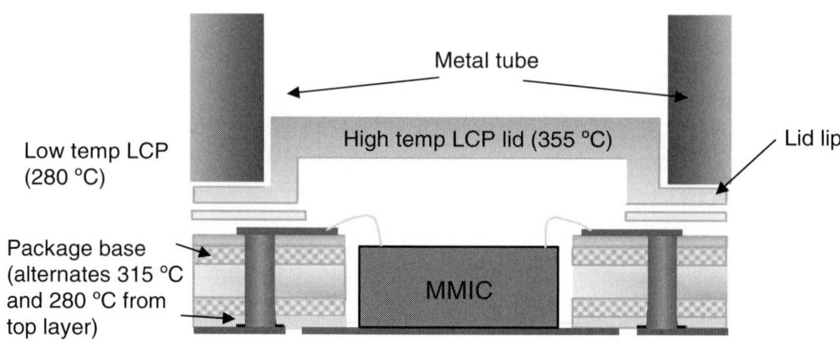

Fig. 3.24 Cross-section showing the process of LCP lid lamination onto a package base [17], [19] (© 2008 IEEE).

must be applied directly to the seal interface. This is due to LCP's low thermal conductivity. Further, as with other types of laser use, this process is relatively expensive.

Alternatively, thermal lamination equipment may be used to apply heat directly to the desired seams. This technique is compatible with standard existing equipment, and it is discussed in this section. A fully homogenous LCP surface mount package was built, without any adhesive films or solders, using this selective sealing technique, to demonstrate the process. We now describe this.

Equipment developed for selective sealing

To demonstrate selective sealing, an air-cavity lid formed of multiple LCP layers was implemented. Figure 3.24 shows a cross-section of the lid lamination stack up with the melting temperature of each LCP layer. The superstrate lids and laminate-base top layers are composed of high-melting-temperature LCP with T_m between 315 °C and 350 °C. A low-melting-temperature LCP film ($T_m = 280$ °C) is inserted between the package lid and the substrate interface. A metal tube with an opening perfectly fitting the package supplies heat and pressure directly onto the perimeter of the lid at the lip. Figure 3.25 shows a metal tool with three bolts used in laminating a lid onto the package base. A cavity is drilled inside a metal block to allow a package base to be inserted and held in place (Fig. 3.25a). Low-melting-temperature LCP film with an internal area punched out is placed above the package substrate to be used as the adhesion layer for the lid. A gas tube is screwed through a side wall of the metal block fixture to allow nitrogen gas flow before and during lid lamination. This fixture is designed to allow nitrogen back-fill and selective heating to occur simultaneously. A lid is placed on top of the LCP stack, and a thick-walled metal tube is pressed down into the hollowed metal block fixture. Within this fixture sits the package substrate (Fig. 3.25b). The three bolts through the metal-block fixture sidewalls hold the tube in place in the XY-plane while conveniently allowing free Z-axis movement. These bolts can be adjusted to allow different metal tubes to be interchanged when different package sizes are being sealed. This lid-lamination

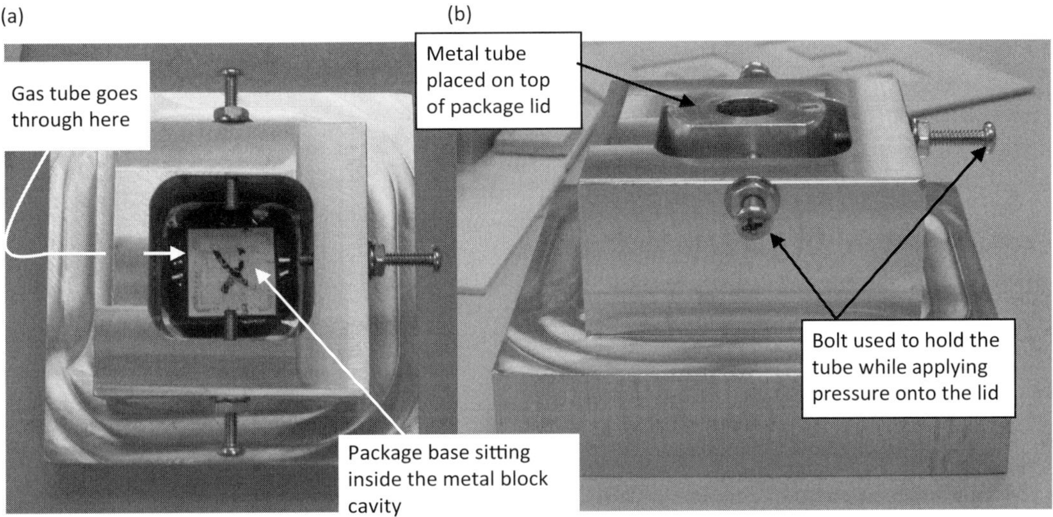

Fig. 3.25 (a) Top view and (b) side view of the lid lamination tool; (c) lid lamination in process.

tool set is placed between the heat plates of the lamination press, as described above in the discussion on uniform planar lamination. In this selective sealing process heat is supplied solely from the top lamination press plate (Fig. 3.25c). Once the top heat plate makes contact with the metal tube, heat is transferred to the package lid perimeter. This allows the low-melting-temperature LCP film to melt and selectively seal the package. The lid and package substrate base are unaffected by this process because they have much higher melting temperatures.

Demonstration prototypes of selective sealing process

Figures 3.26a, 3.26b show prototypes of sealed all-LCP surface mount packages. Figure 3.26c shows a magnified view of a cross-section at a lamination interface.

Fig. 3.26 (a) Top view and (b) bottom view of lid-sealed package and (c) cross-section of the interface between the package base and package lid under microscope.

As can be seen, the low-melting-temperature LCP layer has melted and made a good adhesion to the package lid and the package base layers. The samples were preheated to 150 °C initially for 30 minutes. Then, the press temperature was raised so that the interface temperature between lid and package base reached 280 °C. Finally, the press was shut off and the tool left to cool until it reached room temperature, for handling.

Thermal analysis for selective sealing

Thermal analysis is required in order to properly predict the heat effects during the selective lamination process. For instance, in order to maintain durability, devices fabricated on substrates such as GaAs should not be exposed to temperatures above 300 °C. Figure 3.27 shows a thermal simulation at the cross-section of a package of the temperature distribution during lid lamination [12]. The simulation was performed for a square metal tube placed above the lid lip followed by a metal plate that covers the metal tube. Heat sources are defined at the top of the metal plate, and the temperature is set to provide the necessary level of heat to the interface of the package base and lid. The simulation results show that a

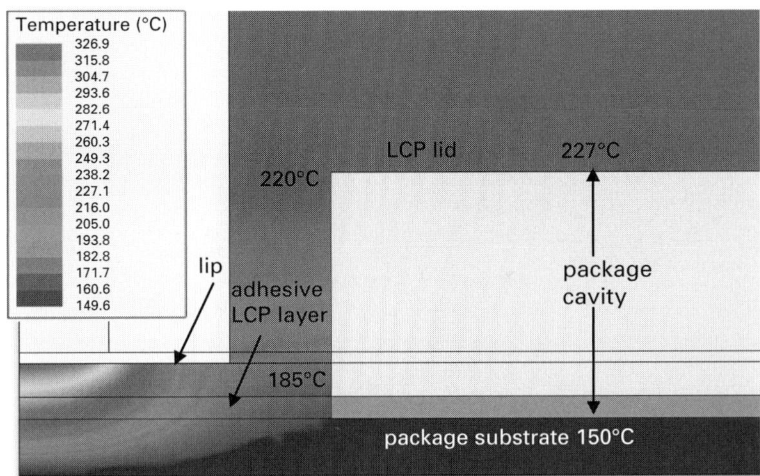

Fig. 3.27 Cross-section of the lid at the edge, showing the and simulated temperature distribution during the lamination process [17] (© 2011 IEEE).

temperature of 280 °C at the interface allows the package base to maintain a temperature below 160 °C. Hence, the simulation predicts that it is safe to package a GaAs device into an LCP surface mount packages with our selective air-cavity seal process.

3.4.5 Evaluating planarity after LCP lid sealing

This section describes a technique to evaluate the planarity of an LCP lid seal. Quad flat no-lead packages constructed with LCP lids were sealed with epoxy to create an air-cavity package. To evaluate their planarity after processing, LCP QFN packages were mounted, using tape, against a mirror surface. This technique allows imperceptible curvatures to be observed easily, by doubling the observable non-planar features and providing a centerline reference. A lack of planarity can increase solder joints in LCP packages, and this can electrically degrade performance. Figure 3.28 presents two air-cavity QFN prototypes simultaneously pressed upwards against a mirror. The real parts are in the lower portion of the photo with the reflected images above. The thin white layers are the LCP laminate and its reflection. The black plastic is a molded LCP cover that provides mechanical protection when pressure is applied to the air-cavity lid. The long-broken line separates two QFNs that have been pressed against the mirror. The QFN on the right shows bowing as observed from the slight curvature of the substrate. A QFN showing reasonable planarity is shown for comparison on the left. This technique allows quick inspection by doubling the observable bowing and providing an easy reference for comparison. Planarity issues observed in the first prototypes have since been resolved on the supplier side.

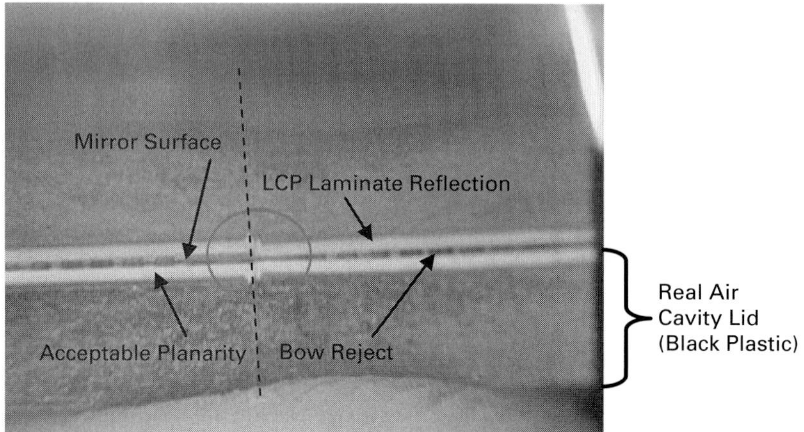

Fig. 3.28 Bowing of an air-cavity part as identified using a mirrored surface, compared side-by-side with a good part (courtesy of Endwave Corporation).

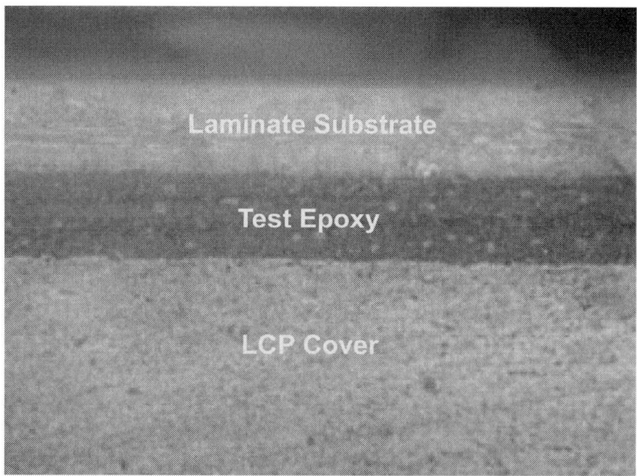

Fig. 3.29 Cross-section view along sawn lid seal epoxy (courtesy of Endwave Corporation).

Cross-section evaluation

A cross-sectional photograph is provided in Fig. 3.29 showing an LCP laminate joined to an LCP cover by means of a sample test epoxy. A diamond saw was used to cut the part along the package-lid wall length. This photograph demonstrates good epoxy coverage and uniform thickness after the part was processed in a tape-mounted frame.

3.4.6 LCP molding techniques

Currently, most connector vendors offer connectors formed from over-molded LCP. For instance, LCP connectors are readily available for high-speed input–output connection at speeds above 25 Gb/s per lane. LCP materials hold out excellent promise for providing low-loss low-dielectric-constant characteristics as an over-mold material. Strong consideration is being given to the use of LCP materials rather than the traditional plastics in over-mold surface mount (SMT) packages. This would allow the elimination, in over-molded SMT packages, of the overcoat materials used to provide a low dielectric constant over the chip. Further, the package would allow near-hermetic operation and provide a moisture sensitivity level 1 (MSL1). This would be of special benefit for air-cavity packages, since these may exhibit a *pop-corning* effect due to internal pressure build-up caused during heating. In this effect, the moisture in a package causes the cavity to pop open. Other current LCP electronic applications include low-cost air-cavity lid formation [14].

A number of plastics manufacturers are able to provide molded LCP parts. LCP for molded parts is available in a number of colors including natural (beige), black, blue, brown, green, and red. Plastic molding typically involves pushing melted material into a mold. The mold is composed of two plates that come together. The lower limit of mold feature size is governed by how well the mold material, i.e. stainless steel, can be machined and by how well small features are tolerated under the pressure exerted by the plastic. Plastic is fed into the mold through a runner that feeds through "gates" which allow material to be distributed. Once the mold is filled and the plastic has solidified, the plates are opened and ejector pins push out the finished part. The ejector pins may leave a small soft impression on the LCP part; one should be careful to place the ejector pins where any residual marking left by ejection action will be acceptable. The sidewalls along the molded part typically need to be angled by 6° in order for the mold to be ejected.

One challenge with LCP is that pellets are typically heated in a hopper and forced into a mold. If the mold itself is not heated, flow analysis has shown difficulties in filling the mold with the LCP. Similarly, this puts a limit on how thin the spaces in the mold can be. Even if the mold is fully filled, knit lines can form where the LCP interfaces have separately cooled and run together. These knit lines greatly compromise the molded part's integrity. In general, this issue is aggravated with larger molds.

Another issue is that the emerging LCP materials intended to be used for such injection molding are often modified with fillers. These fillers allow for simpler workability but also may cause a change in regard to the LCP's electrical, environmental, and physical properties; fillers with either glass or mineral have very different flow and wear characteristics. This can also affect the degree of warpage a molded part may exhibit.

Further, regardless of how flat a mold is machined, warping may occur. As one might expect, the severity of warping appears to increase with size of the part. Fortunately, LCP is especially suitable for molding as it is known for being a highly stable material that may have minimal warpage. Molded parts form consistent dimensions between runs using LCP. For ultra-flat parts, depending on the application, polishing or other post-corrective steps can be taken.

3.5 Chapter summary

This chapter has presented a complete set of fabrication techniques for working with LCP materials suitable for microwave packaging. The standard PCB processes available today allow for the complete processing of LCP. LCP is available in different formats, the laminated format being that most marketed to RF/microwave engineers. Metallization techniques for depositing copper, gold, nickel, palladium, or titanium onto LCP surfaces are possible using different types of equipment, techniques, and adhesion promoters. Specific intermetallic compounds known to form and cause reliability failures in packaging were discussed. Examples of the formation of cavities and the creation of vias through LCP were presented. The techniques used include laser milling, machine milling, stamp punching, and reactive ion etching. Lamination techniques may be implemented to create a multilayer board using standard LCP processes.

Advanced fabrication techniques on LCP were also discussed. Mounting options for LCP were presented for the proper handling of the material. These allow compatibility with existing planar semiconductor processing equipment. Alternative machining topics were discussed, and we demonstrated two techniques for forming dome-shaped LCP lids. Thermal simulations on our developed selective sealing technique were reported that show that the temperature is less than 160 °C inside a package, which allows epoxy, MMICs, and interconnect structures to maintain their integrity during the sealing process. Lastly, LCP injection molding was discussed, and first-hand advice for successful part formation was given.

References

[1] http://www.thefreelibrary.com/Biaxial+oriented+film+technique+exploits+properties+of+LCPs-a010615258

[2] Katsumi Takata, "Manufacturing of liquid crystal polymer flex and its characteristics," IEEE IMS Presentation, Honolulu, June 2007.

[3] http://people.ccmr.cornell.edu/~cober/mse542/page2/files/Schwartz%20R2R%20Processing.pdf

[4] Nanbu, K., Ozawa S., Yoshida K., Saijo K., Suga T., "Low temperature bonded Cu/LCP materials for FPCs and their characteristics," *IEEE Transactions on Components and Packaging Technologies*, vol. **28**, no. 4, pp. 760–764, December 2005.

[5] D. C. Thompson *et al.*, "Characterization of liquid crystal polymer (LCP) material and transmission lines on LCP substrates from 30 to 110 GHz," *IEEE Transactions on Microwave Theory and Techniques*, vol. **52**, no. 4, April 2004.

[6] http://www.semlab.com/goldembrittlementofsolderjoints.pdf

[7] http://tools.ticona.com/tools/mcbasei/product-tools.php?sPolymer=LCP&sProduct=VECTRA

[8] http://www.mlpc.com

[9] http://www.mech.northwestern.edu/medx/web/publications/papers/78.pdf

[10] http://www.mech.northwestern.edu/medx/web/publications/papers/53.pdf
[11] M. J. Rowlands *et al.*, "Manufacture and ultra-high frequency performance of an LCP-based, Z-interconnect, Flip-Chip package," ECTC 2008.
[12] M. Chen, A.-V. Pham, C. Kapusta *et al.*, "Design and development of a hermetic package using LCP for RF/microwave MEMS switches," *IEEE Transactions on Microwave Theory and Techniques*., vol. **54**, no. 11, pp. 4009–4015, November 2006.
[13] http://www.ansoft.com/products/tools/ephysics
[14] K. Gilleo, *MEMS/MOEMS Packaging*, McGraw-Hill, June 2005.
[15] http://www.rjrpolymers.com
[16] K. Takata, A.-V. Pham, "Electrical properties and practical applications of liquid crystal polymer flex", in *Proc. IEEE Polytronic 2007 Conf. Tokyo*, January 2007, pp. 67–72.
[17] K. Aihara, M. J. Chen, A.-V Pham, "Development of thin-film liquid crystal polymer surface mount packages for Ka-band applications," *IEEE Transactions on Microwave Theory and Techniques*, vol. **56**, no. 9, 2008.
[18] A. C. Chen, M. J. Chen, A.-V. Pham, "Design and fabrication of ultra-wideband baluns embedded in multilayer liquid crystal polymer flex," *IEEE Transactions on Advanced Packaging*, vol. **30**, no. 3, August 2007.
[19] K. Aihara, M. J. Chen, A.-V. Pham, "Reliability of liquid crystal polymer air cavity packaging," *IEEE Transactions on Advanced Packaging*, accepted for publication.
[20] http://www2.dupont.com/Plastics/en_US/Products/Zenite/Zenite.html
[21] http://www.solvayplastics.com/sites/solvayplastics/EN/specialty_polymers/Pages/Liquid_Crystal_Polymers.aspx
[22] http://www.kuraray.co.jp/en/products/medical/pdf/vecstar_catalog.pdf
[23] T. Zhang, W. Johnson, B. Farrell, M. St Lawrence, "The processing and assembly of liquid crystalline polymer printed circuits," presented at *Int. Microelectronics Symp.*, 2002.
[24] http://www.rogerscorp.com/acm/products/17/ULTRALAM-3000-Series-Liquid-Crystalline-Polymer-Circuit-Materials.aspx
[25] http://www.nscc.co.jp/english/materials/Circuit/Lseries_h.html
[26] R. J. Ross, "LCP injection molded packages – keys to JEDEC 1 performance," in *Proc. 54th Electronic Components and Technology Conf.*, Las Vegas, 1–4 June 2004, pp. 1807–1811.

4 LCP for wafer-level chip-scale MEMS

As defined in IPC/JEDEC J-STD-012, chip-scale packaging (CSP) refers to a packaging method where the final package area dimensions are no larger than 1.2 times the die. Wafer-level packaging refers to a method where a wafer containing multiple chips is processed for packaging before the individual dies have been sawn-cut for separation [1]. While chip-scale packages have been widely available in high-volume production, hermetic packaging at the wafer level is still either at the research stage or in low-volume manufacturing.

The most popular wafer-level packaging technique is wafer-to-wafer bonding for the packaging of microelectromechanical system (MEMS) devices. Several techniques used by industry to package MEMS devices at the wafer level include epoxy seals, glass frit, glass-to-glass anodic bonding, and gold-to-gold bonding [4–9]. These techniques face two major problems. Firstly, organic materials outgas within the MEMS cavity during bonding processes, owing to the wetting compounds in the glass, gold, or epoxy layers. This contamination detrimentally affects the MEMS reliability. Secondly, most bonding processes utilize high temperatures (300 to 400 °C), and this can degrade MEMS structures [10]. Furthermore, the available hermetic packages and ceramic or glass feed-throughs have significant losses at microwave frequencies, can be expensive, and add considerable weight to a system. Recent advances in low-temperature hermetic wafer-level packaging have shown promise for packaging multiple integrated circuits at wafer level [34]. One such process involves solder bonding, which combines the benefits of low-temperature conditions and thermodynamically stable eutectic bonding. Another viable alternative for packaging is an organic module, in which compact multilayer substrates house active and passive components; however, this presents even more challenges than those mentioned above. Although multilayer chip-on-flex modules using polyimide films are a proven technology for high-density packaging of microwave circuits [11, 12], polyimide is found to be incompatible with RF MEMS switch packaging owing to its high moisture absorption and high outgassing characteristics and the need to use high outgassing epoxies for lamination. In order to improve performance and provide ease of system integration, a hermetic MEMS package must be made with small, light-weight, planar interfaces constructed with MEMS-compatible materials. Thin-film plastic materials using liquid crystal polymer (LCP) are an attractive possibility for this application, owing to the low-temperature processing requirements for forming an interposer package.

This chapter presents developments in LCP-packaged RF MEMS switches in wafer-level chip-scale packages (WL-CSPs). Individual MEMS switch packaging is described below in section 4.1. Package process and environmental robustness will be presented to show the compatibility with MEMS. Electromagnetic design, analysis techniques, and insights will be presented. In section 4.2, research is presented that demonstrates LCP-packaged MEMS devices in an organic multi-chip module (MCM). Using CSP-packaged LCP devices, the hermetic capability in an MCM is presented with a MEMS phase shifter. Phase shifter design, analysis, and measurement will be provided, to demonstrate how LCP enables a hermetic capability for MCM designs.

4.1 Wafer-level chip-scale packaging of an RF MEMS switch

Microelectromechanical system (MEMS) devices, which employ microscopic mechanical motion, require hermetic packaging for proper operation. Until recently, hermetic packaging has been limited to metal and ceramic constructions, which are expensive, bulky, and difficult to design for RF/microwave-frequency operation. LCP, an emerging plastic material with hermetic characteristics, may open up the world of hermetic electronic packaging by offering a solution that balances performance, cost, and environmental robustness.

The MEMS technology has become an attractive alternative to solid-state devices for RF/microwave components; it has advantageous design drivers including miniaturization, portability, performance, and cost [2]. A simple, but common, MEMS structure is the cantilever beam. With appropriate design, a cantilever beam may be used as an RF MEMS switch. A number of companies commercially offer MEMS switches that exhibit performance from DC to 40 GHz [34–38].

MEMS switches have moving parts on a micrometer scale, and so these switches require hermetic, air-tight, packaging in inert environments to prevent contamination by particles or moisture. An ingress of particles may have the following ill-effects on a MEMS switch: wedging switches may become permanently wedged open; switches may be kept closed by stiction (static friction); performance may be degraded so that the switch acts as a resistive material [3].

Packaging provides the environmental shielding required for MEMS devices. By constructing a package using only LCP and copper, MEMS devices are housed in an extremely moisture-resistant enclosure. In general, circuits composed of many chips may be packaged as a single module within LCP. First-level interconnects, which refer to electronic connections through the first, bottom-level, packaging layer, must be analyzed for electrical effects in LCP packaging. In an LCP package, first-level interconnect detuning losses are negligible at X-band using multilayer flex and laser-drilled vias. Microwave measurements demonstrate that an LCP package adds less than 0.2 dB insertion loss and maintains the return loss of a packaged switch at more than 20 dB.

In section 4.1.1 we describe general process and performance aspects for an LCP MEMS package. Section 4.1.2 provides a detailed analysis of an electrical package design. Section 4.1.3 demonstrates measured the electrical performance of a packaged RF MEMS switch in an LCP enclosure.

4.1.1 Packaging steps for LCP-encapsulated RF MEMS

A process has been developed for laminating LCP onto silicon to form an enclosure for packaging an RF MEMS switch without adhesives [13, 14]. One advantage of LCP lamination over conventional packaging techniques is that low-temperature processing can be used. LCP lamination is typically performed at melting temperatures below 315 °C for most commercially available LCP laminates. This processing temperature is considerably lower than the 400 °C minimum temperature required for metallic or glass bonding.

Processing starts with 2 mil thick LCP having single-sided copper cladding. Laser ablation is performed to cut through the 2 mil thick LCP and form an MEMS cavity. Laser ablation is a convenient method to pattern chemically stable LCP. Copper cladding provides a natural laser-stop layer, and this technique is commonly employed for building blind vias. Using laser-ablation techniques the resulting etches are very accurate, with anisotropic vertical sidewalls. Any ash remaining after ablation is removed through a scrub with isopropyl alcohol solvent. By orienting the laminate with the metal upwards and the LCP downwards, the copper acts as a protective roof over an ablated cavity. When bonded to silicon, this cavity acts as a hermetic enclosure with copper above, LCP at the side walls, and silicon below. Lamination can be performed with alignment between an LCP lid and an MEMS die to within 1 μm accuracy. LCP is commercially available as films with melting temperatures between 240 °C and 315 °C, which is low enough not to impact the MEMS switch and allows a robust process.

Once lamination is completed, 100 μm × 100 μm, 50 μm tall, vertical microvias are formed in LCP using laser processes. Lastly, electrical contacts are formed on the package surface. The contact pads have a 200 μm × 250 μm area. Fifty-six packaged parts are individually separated with a dicing saw. Each packaged RF MEMS switch is 1.3 mm × 1.3 mm wide and 0.7 mm tall. Figure 4.1 demonstrates the process flow for laminating LCP on silicon. The final packaged dimensions are designed to be larger than the actual 100 μm long switch, for handling and packaging considerations. In the final package cross-section shown in Fig. 4.2, RF signals enter a package from the left along a metal trace, traverse a microvia in the LCP, pass through a closed MEMS switch, go up another interconnect, and exit at the right-hand side. Figure 4.3 shows a three-dimensional diagram of an LCP-packaged RF MEMS switch with the three layers separated for clarity. Figure 4.4 shows an actual packaged RF MEMS switch prototype, and Fig. 4.5 shows a well-defined via on a MEMS package.

Tests are performed to ensure that the LCP processing is compatible with MEMS. In Chapter 2 we explained that the low outgassing of LCP meets space-level

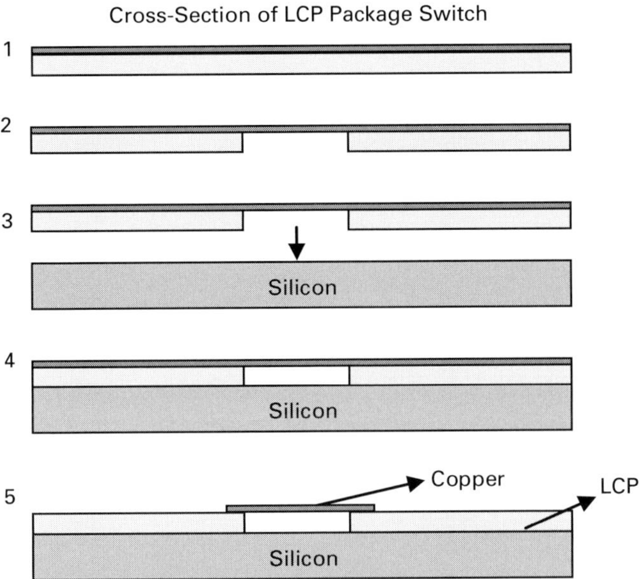

Fig. 4.1 The lamination of LCP onto Si [14] (© 2006 IEEE).

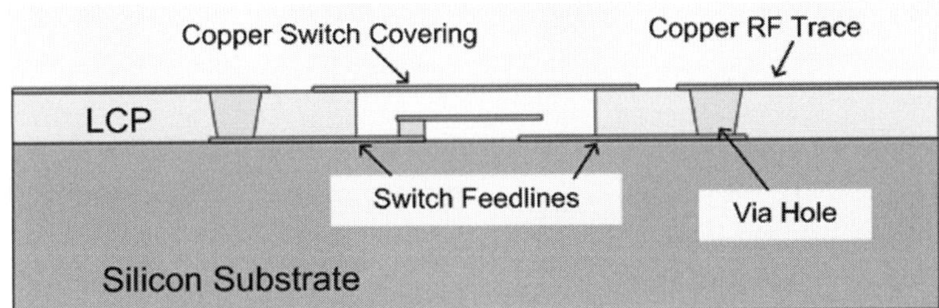

Fig. 4.2 An LCP-packaged MEMS device, showing the final cross-section [32] (© 2006 IEEE).

standards. Adhesion strengths were also discussed in chapter 2, and it was shown that LCP provides adequate adhesion between the copper and silicon. In regard to structural integrity, LCP exhibits some risk of deformation during processing owing to its decreased viscosity when melted. This is akin to when an ice cube is left in a room and begins to melt away at its exterior surfaces to form a puddle. Just as for an ice cube, the LCP outer surfaces are at risk of becoming fluid and freezing into a different shape when re-cooled. The amount of LCP reflow is measured by the changes in cavity dimensions after processing. In this subsection we study how process variations affect the physical cavity structure.

Throughout lamination, laser-formed cavities maintain well-defined sidewalls. Figure 4.6 depicts an intact open rectangular cavity in LCP laminated onto a silicon substrate. A cross-section is shown in Fig. 4.7; a rectangular hole 200 μm wide reveals the interdigitated fingers printed on silicon seen in Fig. 4.6. This rectangular

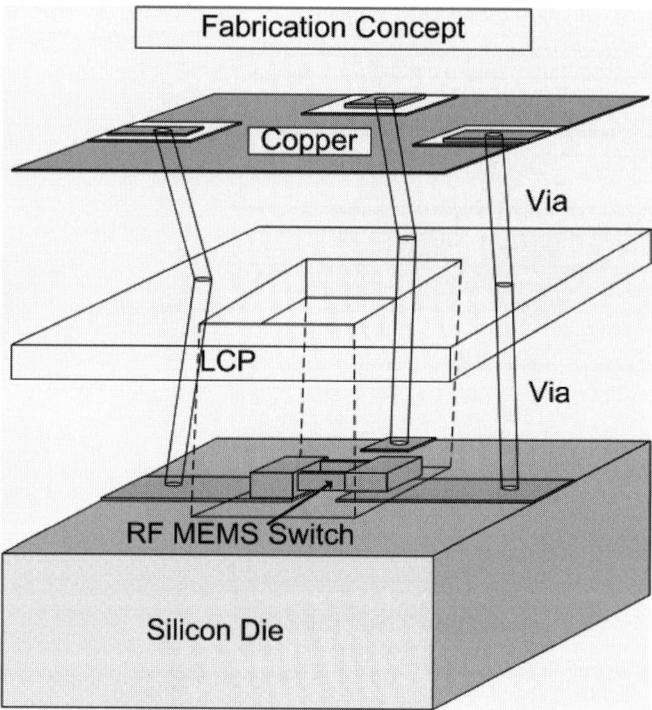

Fig. 4.3 A packaged RF MEMS switch in an LCP enclosure [14] (© 2006 IEEE).

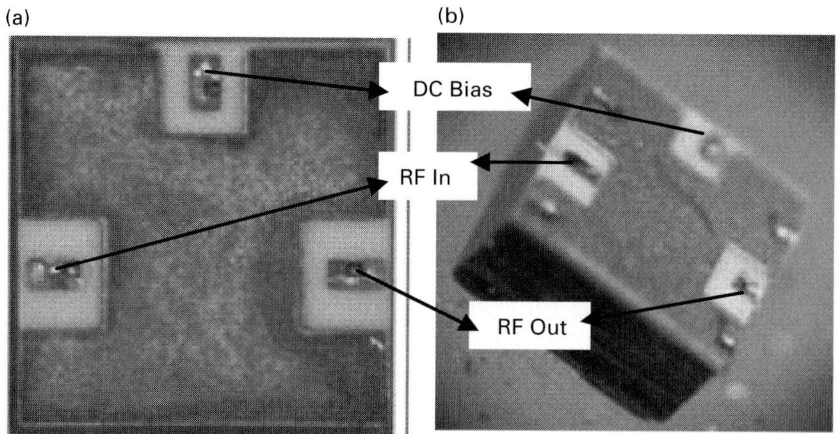

Fig. 4.4 Prototypes of packaged RF MEMS switch using LCP capping. (a) Plan view, (b) three-dimensional view [13, 14, 32] (© 2006 IEEE).

test cavity is the same size as a MEMS switch enclosure. The reflow is measured to be less than 5 μm at the cavity sidewall midpoints, where the LCP cavity walls are nearest to the enclosed MEMS devices. A worst-case reflow of 25 μm was exhibited at the corners that lacked sensitive device features.

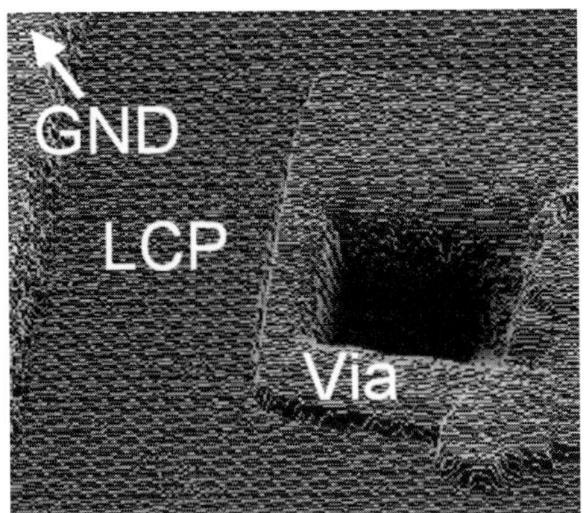

Fig. 4.5 Close-up of a MEMS part showing via formation and metallization under a confocal microscope [33] (© 2006 IEEE).

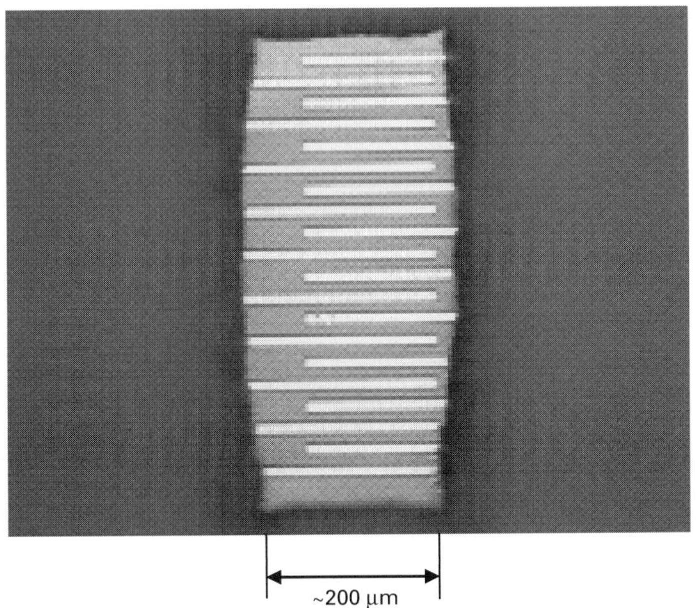

Fig. 4.6 Top-down photograph of LCP cavity on silicon [13, 14] (© 2006 IEEE).

4.1.2 Package electrical design

In this subsection, the impact of packaging on electrical performance is considered. We explain how simulations on electromagnetic models and circuits are conducted and applied to the design process.

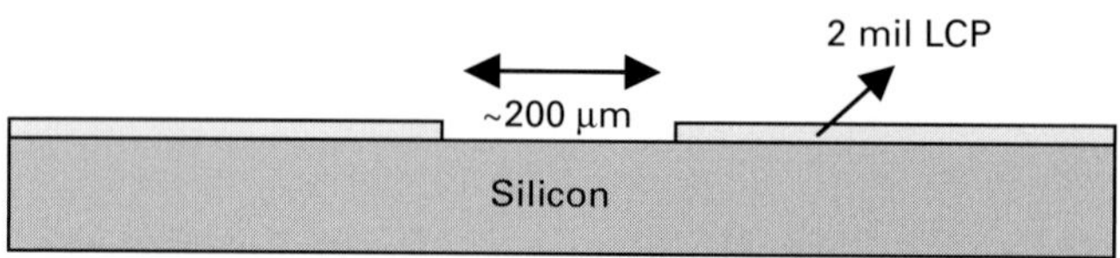

Fig. 4.7 Cross-section of LCP laminated on Si [14] (© 2006 IEEE).

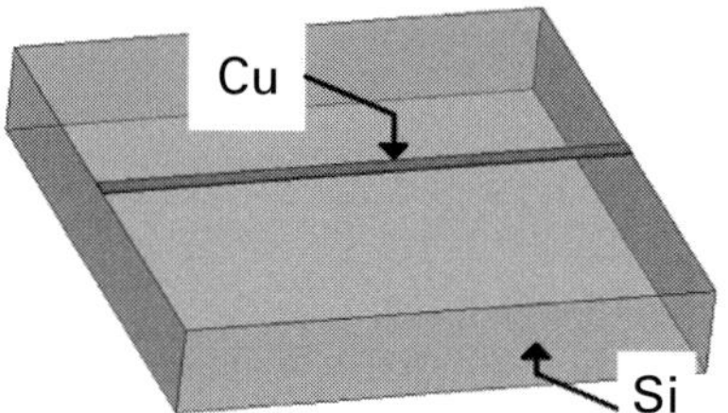

Fig. 4.8 Finite element method simulation of an unpackaged through line in LCP with thickness *H* [14] (© 2006 IEEE).

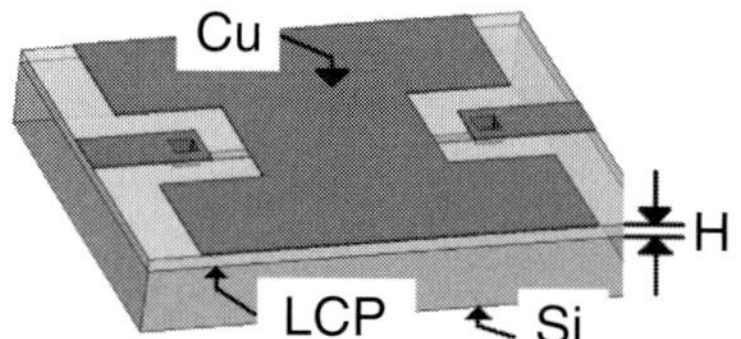

Fig. 4.9 Finite element method simulation of a packaged through line in LCP with thickness *H* [14] (© 2006 IEEE).

Electromagnetic modeling

In order to evaluate packaging effects on RF MEMS switches, full-wave electromagnetic simulations were conducted using a finite element method (FEM) solver. The basic structures for studying insertion and return losses include a bare microstrip transmission line on silicon with bulk conductivity 1 S/m. A bare microstrip line is used to model the unpackaged devices, as shown in Fig. 4.8. The bare microstrip line is then packaged in a copper-capped, 2 mil tall, LCP cavity, as shown in Fig. 4.9. Unpackaged MEMS switches have been designed to have high 80 Ω characteristic impedances; adding LCP packaging actually provides the circuitry for an improved 50 Ω match. For mechanical robustness, a 2 mil high cavity is chosen. Values of 2.9 for the dielectric constant (abbreviated as either ε_r or DK) and of 0.0025 for the loss tangent (abbreviated as either tan d, tan θ, or *DF*) were used to model the LCP, as provided in material data sheets. The copper vias are 100 μm × 100 μm with 5 μm thick walls and form the first-level interconnects. Each metal layer is also 5 μm thick. The final chip is 3000 μm by 3000 μm.

An analytic line impedance design tool is employed to determine the microstrip width on 254 μm thick silicon. The trace widths corresponding to 50 Ω and 80 Ω

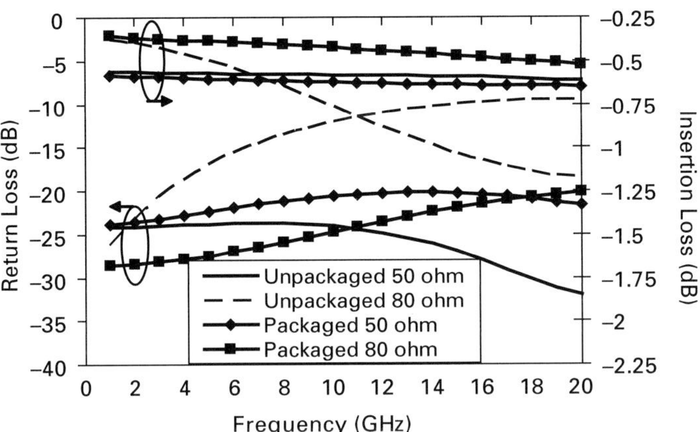

Fig. 4.10 Return and insertion losses of 50 Ω and 80 Ω through lines and an unpackaged 50 Ω line [14] (© 2006 IEEE).

microstrip lines (unpackaged) are 197 μm and 50.7 μm, respectively. In package simulations, microstrip sections feeding coplanar waveguides are de-embedded at each port. De-embedding removes all RF characteristics up to the port's reference plane so that only the device under test (DUT) is simulated.

Figure 4.10 shows simulation results for unpackaged and packaged microstrip lines over a frequency range from 0 GHz to 20 GHz. The return loss is plotted on the left-hand scale and the insertion loss on the right-hand scale. When an 80 Ω microstrip line is packaged with a 2 mil high metal lid, its characteristic impedance is tuned down closer to 50 Ω. This means that a lid addition to the 80 Ω switch allows better matching to 50 Ω terminations. In this case, the return loss of a packaged 80 Ω microstrip line improves from 13 dB to 25 dB at 10 GHz. Similarly, the insertion loss improves from 0.76 dB to 0.42 dB at 10 GHz. However, the insertion loss of a 50 Ω microstrip line when LCP packaged increases from 0.581 dB unpackaged to 0.624 dB packaged, and this difference means that the LCP packaging adds a less than 0.05 dB parasitic effect at 10 GHz. The return loss for a packaged 50 Ω microstrip line changes from 24.1 dB unpackaged to 20.6 dB packaged at 10 GHz. Note that a good match is maintained after packaging, with return loss better than 20 dB.

Table 4.1 gives simulation results for unpackaged and packaged 50 Ω and 80 Ω microstrip lines in 1 mil and 2 mil height metal lids. High-characteristic-impedance microstrip lines are tuned closer to 50 Ω transmission lines when they become striplines with 1 mil high or 2 mil high metal lids. In this package the capacitance per unit length of the striplines increases, which in turn decreases the characteristic impedance. This phenomenon is described by the characteristic impedance equation

$$Z_0 = \sqrt{\frac{R + j\omega L}{G + j\omega C}}, \tag{4.1}$$

Table 4.1. Return and insertion losses as simulated with the finite element method [14] (© 2006 IEEE)

	Unpackaged		Packaged (H = 1 mil)		Packaged (H = 2 mil)	
Width (μm)	S11 (dB)	S21 (dB)	S11 (dB)	S21 (dB)	S11 (dB)	S21 (dB)
197.4 (50 Ω)	−24.07	−0.58	−7.37	−1.42	−11.16	−0.88
50.7 (80 Ω)	−12.03	−0.76	−21.39	−0.46	−24.72	−0.42

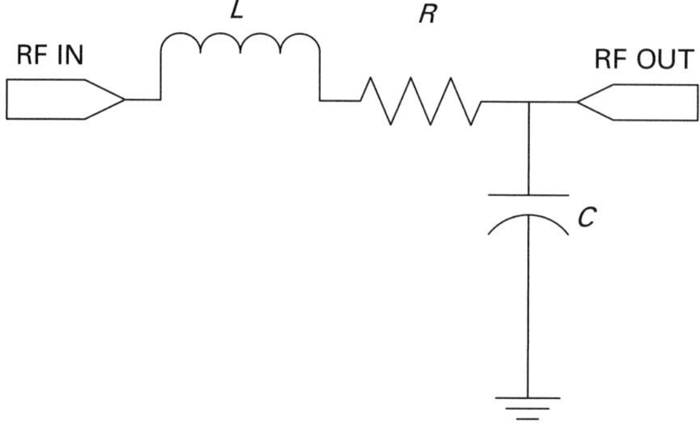

Fig. 4.11 Electrical via model for an LCP-packaged switch with $C = 120$ fF, $L = 124$ pH, and $R = 0.05\ \Omega$ [14] (© 2006 IEEE).

where R is the series resistance per unit length (in ohm/m), L is the series inductance per unit length (in H/m), G is the shunt conductance per unit length (in mho/m), and C is the shunt capacitance per unit length (in F/m).

Circuit modeling

An equivalent circuit model for microvia interconnects has been developed from simulations using method-of-moments (MoM) software. Further, an equivalent-circuit model has been developed to facilitate the understanding of switch performance.

When a device is packaged, the interconnects and the nearby copper can together cause additional tuning effects. An interconnect model is shown in Fig. 4.11 that has $C = 120$ fF, $L = 124$ pH, and $R = 0.05\ \Omega$. This lumped circuit models the via interconnect strictly: C models the capacitance between an interconnect via and the surrounding ground and L models the inductance associated with a narrow via constructed through an LCP thin film from the package exterior to the metal trace on the chip.

The S-parameters are measured for a packaged through line. An analytical method employing a circuit simulator is used to de-embed all path elements other than interconnect transitions, using techniques given in [15]. Figure 4.12 gives the modeled and measured transition S-parameters. This shows agreement to 0.02 dB

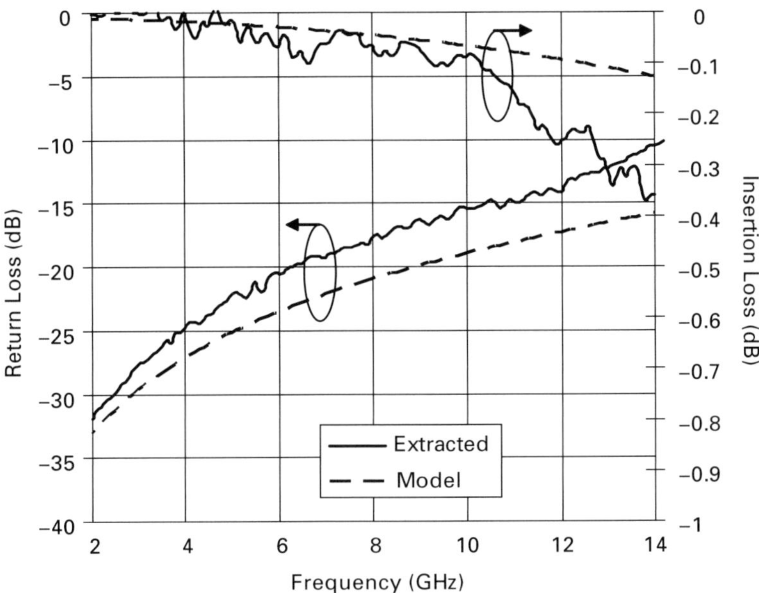

Fig. 4.12 Modeled and measured package interconnect losses [14] (© 2006 IEEE).

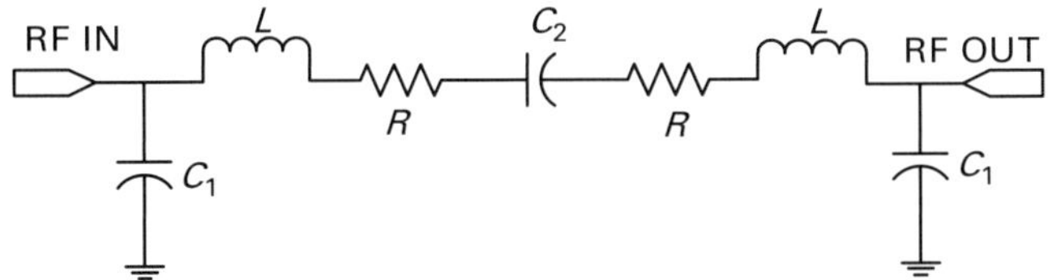

Fig. 4.13 Open-switch model with $L = 200$ pF, $R = 2\ \Omega$, $C_1 = 90$ fF, and $C_2 = 34$ fF [33] (© 2008 IEEE).

between the model and measurement insertion losses at 10 GHz, the frequency of interest. The model and the measurements both show less than 0.07 dB insertion losses per package transition at 10 GHz. The return loss shows agreement to less than 4 dB between the modeled and extracted measurement.

Furthermore, lumped-element LCP-capped switch models have been developed using measured S-parameters. For an LCP-capped switch in an open position, two *RLC* sections with $R = 2\ \Omega$, $L = 200$ pH, and $C_1 = 90$ fF are separated by a $C_2 = 34$ fF series capacitor, which models the capacitance between RF electrodes, as shown in Fig. 4.13. A closed-state LCP-capped switch model consists of two sections each with a $C_1 = 105$ fF shunt capacitance, a 420 pH series inductance, and a 4 Ω series resistance, as in Fig. 4.14. The differences between the two models are due to physical changes between the switch states, which cause variations of 20 pH in inductance and an increase in capacitance in the closed state of 15 fF over the lumped-element value.

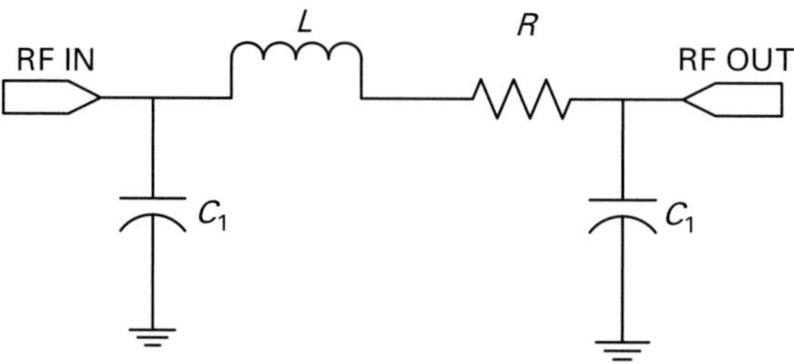

Fig. 4.14 Closed-switch model with $L = 420$ pH, $R = 4\ \Omega$, and $C = 105$ fF [33] (© 2008 IEEE).

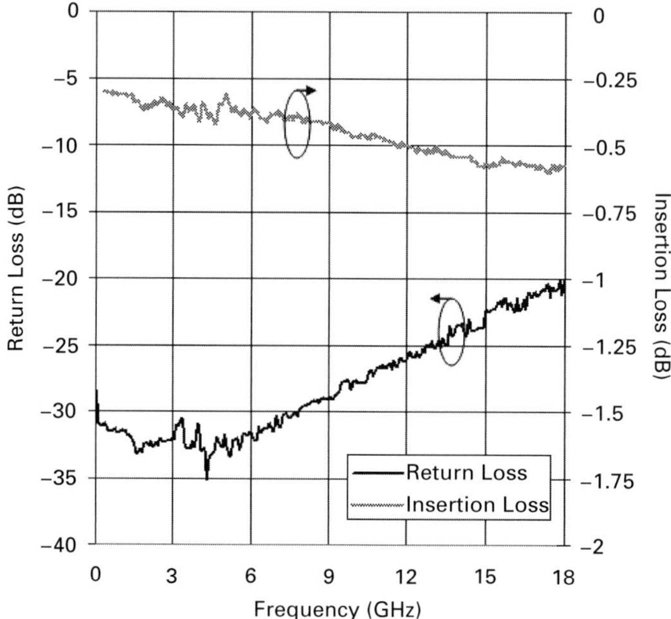

Fig. 4.15 Measured insertion loss and return loss versus frequency when the switch is closed [14] (© 2006 IEEE).

4.1.3 Package electrical measurements

Measurements of S-parameters can be performed with a probe station, a network analyzer, and coplanar waveguide probes with a 350 μm pitch. A load–reflect–match (LRM) calibration is performed to establish reference planes at the RF probe tips. This calibration involves probing known structures in order to calculate the added RF effects due to the test setup. After calibration, measurements no longer "see" the losses contributed by probes, cables, and connectors, which typically could be 6 dB or higher.

A DC probe is used to electrostatically bias a switch between the on-state and the off-state, with 0 V and 90 V applied, respectively. Measured results on an LCP-packaged switch in a closed state are provided in Fig. 4.15 for the X-band region

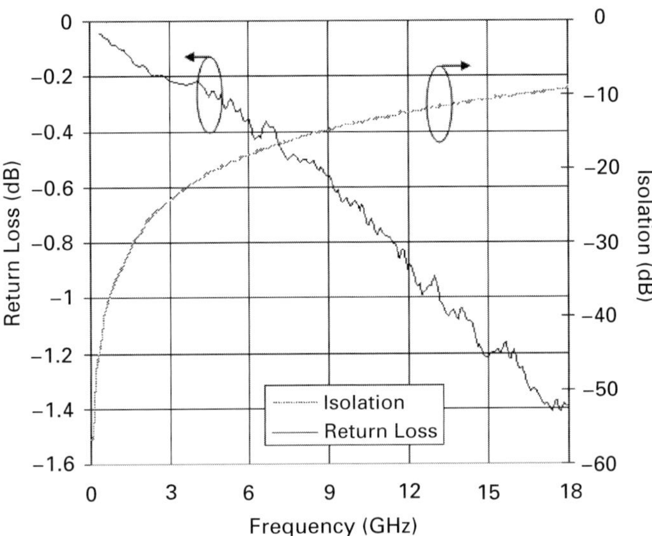

Fig. 4.16 Measured isolation and return loss versus frequency when the switch is open [14] (© 2006 IEEE).

and up to 18 GHz. The packaged switches show only ~0.45 dB total insertion loss at the X-band owing to the use of low-loss LCP material, microvias, and excellent shielding. This insertion loss measurement includes an additional 0.07 dB loss per interconnection at input and output, ~0.3 dB being attributed to the unpackaged MEMS switch at the X-band. In addition, the measured return loss is better than 25 dB at the X-band. This excellent return loss is achieved because a metal package cap tunes the switch's characteristic impedance close to 50 Ω and also limits radiation losses.

The S-parameters of a packaged MEMS switch were also measured in the open and off state (0 V bias). Figure 4.16 shows the measured S-parameters for an off-state switch. The measured isolation of the packaged switch is 14 dB and is the same as that for an unpackaged switch to within 1 dB. Since the particular switches had been optimized for an 80 Ω characteristic impedance system rather than a 50 Ω system, isolation is an especially relevant metric, and in this case it shows that minimal packaging causes negligible performance degradation.

4.2 Hybrid phase shifter on multilayer organic multi-chip module

Organic modules for RF/microwave components and subsystems have continued to be an attractive packaging platform for microwave and millimeter-wave frequency applications for the last 20 years [16–21]. These organic modules house chips under polymer films to form a complete package. Researchers have attempted to integrate

Table 4.2. The hybrid MEMS design and, for comparison, previous TTD phase shifters [23–29]

Phase shifter	Year	Freq. (GHz)	Bits	RL (dB)	IL (dB)	IL (dB/bit)
Hybrid design	2007	10	2	~14	4.9 ± 0.24	2.45 ± 0.12
[25]	2006	14	4	19.7	0.96	0.24
[26]	2003	12	3	10	2	0.67
[27]	2003	10	2	~20	0.6 ± 0.3	0.3
[28]	2003	10	4	~15	1.2 ± 0.5	0.3
[29]	2001	10	4	~20	2.4 ± 0.2	0.6
[24]	1998	34	4	15	2.25	~0.56
[23]	1989	20	4	n/a	10.4 ± 0.8	~2.6

microelectromechanical systems (MEMS) components into an organic module [3, 22]. Although organic modules have demonstrated excellent signal-integrity capabilities up to 110 GHz, they have the drawback of using moisture-absorbing polyimide. It has been shown that MEMS devices are not compatible with modules constructed with polyimide, owing to their high gas-permeability rates. Further, the polyimide inside an MEMS cavity outgasses during bonding processes using glass, gold, or epoxy layers. If these types of contamination occur, MEMS switches may be detrimentally affected.

This new integration scheme using LCP extends the existing chip-on-flex (CoF) technology, where chips are directly attached to a flexible board. Initial thin-film LCP packaging enables physical hardware abstraction by providing the required moisture protection for an MCM. In the present case, hardware abstraction is possible because MEMS devices are first packaged into LCP before being integrated into flexible polyimide modules. The resulting phase shifter is believed to be the first hybrid MEMS circuit on a CoF platform using commercially available MEMS switches.

The LCP-packaged MEMS devices discussed in section 4.1 were demonstrated in an organic multi-chip module (MCM). A hybrid MEMS phase-shifter circuit was built using this new organic integration technology [32, 33]. The realized phase shifter was measured to have 2.45 ± 0.12 dB insertion loss per bit at 10 GHz. The worst-case phase variation at 10 GHz was measured to be less than 5°. For comparison, Table 4.2 highlights how this hybrid design compares with previously published true-time delay (TTD) phase shifters. The losses in this circuit are primarily attributed to the switch-device loss characteristics, with a minimal contribution due to the packaging. Further discussion on this is found later in this chapter.

Fabrication processes for attaching MEMS devices to an organic motherboard will be described in section 4.2.1. The electrical design of a packaged phase shifter will be considered in section 4.2.2. Measured results are presented and discussed in section 4.2.3.

4.2.1 Description of processes

Processing begins with individual fully formed LCP-packaged MEMS single-pole single-throw (SPST) switches. Recall that the silicon MEMS parts are first packaged into a cavity composed of LCP and copper for hermetic protection, as described above in section 4.1. The LCP-capped MEMS parts are then integrated into a fully packaged phase shifter using standard CoF processing. In this scheme, many MEMS devices are encapsulated in a stack and form interconnections to a motherboard dielectric.

Processing begins by tension-mounting 2 mil thick polyimide film on a steel ring for planar process handling [33]. Next, bare polyimide dielectric is metallized; this involves sputtering titanium and then tungsten seed layers and electroplating copper to a thickness of 4 μm. The metal is then etched using lithography steps involving spun-on photoresist, ultraviolet mask exposure, photoresist development, copper etch, and photoresist strip. This forms signal traces along the top side of the dielectric and a ground plane along the bottom surface. A minimum feature size on the copper of 25 μm for both lines and spaces (the distances between lines) is required for reliable production. Next, the LCP-capped MEMS switches are aligned to make electrical connections between the MEMS and the polyimide film pads with the aid of fiducial markings. The MEMS parts are then affixed to the polyimide with a dam-and-fill step. Dam-and-fill processing consists of application of the dam plastic along the periphery to form a well. The wall formed is cured for 30 minutes at 125 °C and then for 90 minutes at 165 °C. The well is subsequently filled in with lower-viscosity fill plastic at under 80 °C to minimize air entrapment. The fill plastic is cured at 125 °C for 60 minutes and 165 °C for 90 minutes. The molding compounds form a 1 mm thick layer. Lastly, 60 μm × 60 μm square via interconnections are formed through the polyimide to reach the LCP-capped MEMS package by steps including laser ablation, copper sputtering, and electroplating. Additional laminates may be stacked using conventional flex technology. A fully packaged MEMS device cross-section in a film stack is illustrated in Fig. 4.17.

4.2.2 Design of a multi-chip module MEMS phase shifter

A two-bit MEMS phase shifter was designed and implemented in an organic module to demonstrate hybrid MEMS integration. The design takes into account the electrical and integration aspects of the delay lines, junction matching, and a fully composed phase shifter.

Design of true-time delay lines

The true-time delay refers to the situation where there is a constant transient response across a transmission line length with respect to frequency. Hence the phase shifts across the transmission line are directly proportional to the time taken for a signal to propagate from one end to the another. True-time delay lines are built

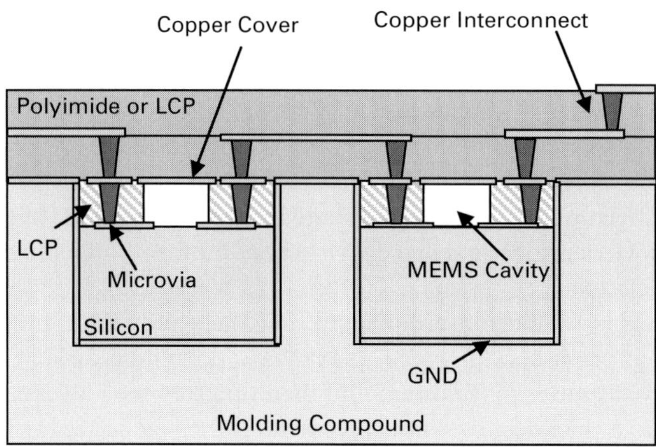

Fig. 4.17 Cross-sectional diagram of LCP and polyimide MCM for MEMS [33] (© 2008 IEEE).

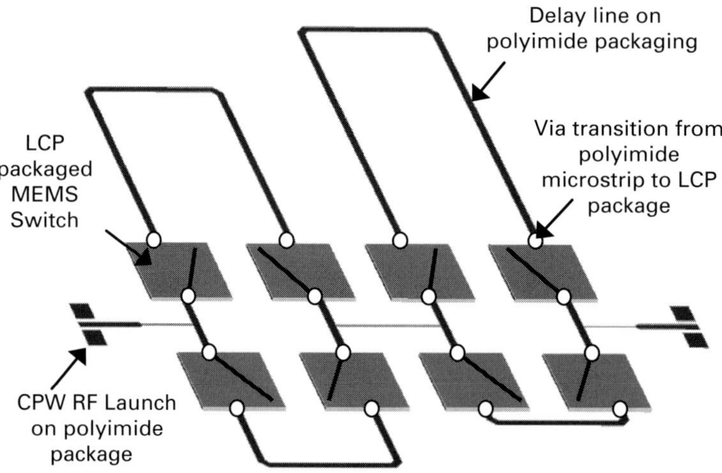

Fig. 4.18 Circuit topology in a three-dimensional view of a TTD switched-line phase shifter packaged in thin-film organic LCP and polyimide [33] (© 2008 IEEE).

into packaging with 4 μm thick copper on 2 mil thick polyimide films. A 4 μm thick copper package ground plane is located at the interface between the polyimide and the LCP-capped switches. The signal lines are printed on the opposite polyimide surface. The phase shifting is controlled by directing the RF signals through delay lines of varying physical length. At a specific frequency, 0°, 90°, 180°, or 270° output phase shifts may be chosen by closing switch combinations. A packaged true-time delay phase-shifter circuit diagram is illustrated in Fig. 4.18.

Fifty-ohm characteristic impedance transmission lines on 2 mil thick polyimide film were analytically simulated at lengths corresponding to the desired X-band phase delays. The line length l may first be calculated by using

$$l = \frac{\theta}{360}\frac{v_p}{f}, \tag{4.2}$$

where θ is the phase shift (in degrees), v_p is the propagation velocity (in m/s), and f is the frequency (in Hz).

The phase velocity v_p is given by

$$v_p = \frac{\omega}{k} = \frac{2\pi f}{2\pi / \lambda} = \frac{C}{\sqrt{\varepsilon_{reff}}}, \tag{4.3}$$

where ω is the angular frequency (in rad/s), k is the wavenumber (in m^{-1}), λ is the wavelength (in m), c is the speed of light in vacuum (in m/s), and ε_{reff} is the relative effective dielectric constant (dimensionless).

The lengths calculated from (4.2) were applied at $\theta = 90°$ and 180° and added to the reference line lengths. By rearranging (4.2), the relative phase shift p may be given as

$$p = 360\frac{f}{v_p}(l_1 - l_2) \tag{4.4}$$

where l_1 is the delay line length (in m) and l_2 is the reference line length (in m).

Junction matching section

A matching section is required at SPST junctions. The reason is that off-state tuning stubs add an associated detuning capacitance. Matching is performed by placing a high-impedance transmission line at the junction input to provide inductance cancellation. In realized form, the inductive line width is set to 50 μm, a minimum design rule. The inductive length is calculated from transmission line theory by solving for equation (4.5) below [31]:

$$jZ_0\beta l = j\omega L, \tag{4.5}$$

where j is the imaginary number (dimensionless), Z_0 is the inductive line characteristic impedance (in Ω), β is the propagation phase constant (in Np/m), l is the line length (in m), ω is the angular frequency (in rad/s), and L is the inductance value (in H).

The desired inductance L is given as

$$L = \frac{Z_0 l \sqrt{\varepsilon_{reff}}}{c}, \tag{4.6}$$

where ε_{reff} is the effective dielectric constant (unitless) and c is the speed of light (in m/s).

This equation gives an inductive line length of about 0.8 mm, built onto the polyimide package and located at the switch junctions. This inductor compensates the 230 fF capacitance presented by the off-state switch and stub section. The inductive

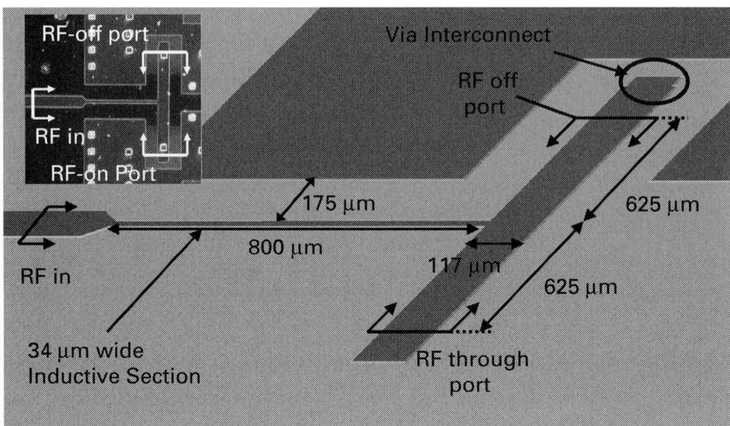

Fig. 4.19 Illustration of a T-junction in an organic package, overlaid with a photograph of its realization [33] (© 2008 IEEE).

section is 34 μm wide, as fabricated on polyimide. A distance of 175 μm separates the inductive sections from the nearby ground planes; this is more than five times the 34 μm inductor width and allows only minimal signal–ground coupling. The realized T-junction positioned in the package is shown in Fig. 4.19, along with the corresponding circuit diagram.

An analysis was performed on a three-port T-junction with impedance terminations based on empirical LCP-capped switch measurements. A majority of input power is transferred to RF-through (see Fig. 4.19), but a small percentage is transferred to the RF-off path owing to the 14 dB isolation path at 10 GHz in MEMS switches. Hence, such a T-junction may be considered as an asymmetric power divider with outputs having ~91.2% and ~5.9% of the original input power. Thus, the paths from the junction inputs to the on-state RF-through paths have ~0.4 dB insertion loss. The remaining unaccounted 2.9% input power is attenuated in conductor and dielectric losses. Specifically, it comes from ~1.6% at the package-matching sections and ~1.3% at the packaged T-stub sections. Simulated results for the insertion loss and the return loss between the input and through paths are plotted in Fig. 4.20. Over the 1 to 10 GHz band, the insertion loss varies from 0.42 to 0.74 dB, and the return loss from the RF-in port to the RF-through port is better than 10 dB. Over the narrow 9.5 to 10 GHz band, the plots show less than 0.42 dB insertion loss and better than 20 dB return loss through the matching and junction sections.

Fully composed phase shifter

A complete, fully composed, phase shifter was simulated in a microwave circuit simulator to account for the package effects on the overall design. The LCP-capped MEMS switches were modeled with two-port S-parameter measurements. Each phase shifter included four LCP-capped MEMS switches packaged into an

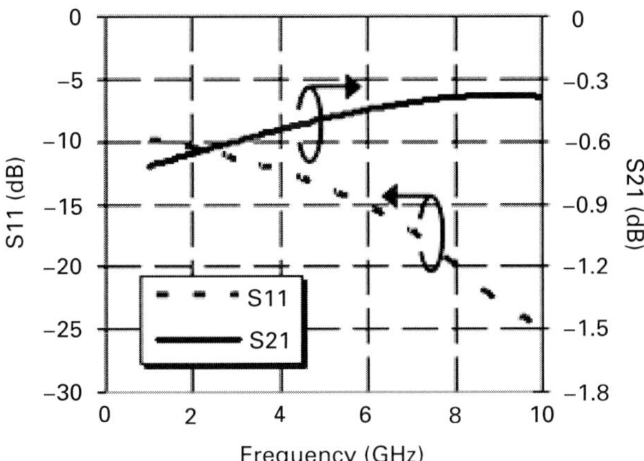

Fig. 4.20 Simulated insertion loss S21 and return loss S11 of through-path packaged junctions [33] (© 2008 IEEE).

RF path. The LCP-capped switches add a 0.55 dB additional loss per switch, or a 2.2 dB total loss. An additional 0.015 dB loss is incurred at each package via used to connect the delay lines on the polyimide to the LCP-capped MEMS switches. The MEMS SPST components are roughly 1 mm × 1 mm in this hybrid design, and this dimension largely determines the overall circuit size. The SPST modules are separated from each other by 1.25 mm, the minimum allowable spacing in chip-on-flex technology. The package vias were modeled using method of moments (MoM) analysis on the polyimide structures. The MoM numerical technique utilizes a rectangular mesh to solve for the electromagnetic fields using Maxwell's equations. The remaining transmission lines on the polyimide packaging were simulated using the built-in microstripline and waveguide models provided in the microwave circuit simulation.

A fully packaged two-bit phase shifter was simulated, and the results in Fig. 4.25 show 3.884, 3.759, 4.140, and 4.128 dB insertion losses across 0°, 90°, 180°, and 270° phase shifts, respectively, at 10 GHz. The average packaged-phase-shifter loss at 10 GHz is found to be 3.978 ± 0.219 dB, or 1.989 ± 0.11 dB/bit. The return losses of a packaged phase shifter are simulated as being better than 14.7 dB across all phases at 10 GHz. The simulated phase-shifts show less than 5° offset from the desired values at 10 GHz. The simulated phase-shifter losses are summarized, and the contributions from each stage are provided in Table 4.3. The losses are roughly identified as 0.4 dB from each T-junction, 0.015 dB from each via through the polyimide, 0.55 dB per MEMS switch, and between 0.077 and 0.296 dB per delay line.

4.2.3 Packaged phase shifter measurements

A realized hybrid MEMS phase shifter packaged in thin-film organics as measured with four DC probes and two RF probes is shown in Fig. 4.21. Four DC pads

Table 4.3. Loss analysis of the phase shifter in dB [33] (© 2008 IEEE)

		0°	90°	180°	270°
Bit 1	T-junction	0.4	0.4	0.4	0.4
	Via	0.015	0.015	0.015	0.015
	MEMS	0.55	0.55	0.55	0.55
	Via	0.015	0.015	0.015	0.015
	Delay	0.123	0.229	0.123	0.229
	Via	0.015	0.015	0.015	0.015
	MEMS	0.55	0.55	0.55	0.55
	Via	0.015	0.015	0.015	0.015
	T-junction	0.4	0.4	0.4	0.4
Bit 2	T-junction	0.4	0.4	0.4	0.4
	Via	0.015	0.015	0.015	0.015
	MEMS	0.55	0.55	0.55	0.55
	Via	0.015	0.015	0.015	0.015
	Delay	0.077	0.077	0.296	0.296
	Via	0.015	0.015	0.015	0.015
	MEMS	0.55	0.55	0.55	0.55
	Via	0.015	0.015	0.015	0.015
	T-junction	0.4	0.4	0.4	0.4
Total		4.12	4.23	4.34	4.45
Total/bit		2.06	2.12	2.17	2.23

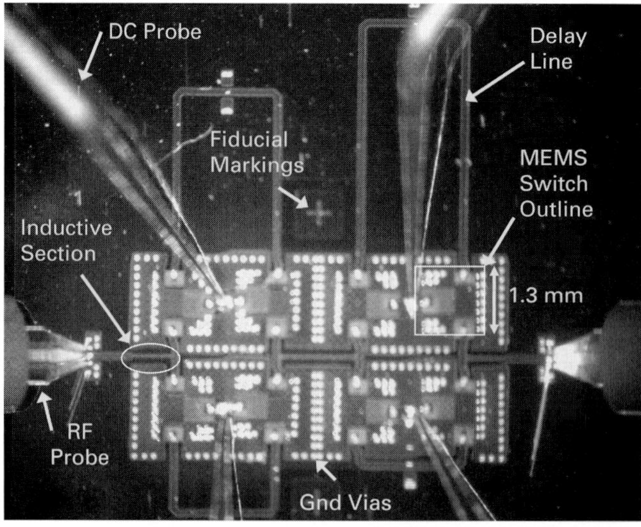

Fig. 4.21 Realized MEMS phase shifter [33] (© 2008 IEEE).

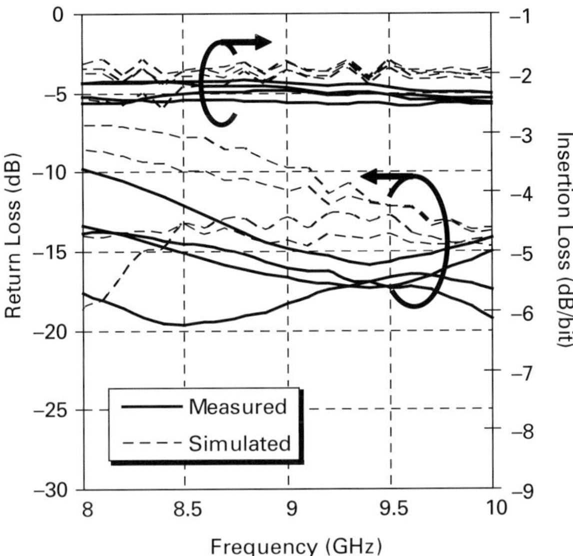

Fig. 4.22 Phase shifter measurements of return loss and insertion loss and the values obtained from a simulation, plotted over the intended operation band [33] (© 2008 IEEE).

control a total of eight switches; each DC pad controls two switches. Measurement involves applying 90 V DC to turn on the in-path switches, and 0 V DC is applied to the remaining switches. No DC current is drawn for actuation during this process, and hence the power drawn is near-zero.

The S-parameter measurements were obtained with a network analyzer (PNA) on a probe station. Probes with a 150 μm pitch contact the phase shifter directly at coplanar waveguide (CPW) contacts. Calibration was performed using load–reflect–match (LRM) on an impedance standard substrate (ISS) to allow the measurements to be referenced to the probe tips. The measured phase-shifter S-parameters in the 0°, 90°, 180°, and 270° configurations are shown in Fig. 4.22. Wideband S-parameter measurements are also provided, as shown in Fig. 4.23. At 10 GHz operation frequency, the packaged phase shifter insertion losses were measured to be 5.00, 5.06, 4.66, and 4.85 dB for 0°, 90°, 180°, and 270° phase shifts, respectively. Hence, the average insertion loss at 10 GHz was 4.89 ± 0.23 dB or 2.45 ± 0.12 dB/bit. On average, this result is 0.46 dB/bit more than the simulated value. The return loss was 14.2 dB at 10 GHz, which matches the simulation to within 0.5 dB. The wideband data indicate minor deviations from an ideal switched- line phase shifter at 6 GHz and 7.5 GHz due to ringing in the off-state transmission lines arising from low-isolation switch implementation.

The wideband insertion phases and group delays are plotted in Figs. 4.24 and 4.25, respectively. The phase shift errors are found to be less than 5° at 10 GHz. The relative phase shifts at 10 GHz are 0°, 90.2°, 181.8°, and 274.4° degrees. Across the entire band, the phase shifts were found to be accurate to within 15°. The group delays were measured as ~165, ~190, ~215, and ~240 ps. They are distinct and are

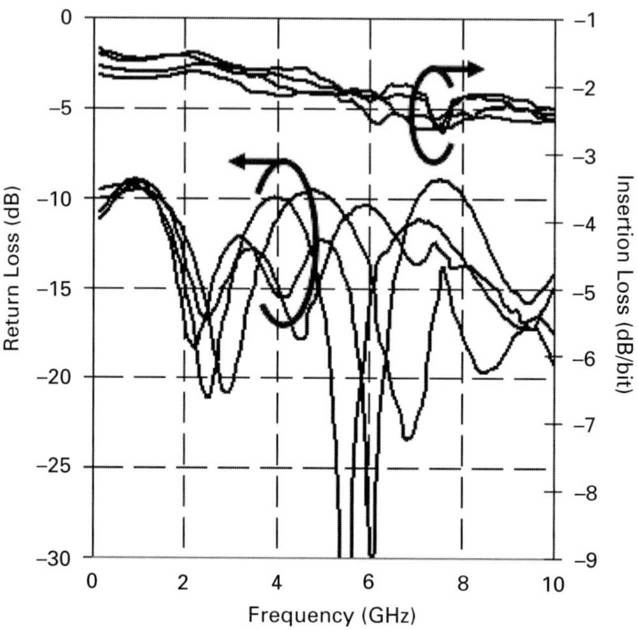

Fig. 4.23 Wideband phase shifter measurements of return loss and insertion loss [33] (© 2008 IEEE).

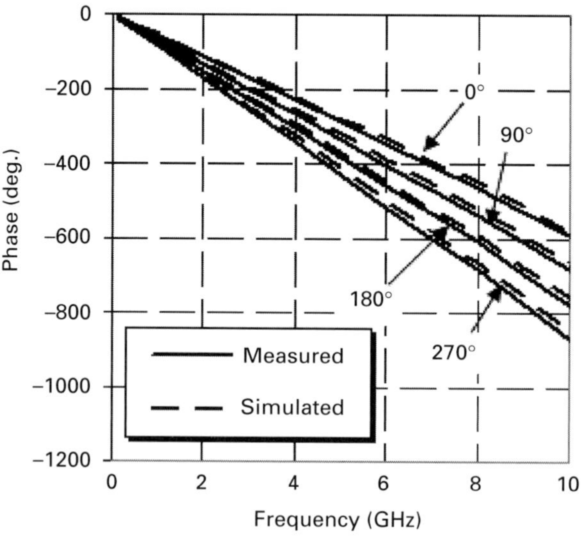

Fig. 4.24 Wideband phase shifter measurement of insertion phase shift in degrees [33] (© 2008 IEEE).

Table 4.4. Results of the simulation with the measured results for comparison, at 10 GHz [33] (© 2008 IEEE)

	Sim. IL (dB/bit)	Measured IL(dB/bit)	Error
0°	1.942	2.5	22.3%
90°	1.880	2.53	25.7%
180°	2.070	2.33	11.2%
270°	2.064	2.43	15.1%

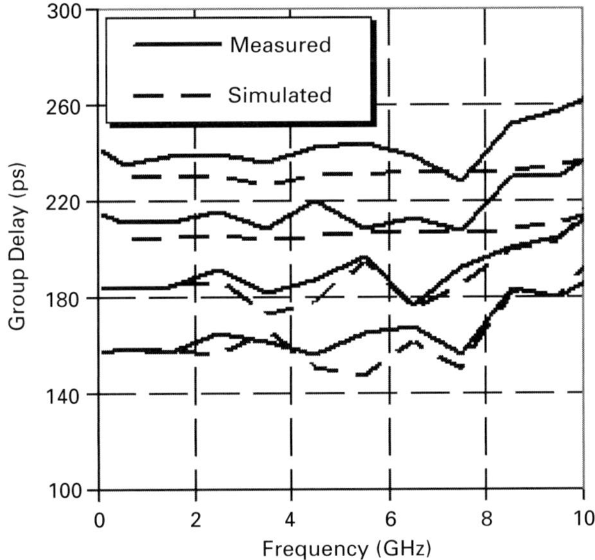

Fig. 4.25 Wideband phase shifter measurement of group delay [33] (© 2008 IEEE).

separated by ~25 ps across all phase shifts from DC to 10 GHz frequencies. The errors were caused by off-state resonances at 6 and 7.5 GHz. The low group-delay variations across other frequencies indicate excellent linearity in the phase shifter. The delay times, of ~200 ps, are slightly higher in this hybrid design as compared with those for MEMS phase shifters built on-wafer, since fully packaged components were used.

Table 4.4 provides a detailed comparison between the simulations and the measurements for each phase shift. This table indicates that the measured losses approximately match the simulated results shown in Fig. 4.22. In comparison with the simulation the insertion losses agree to within 25.7%, and the differences may be attributed to variations in the LCP-capped MEMS performance. The average error between simulation and measurement is 18.6%. Most losses are due to the LCP-capped MEMS switches, and the losses total 2.2 dB across the two-bit design.

About 0.12 dB loss is due to the combined package vias. Note that the electrical measurements are made on a fully packaged phase shifter designed using commercially available MEMS devices. This design would exhibit less loss with the following two changes to the packaged switch: (1) a lower through loss, and (2) isolation characteristics greater than 20 dB for good off-state matching.

4.3 Chapter summary

In this chapter, we have presented a novel concept and process for packaging MEMS devices into an organic laminate package to provide near-hermetic protection. A packaging technique has been developed that employs minimal length transitions through 2 mil thick LCP. This LCP layer provides MEMS chips with low-weight, low-bulk, near-hermetic, planar interfacing. Further, the package was designed to provide excellent high-frequency matching while introducing minimal parasitic loss. Simulations show that the entire package introduces miniscule electrical degradation to the overall circuit performance. At X-band, the measured LCP-packaged switches exhibit less than 0.5 dB insertion loss, greater than 25 dB return loss, and 14 dB isolation.

Furthermore, the design and realization of a two-bit MEMS phase shifter implemented in multilayer organic module technology was demonstrated. Such MEMS switch packages designed with LCP are integrated into this design with chip-on-flex technology. We devised a novel system-in-package module using thin-film technology for MEMS in a hybrid LCP–polyimide stack configuration. A realized switched line true-time delay phase shifter was measured to have an average loss of 2.45 ± 0.12 dB/bit with less than 5° phase variations at 10 GHz.

References

[1] http://www.siliconfareast.com/wl_package.htm
[2] http://wireless.fcc.gov/outreach/2004broadbandforum/comments/YDI_benefits60GHz.pdf
[3] G. Rebeiz, *RF MEMS: Theory, Design, and Technology,* John Wiley and Sons, 2003.
[4] http://flipchips.com
[5] S.-A. Kim, Y.-H. Seo, Y.-H. Cho, G. H. Kim, J.-U. Bu, "Fabrication and characterization of a low-temperature hermetic MEMS package bonded by a closed loop AuSn solder-line," in *Proc. IEEE 16th Annual Int. Conf. on MEMS* '03, Kyoto, January 2003, pp. 614–617.
[6] S.-J. Kim, Y.-S. Kwon, H.-Y. Lee, "Silicon MEMS packages for coplanar MMICs," In *Proc. Asia-Pacific Microwave Conf.*, Sydney, December 2000, pp. 664–667.
[7] A. Jourdain, P. De Moor, S. Pamidighantam, H. A. C. Tilmans, "Investigation of the hermeticity of BCB-sealed cavities for housing (RF-) MEMS devices," in *Proc. 15th IEEE Conf. on Micro Electro Mechanical Systems*, Las Vegas, 2002, pp. 677–680.

[8] A. Margomenos, L. P. B. Katehi, "Fabrication and accelerated hermeticity testing of an on-wafer package for RF MEMS," *Transactions on Microwave Theory and Techniques*, vol. **52**, no. 6, pp. 1626–1636, June 2004.
[9] R. M. Henderson, L. P. B. Katehi, "Silicon-based micromachined packages for high-frequency applications," *IEEE Transactions on Microwave Theory and Techniques*, vol. **47**, no. 8, pp. 1563–1569, August 1999.
[10] R. Tummala, V. Madisetti, "System on chip or system on package?" *IEEE Design & Test of Computers*, vol. **16**, no. 2, pp. 48–56, April–June 1999.
[11] http://www.kapton-dupont.com
[12] http://www.gore.com/electronics
[13] M. Chen, N. Evers, C. Kapusta *et al.* "Development of a hermetically sealed enclosure for MEMS in chip-on-flex modules using liquid crystal polymer (LCP)," in *Proc. Conf. on ASME Interpack, part C*, San Francisco, July 2005, pp. 2057–2060.
[14] M. J. Chen, A-V. Pham, C. Kapusta *et al.* "Design and development of a package using LCP for RF/microwave MEMS switches," *IEEE Transactions on Microwave Theory and Techniques*, vol. **54**, no. 11, pp. 4009–4015, November 2006.
[15] A. Pham, J. Laskar, V. Krishnamurthy, H. S. Cole, T. Sitnik-Nieters, "Ultra-low loss millimeter wave multichip module interconnects," *IEEE Transactions on Components, Packaging, and Manufacturing Technology, Part B*, vol. **21**, no. 3, pp. 302–307, August 1998.
[16] http://dynacocorp.com/Documents/LCPInfo.pdf
[17] J. W. Balde, "Crisis in technology: the questionable U.S. ability to manufacture thin-film multichip modules," *IEEE Proceedings*, vol. **80**, no. 12, December 1992.
[18] M. Pecht, "Characterization of polymides used in high density interconnects," *IEEE Transactions on Components, Packaging, and Manufacturing Technology – Part B*, vol. **17**, no. 4, November 1994.
[19] W. Daum, W. E. Burdick, R. A. Fillion, "Overlay high-density interconnect: a chips first multichip module technology," *IEEE Computer*, vol. **26**, no. 4, April 1993.
[20] T. R. Haller, B. S. Whitmore, P. J. Zabinski, B. K. Gilbert, "High frequency performance of GE high density interconnect modules," *IEEE Transactions on Components, Hybrids, and Manufacturing Technology*, vol. **16**, no. 1, February 1993.
[21] F. Liu, V. Sundaram, B. Wiedenman, R. Tummala, "Advances in high density interconnect substrate and printed wiring board technology," in *Proc. IEEE 6th Int. Conf. on Electronic Packaging Technology*, Shenzhen, September 2005, pp. 307–313.
[22] J. T. Butler, P. B. Chu, V. M. Bright, R. J. Saia, "Adapting multichip module foundries for MEMS packaging," *Int. J. Microcircuits and Electronic Packaging*, vol. **21**, no. 2, pp. 212–218, 1998.
[23] V. E. Dunn, N. E. Hodges, O. A. Sy, W. Alyassini, M. Feng, Y. C. Chang, "MMIC phase shifters and amplifiers for millimeter-wavelength active arrays," *IEEE MTT-S Int. Microwave Symposium Dig.*, vol. **1**, no. 1, pp. 127–130, June 1989.
[24] N. S. Barker, G. M. Rebeiz, "Distributed MEMS true-time delay phase shifters and wide-band switches," *IEEE Transactions on Microwave Theory and Techniques*, vol. **46**, no. 11, part 2, pp. 1881–1890, November 1998.
[25] N. Kingsley, J. Papapolymerou, "Organic 'wafer-scale' packaged miniature 4-bit RF MEMS phase shifter," *IEEE Transactions on Microwave Theory and Techniques*," vol. **54**, no. 3, pp. 1229–1236, March 2006.

[26] Y. J. Ko, J. Y. Park, J. U. Bu, "Integrated RF MEMS phase shifters with constant phase shift," in *Proc. IEEE MTT-S Int. Microwave Symp. Dig.*, Philadelphia, vol. **3**, no. 1, June 2003, pp. 1489–1492.

[27] G.-L. Tan, R. E. Mihailovich, J. B. Hacker, J. F. DeNatale, G. M. Rebeiz, "Low-loss 2- and 4-bit TTD MEMS phase shifters based on SP4T switches," *IEEE Transactions on Microwave Theory and Techniques*, vol. **51**, no. 1, part 2, pp. 297–304, January 2003.

[28] G. L. Tan, R. E. Mihailovich, J. B. Hacker, J. F. DeNatale, G. M. Rebeiz, "A 4-bit miniature X-band MEMS phase shifter using switched-LC networks," in *Proc. IEEE MTT-S Int. Microwave Symp. Dig.*, Philadelphia, vol. **3**, June 2003, pp. 1477–1480.

[29] M. Kim, J. B. Hacker, R. E. Milailovich, J. F. DeNatale "A DC-to-40 GHz Four-Bit RF MEMS True-Time-Delay Network," *IEEE Microwave and Wireless Components Letters*, vol. **11**, no. 2, February 2001.

[30] G. D. Lynes, "Ultra broadband phase shifters," in *Proc. IEEE MTT-S Int. Microwave Symp. Dig.*, vol. **73**, June 1973, pp. 104–106.

[31] D. M. Pozar, *Microwave Engineering,* 2nd edition, John Wiley and Sons, 1999.

[32] M. Chen, A. Pham, C. Kapusta *et al.* "Development of multilayer organic modules for hermetic packaging of RF MEMS circuits," in *Proc. IEEE MTT-S Int. Microwave Symp. Dig.*, San Francisco, June 2006, pp. 271–274.

[33] M. J. Chen, A-V. Pham, C. Kapusta *et al.* "Multilayer organic multi-chip module implementing hybrid microelectromechanical systems" *IEEE Transactions on Microwave Theory and Techniques*, vol. **56**, no. 4, pp. 952–958, April 2008.

[34] P. P. Chang-Chien, K. J. Tornquist, M. Y. Nishimoto, *et al.* "Low-temperature, hermetic, high-yield wafer-level packaging technology," *Northrup Grumman Technology Review Journal*, pp. 57–78, Spring/Summer 2006.

[35] http://www.radantmems.com

[36] http://www.xcomwireless.net

[37] http://www.memtronics.com

[38] http://www.delfmems.com

5 LCP for surface mount interconnects, packages, and modules

This chapter presents the design and development of thin-film LCP surface mount (SMT) package feed-throughs for DC to Ka-band applications. Three types of feed-through design will be introduced, via feed, bandpass feed, and lumped element feed, that interface signals from the outside world to a component inside a package. The packages are constructed using multilayer LCP films and are surface mounted on a printed circuit board (PCB) for use. The utilization of an all-LCP enclosure provides a hermetic environment for microwave and millimetre-wave monolithic integrated circuits (MMICs). In addition, mounting MMICs inside a package cavity allows enhanced thermal dissipation because the metal submounts can make direct contact with the PCB or motherboard ground through solder or epoxy. Applications that require lightweight, hermetic, and low-loss modules, which can be developed using LCP, include, but are not limited to, vehicular-collision-warning short-range radar, radar for ground-moving vehicles, point-to-point communication, ground–satellite communication, intersatellite links, and airborne radar. A phased-array system for ground-vehicle or airborne applications may require thousands of modules in the RF link. If each module, typically cased in ceramic and metal, were replaced with LCP packages, then the total weight of a system could be reduced by more than 66%, because ceramic [1] is three times denser than LCP [2]; this can lead to improvements in fuel efficiency.

Section 5.1 shows the design, the modeling, and measurement process of a Ka-band package feed-through using vias. The experimental results demonstrate that a package via feed-through including a PCB signal launch structure and bond wires achieves a return loss of better than 20 dB and an insertion loss of less than 0.4 dB at the Ka-band. The package has a measured port-to-port isolation greater than 45 dB up to 40 GHz. An amplifier packaged inside an LCP SMT package is characterized. The measured data are then compared with a circuit-model simulation of the package feed-through, using the on-wafer data of an amplifier to validate the lumped-circuit model.

Section 5.2 introduces a bandpass feed-through concept that aims to ultimately remove vias from package feed-throughs in order to eliminate potential moisture-leak paths into a package cavity and to overcome the Z-expansion disadvantage of LCP in a temperature cycle. Excellent performance is measured from 15 GHz to 35 GHz.

In section 5.3 we improve the concept introduced in section 5.2: the feed-through size is reduced by utilizing a lumped element filter. Excellent interconnect performance is measured from 5 to 20 GHz.

In section 5.4 we take the Ka-band package interface developed in the first section of this chapter and incorporate the design into a multi-chip Ka-band receiver down-converter module using thin-film LCP SMT packages. The module, housing a buffer amplifier, LNA, mixer, attenuator, seven bypass capacitors, and resistors, down-converts a Ka-band input signal to a 1.2 GHz intermediate-frequency (IF) output. In the receiver system, the detected RF signal is down-converted to the IF range and then digitized through an analog-to-digital converter (ADC) to process and extract the incoming message.

5.1 Design process for a thin-film LCP surface mount package and feed-through

At millimeter-wave frequencies, a package feed-through requires careful design, matching, and optimization so that it maintains a 50 Ω characteristic impedance throughout a desired band. Figure 5.1 demonstrates schematic diagrams of a thin-film LCP surface mount package. Looking at the cross-section of the package in Fig. 5.1c, an electrical signal enters the package through an underside metal trace and traverses through a plated via to the top of the package base. A cavity has been drilled through the base LCP layers by laser ablation [3], and the cavity bottom consists of package-base backside copper. A monolithic microwave integrated circuit (MMIC) is mounted inside the base cavity using high-thermal-conductivity epoxy or solder for enhanced thermal dissipation. Mounting the chip inside the package base cavity further enhances the thermal dissipation, in comparison with that for a non-cavity package using thermal vias to connect the chip's ground to the motherboard. Die attachment and wire bonding are followed by lid lamination, which hermetically seals the package cavity. Table 5.1 shows the dimensions *A*–*G* of the single-chip package prototypes. In this section, the design of a low-loss organic SMT package feed-through is investigated.

Package feed-through design overview

A vertical via package feed-through was designed and developed using full-wave finite element method analysis and quasi-static approaches [4]. The package base was constructed with a 254 μm (10 mil) thick LCP multilayer substrate and a 254 μm diameter via process for mechanical drilling compatibility. Figure 5.2a shows a complete package feed-through model. Initially, the feed-through is divided into four microwave transition subsections to be designed, simulated, and fine-tuned individually to the desired performance. Typically, each design aims to achieve 30 dB return loss in simulation. Owing to uncertainties in the fabrication and assembly process, up to 10 dB return loss degradation is expected during characterization

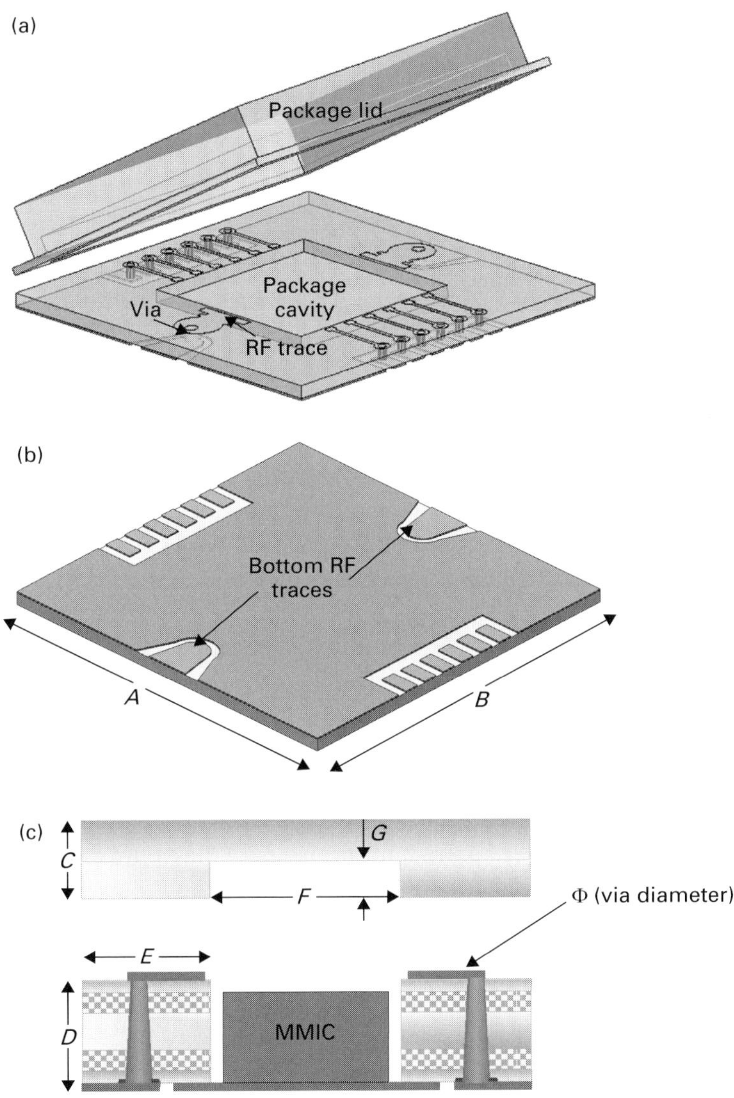

Fig. 5.1 (a) Schematic diagram of a multilayer thin-film LCP package, [32] (© 2008 IEEE); (b) the bottom side of a multilayer thin-film LCP package [32] (© 2008 IEEE); and (c) a cross-section of an LCP surface mount package [23] (© 2006 IEEE). For the dimensions *A*–*G*, see Table 5.1.

in the microwave and millimetre-wave range. One can perform a monte-carlo-type simulation using various software packages [5], but the predictions from it may not be accurate in these frequency ranges; greater accuracy can be obtained using three-dimensional electromagnetic software. However, with more than just a few variables, the run time for a three-dimensional electromagnetic simulation can blow up exponentially.

Table 5.1. Package dimensions from Fig. 5.1b, c

Var.	Dim. (mm)
A	6~8
B	6~8
C	2
D	0.254
E	3.1
F	>1 mm + chip width
G	1
Φ	0.254

The finely tuned subsections are combined, simulated, and optimized once again. The subsections comprise a via transition that includes a via barrel and via pads above and below the via, a microstrip line-to-bond-wire transition on top of a package base, a ground-backed coplanar waveguide (CPWG) below the package base, and a CPWG launch on a PCB test board, which makes a smooth transition from the package edge to the probe pads. The design of the last subsection is for characterization purposes only and is not required when the package is in use in a chain of RF components. In actual use, the package edge would make a transition to a 50 Ω characteristic impedance microstrip line or a CPWG. One can also measure a back-to-back package feed-through with a through–reflect–line (TRL) calibration that defines a reference point at the package edges and use time-gating to extract the data for a single-feed transition. The reader should keep in mind, however, that this method would give only an estimate. Figure 5.2b illustrates each subsection; the subsections are discussed separately below.

5.1.1 Via transition

A via transition through a multilayer package base includes a via barrel and via pads placed above and below the via. The via height and diameter are fixed at 254 μm, so the via pad diameters and their locations with respect to the via can be varied in such a way that the square root of the ratio of the via transition inductance and capacitance with respect to ground,

$$Z_0 = \sqrt{\frac{\text{via inductance}}{\text{via capacitance}}}, \tag{5.1}$$

is maintained at 50 Ω. This design was performed using commercially available model extraction software that computes the via transition inductance and capacitance. Figure 5.3 shows a simulated structure on Q3D. The initial design dimensions show diameters of 800 μm for the bottom via pad and 900 μm for the top via pad, and the two via pads are offset by 280 μm in the direction of signal propagation.

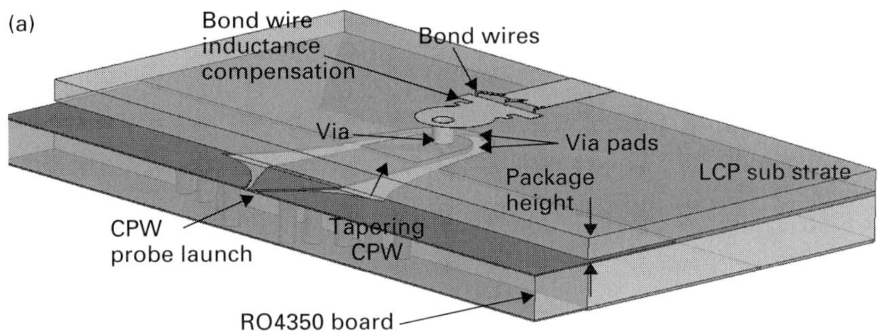

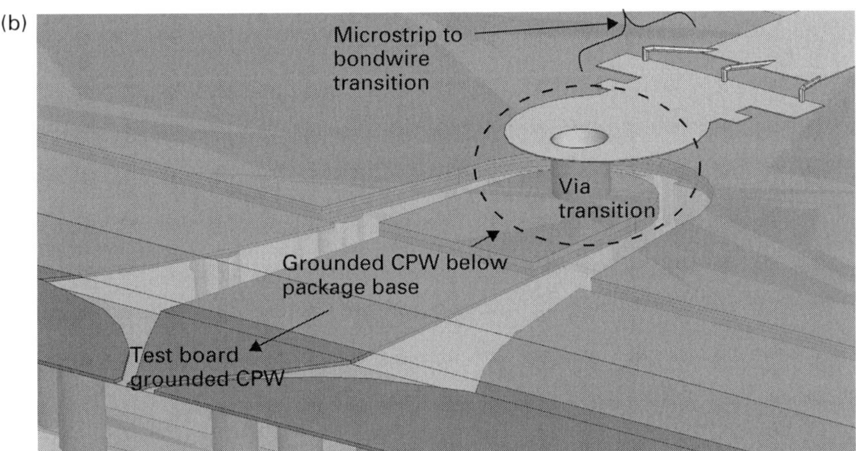

Fig. 5.2 (a) Ka-band thin-film LCP package feed-through transition design structures [32] (© 2008 IEEE) and (b) package feed-through transition showing details of the four subsections.

A simulation predicted that our optimized via pads and barrel inductance would be 397.68 pH and that the via pads and barrel capacitance to ground would be 156.6 fF. Taking the square root of the via transition inductance to capacitance ratio, (5.1), yields 50.38 Ω, which has a less than 0.8% variation from 50 Ω. Since the reflection coefficient Γ is given by

$$\Gamma = \frac{Z_L - Z_0}{Z_L + Z_0}, \tag{5.2}$$

the above optimized values give 0.003785. Thus the return loss, which is given by

$$-20 \log |\Gamma|, \tag{5.3}$$

is found to be 48 dB using (5.2) and (5.3). The subsection design of a via transition gives only an estimate, and the actual performance may vary once simulation on a three-dimensional electromagnetic simulator is carried out. Hence, once all four subsection structures are merged in an EM simulation, the dimensions of the via pads and their locations with respect to the via may vary.

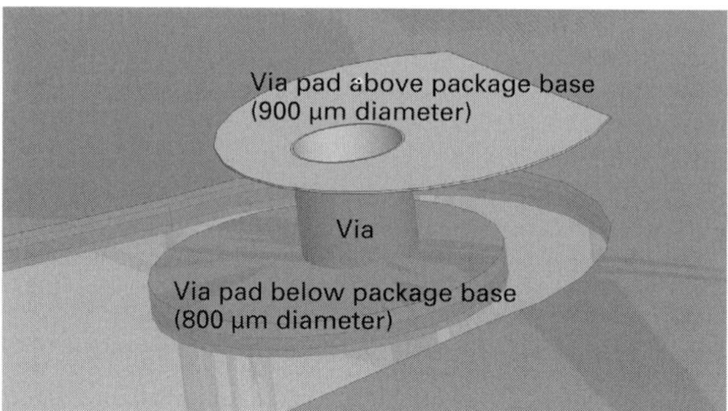

Fig. 5.3 Simulation structure of the package via transition.

5.1.2 Microstrip line to bond wire transition

The next subsection of the package feed-through includes a microstrip line on the package base and bond wires that connect the microstrip line to an MMIC inside the package. Figure 5.4 shows the structure of this subsection. A 50 Ω microstrip line on a 10 mil LCP package base is connected to a via pad on top of the base. Then, a bond wire connection is made from the package base microstrip line to the packaged MMIC. Because the bond wire connection is almost purely inductive, its characteristic impedance at the bond wires becomes very high, causing a discontinuity and high reflection at millimeter-wave frequencies. Thus, it is necessary to tune out the bond wires' inductance: the end of the microstrip line was widened to compensate for it. The pad size was tuned using FEM software. Three parallel wires were also used to minimize the parasitic inductance at the Ka-band. The three bond wires were connected to another 50 Ω microstrip line to establish a two-port simulation. Note that the second microstrip line does not have wide pads. On an actual package, the bond wires connect the microstrip line on a package base to an MMIC inside a cavity. Hence, bond-wire inductance compensation has to be achieved for a package base microstrip line. It is assumed that the MMIC's RF in/out pads are matched to 50 Ω internally. It becomes more challenging to design a feed-through that serves as a matching network to compensate an unmatched impedance coming out of an MMIC to 50 Ω. The package feed needs at the same time to match the impedance coming out of the MMIC. This requires more precision than a 20 dB S11 design. Also, the package fabrication tolerance may not allow for such a precise requirement although this is more feasible in an IC fabrication environment, where the tolerance is in the submicron range. Additionally, from a mass production perspective, a generic 50 Ω package-feed design, instead of MMIC-specific package-feed designs, makes much more sense as it is cost effective.

Initially, the capacitance and inductance of the bond wire transition was extracted using software. The bond wire length in simulation was set to 330 μm,

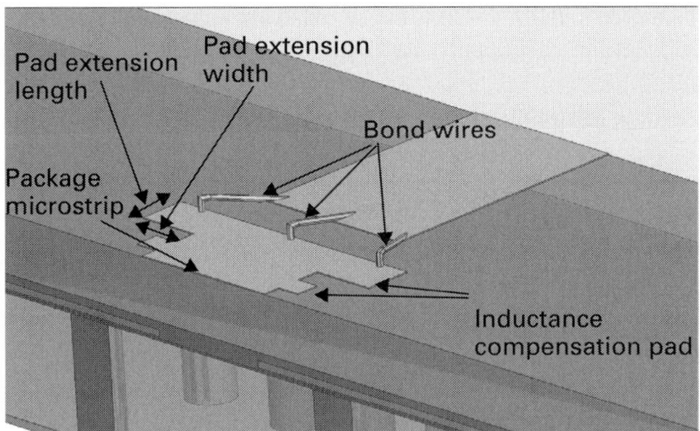

Fig. 5.4 Simulation structure of the package base microstrip to bond wire transition.

(~13 mil) in the simulation. This assumes a 100 μm (4 mil) spacing between the package trace and the chip's ground pad, a distance of 150 μm (6 mil) to the RF signal pad on the chip, a 100 μm chip thickness, and 50 μm epoxy. Typically, the datasheet for an MMIC from a major chip supplier may recommend a similar wire length for assembly. Different sizes of bond-wire inductance-compensation pads were simulated, and the characteristic impedance was approximated using (5.1). The compensation pad width was kept constant at 200 μm to reduce the number of variables. The extension pad length is defined as the distance a compensation pad extends from the 50 Ω microstrip line width (see Fig. 5.4). For example, a pad with 170 μm length and 200 μm width shows a 208 fF capacitance and a 650 pH inductance throughout the bond wire transition, according to Q3D. Estimating its characteristic impedance using (5.1) yields 55.9 Ω. The pad extension length was varied from no extension to 970 μm and the corresponding characteristic impedance was estimated. Table 5.2 shows the simulated pad dimensions in column 2, the extracted capacitance and inductance in columns 3 and 4, and the estimated characteristic impedance in column 5. It can be seen from Table 5.2 that a pad length between 670 and 770 μm achieves closest to the required 50 Ω characteristic impedance.

Looking closely at Table 5.2, one may see that total transition capacitance increases by 10 fF for every 100 μm increase in pad extension length on each side of the 50 Ω microstrip line. This constant pattern of capacitance increase can be verified using a capacitance approximation equation modified from Bogatin [7], Kaupp [6], and the Technick webpage [8]:

$$C = \frac{2.986\varepsilon_0\left(\varepsilon_r + 1.41\right)}{\ln\left(\dfrac{5.98H}{0.8W + T}\right)}, \tag{5.4}$$

where ε_r is the substrate dielectric constant, ε_0 is the permittivity, H is the substrate thickness, W is the extension pad width, and T is the extension pad thickness.

Table 5.2. Various pad extension lengths (1–12) with estimated impedance and simulation prediction

	Pad extension (length in μm × width in μm)	Capacitance (fF)	Inductance (pH)	Est. Z_0 (Ω)	S11/S21 from simulation at 40 GHz (dB)
1	0 × 0 (1 bond wire)	194	812	64.7	−7.3/−1.1
2	0 × 0 (3 bond wires)	194	650	57.9	−18.2/−0.099
3	70 × 200	198	650	57.3	−22/−0.059
4	170 × 200	208	650	55.9	−31.9/−0.035
5	270 × 200	216	650	54.9	−21.8/−0.062
6	370 × 200	226	650	53.6	−17/−0.123
7	470 × 200	236	650	52.5	−14.2/−0.203
8	570 × 200	246	650	51.4	−11.6/−0.382
9	670 × 200	252.5	650	50.7	−9.0/−0.618
10	770 × 200	262.5	650	49.8	−6.9/−1.052
11	870 × 200	272.5	650	48.8	−5.1/−1.663
12	970 × 200	282.5	650	48	−3.5/−2.656

Using (5.4) with a 100 μm pad extension length, 200 μm pad width, and 9 μm pad thickness, each extension pad adds 5.13 f F on each side, which explains the increase of 10 f F that is found every time extension pads are increased by 100 μm on each side of the microstrip line.

Figure 5.5 shows rough FEM-based predictions for a microstrip line to bond wire transition with and without inductance compensation pads at the edge of the package-base microstrip line. Simulation without compensation includes two 50 Ω microstrip lines connected with one 330 μm long bond wire. The return loss improves to greater than 20 dB, and the insertion loss is less than 0.06 dB for the transition with bond wire compensation; this may be compared with the 7 dB return loss and the more than 1 dB insertion loss for a non-compensated transition up to 40 GHz.

Figure 5.6a shows a set of return losses, and Fig 5.6b shows a set of insertion losses for this bond-wire transition subsection for various pad extension lengths. Three values of the pad extension length showed better performances than that with no pad extension, and the best results are those for the 170 μm pad length. The simulations showed a better than −30 dB return loss and a 0.05 dB insertion loss at 40 GHz. Column 6 in Table 5.2 summarizes the insertion and return loss simulated at 40 GHz on the basis of the pad size dimensions shown in column 2. It seems that the estimated characteristic impedances (5.1) have a discrepancy of approximately 21% from the three-dimensional electromagnetic simulations in terms of the capacitances extracted from the software, and of about 10% in terms of the characteristic impedance. With a 10% error in the characteristic impedance, the return loss degrades to 9 dB, and the insertion loss degrades to 0.62 dB at 40 GHz from the best simulation result achieved. This shows that optimization

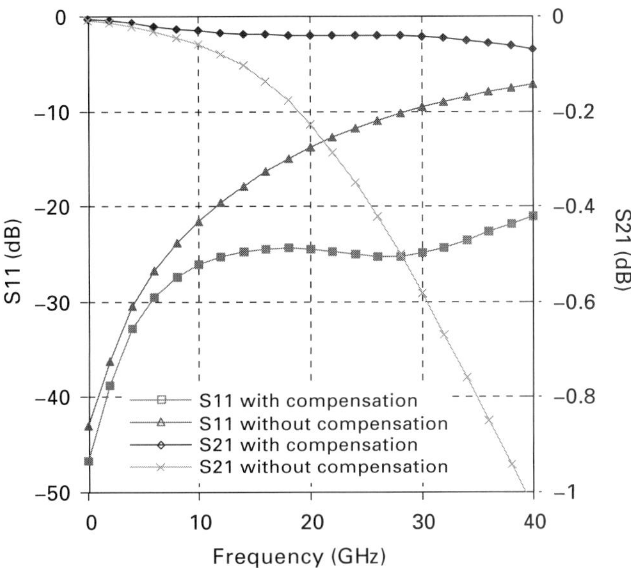

Fig. 5.5 Simulation data of package base microstrip to bond wire transition with and without bond wire inductance compensation: insertion loss with and without compensation (diamonds and crosses); return loss with and without compensation (squares and triangles).

is necessary after all four subsections of the package feed-through have been designed and merged together.

5.1.3 Grounded coplanar waveguide (CPWG) below package base

The third subsection consists of the back side (Fig. 5.1b) of a surface mount package, which makes contact with the PC board. Figure 5.7 illustrates this subsection. A 50 Ω CPWG was designed at the bottom of a package. It extends from the package edge to its bottom via pad. The signal trace width of the CPWG is linearly tapered down towards the bottom via pad while the 50 Ω characteristic impedance is maintained by varying the spacing between the signal line and ground plane. Table 5.3 shows the relevant dimensions of this subsection structure. The total signal line length, *Sig*_L, is roughly 2 mm and is sliced eight times; this determines the signal-to-ground spacing dimensions from the signal line width at each sliced section; *Gap_B* is the first slice, and *Gap_A* is the last slice (See Fig. 5.7). Curved ground planes are drawn on the basis of the eight spacing dimensions just determined. The tapered CPWG is designed on a 508 μm (20 mil) microwave laminate board, with dielectric constant 3.48. The simulation assumes that the package is mounted on a test PC board. Our design also takes into account the 254 μm thick dielectric material that is part of the LCP package base, above the signal trace. This

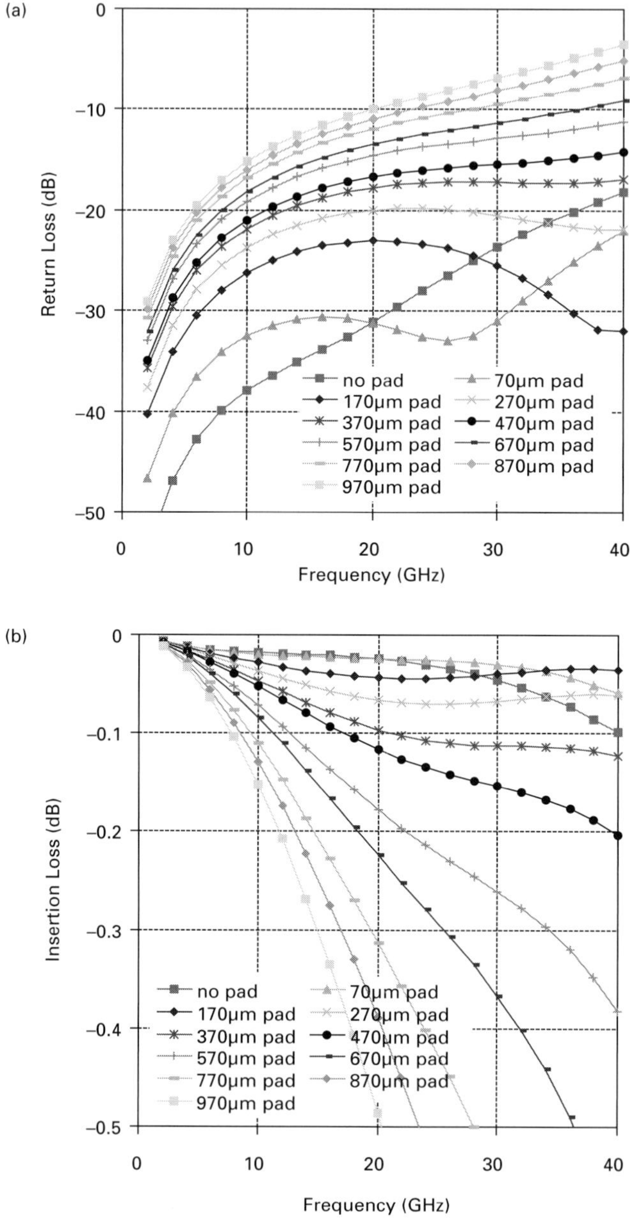

Fig. 5.6 (a) Insertion loss and (b) return loss simulation data for package base microstrip to bond wire transition with various inductance compensation pad lengths.

is necessary because having a dielectric material rather than air above the conductor changes its effective dielectric constant. Figure 5.8 shows simulated predictions for our CPWG design laid out below the package base. The simulation shows a better than 20 dB return loss and a less than 0.3 dB insertion loss up to 40 GHz.

Table 5.3. Dimensions for CPWG below package base

Dimension variable	(μm)
Gap_A	723
Sig_A	1117
Gap_B	240
Sig_B	800
Sig_L	2025
Test board	508
Package substrate	254

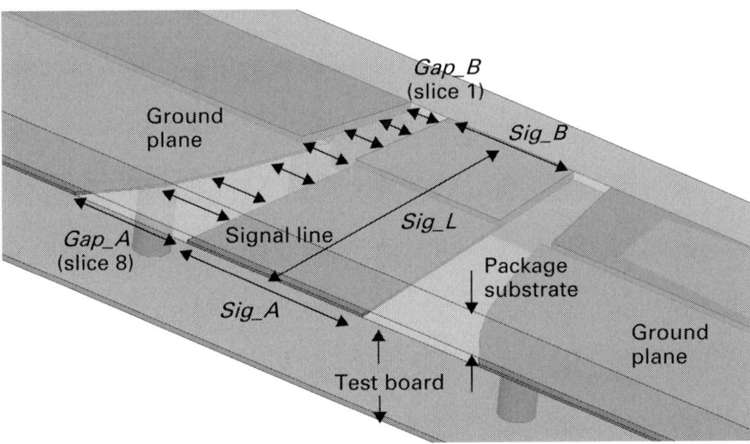

Fig. 5.7 Simulation structure of the package-base grounded CPW to test-board grounded CPW transition.

5.1.4 Coplanar waveguide probe launch on test board

The fourth subsection includes a CPWG design on a PC board, which connects the signal trace at the package-base edge to the probe signal launch pads on a test board. Figure 5.9 illustrates this transition. The CPWG signal trace width at the package-base edge is 1.1 mm, and the spacing to the ground plane is 0.7 mm. However, the ground–signal–ground RF probes used to take measurements have only a 150 μm pitch between the signal and each ground probe. The probe pitch is chosen to be narrow for high-frequency-measurement accuracy. The probe pads on the test board were designed to have a 76.2 μm (3 mil) signal pad width and a 76.2 μm spacing between the signal pad and the ground planes. Therefore, a tapering design is necessary to connect the probe pads and the CPWG located below the package base. This is achieved by linearly increasing the signal trace width from a 3 mil probe pad width to the signal width at the package-base edge and designing the spacing between the signal trace and ground planes to maintain a 50 Ω

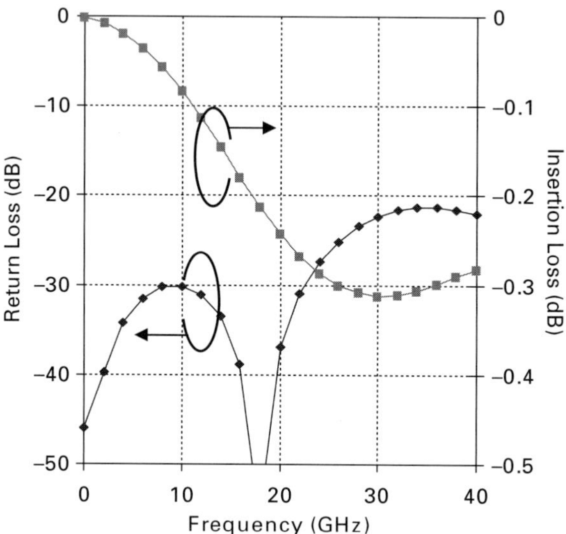

Fig. 5.8 Simulation data for the package-base grounded CPW to test-board grounded CPW transition.

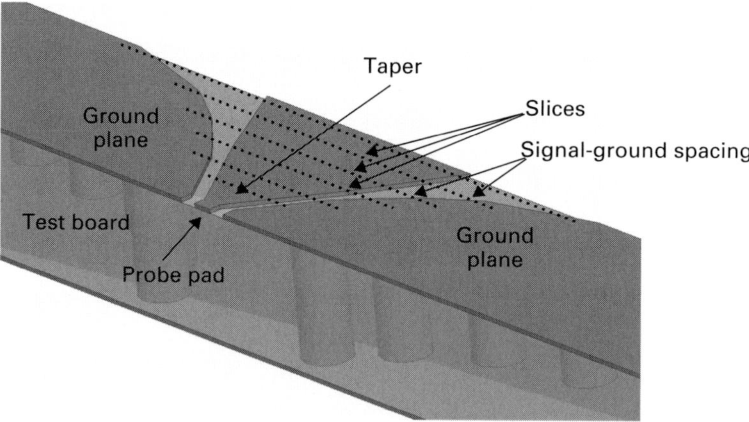

Fig. 5.9 Simulation structure of the probe launch to ground-backed CPW transition on the ground test board.

characteristic impedance throughout the taper. The spacing was roughly estimated by slicing the taper into five sections and computing the corresponding 50 Ω signal-to-ground spacing from the signal width measured at each slice (See Fig. 5.9). The tapering was designed to the best possible fabrication capability by an express prototype PCB fabrication service. Figure 5.10 shows an FEM-based prediction for a CPWG transition from RF probe pads to the edge of the package base on a PC board. The predictions show that the transition would incur 0.6 dB of insertion loss and 17 dB of return loss at 40 GHz.

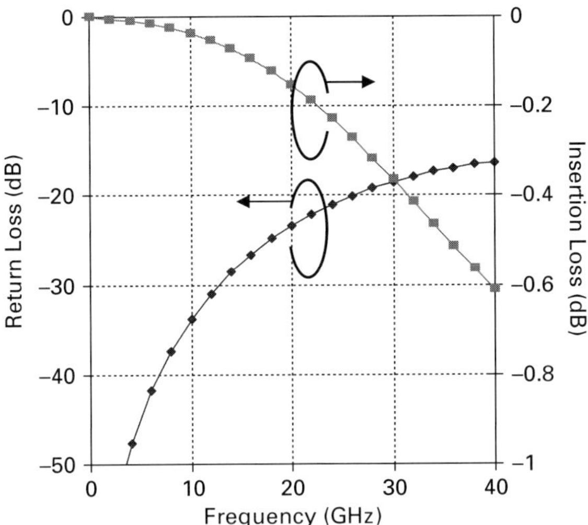

Fig. 5.10 Simulation data of the probe launch to ground-backed CPW transition on the test board.

Fig. 5.11 Cross-section of the package showing its copper ground and solder thicknesses [32] (© 2008 IEEE).

5.1.5 Electrical performance sensitivity of package trace thickness

Once the initial design is optimized, the four subsections comprising the via transition, the microstrip line to bond wire transition on the package base, the package base CPWG design, and the RF-pad to package-pad CPWG transition are combined together to be simulated in one file. The design process is followed by a fine-tuning of the combined package feed-through structure. Once the complete feed-through structure is optimized, the electrical performance sensitivity is investigated. Two parameters will be analyzed here, the variation in the total metal thickness at the package ground and test board interface and the variation in the bond wire length. Figure 5.11 illustrates the copper and solder thickness parameters that together comprise the total metal thickness at the package and test board interface. Figure 5.12 shows the package feed-through's electrical sensitivity to the ground copper and solder thicknesses. The total interface metal thickness varies from 64 μm, 83 μm, 98 μm to 118 μm in simulations. As can be seen, the total metal thickness of the copper ground and solder does not have a dramatic effect on the electrical performance. Up to 72 μm of copper ground can be mechanically supported by the available cladding thickness of LCP films. Figure 5.13 shows how the simulated

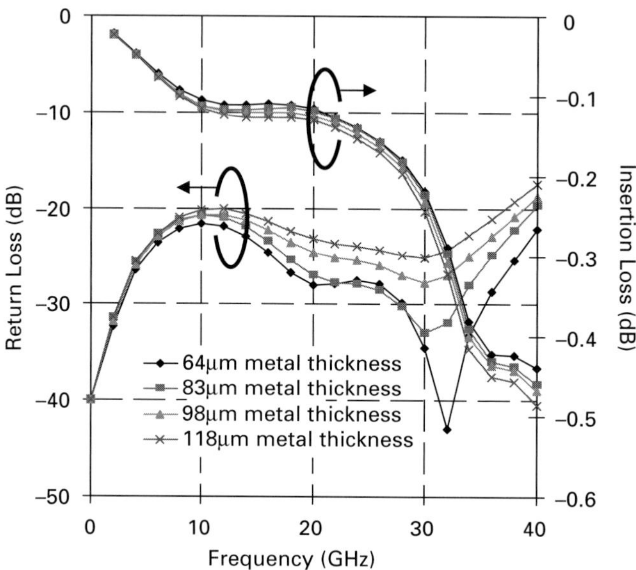

Fig. 5.12 Simulated S-parameters of the package feed-through for various package copper ground thicknesses (the solder thickness is fixed at 30 μm) [32] (© 2008 IEEE).

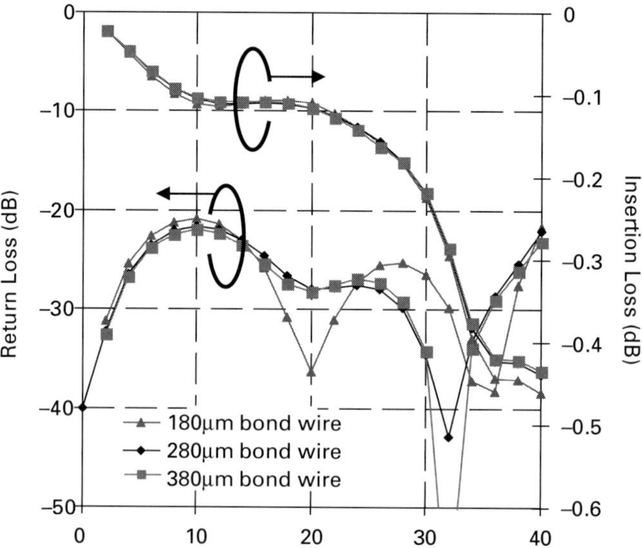

Fig. 5.13 Simulated S-parameters of the package feed-through for various lengths of the three bond wires [32] (© 2008 IEEE).

electrical performance of a feed-through changes with bond wire length. As the bond wire length varies from 180 to 380 μm, the insertion and return losses are maintained to within 0.03 and 8 dB, respectively. Note that the bond wire length represents the distance between the two contacts of the wire. Because bond wires become stretched vertically after the first bond, the total length of each wire is slightly more than that mentioned on the plot.

5.1.6 Theoretical attenuation of a package feed-through

In order to investigate a design limit, the theoretical attenuation of a package feed-through must be explored and compared with simulation data. This comparison is intended to assess the performance of an optimized feed-through design in comparison with the theoretical limit. The theoretical attenuation consists of the loss due to the signal trace metal, the loss due to the dielectric loss tangent of the test board and package base, the attenuation due to the dielectric's conductivity, and the attenuation due to radiation. In the present subsection, the attenuation due to the metal trace and the dielectric loss tangent are computed to approximate the theoretical minimum loss of our LCP package feed-through. The attenuation due to radiation is ignored because it introduces a minimal loss effect and so is smaller by orders of magnitude compared with the attenuation caused by the metal's conductivity, surface roughness, and the loss tangent. A detailed set of equations is presented in the chapter appendix.

Equations (5.5)–(5.8) [9] below are used to convert the S-parameter data obtained from FEM simulation to the loss, α, in dB. Defining

$$A = \frac{(1+\mathrm{S11})(1-\mathrm{S22})+\mathrm{S12}\times\mathrm{S21}}{2\times\mathrm{S21}}, \tag{5.5}$$

$$D = \frac{(1-\mathrm{S11})(1+\mathrm{S22})+\mathrm{S12}\times\mathrm{S21}}{2\times\mathrm{S21}}, \tag{5.6}$$

we write

$$\gamma = \frac{1}{l}\ln\left(A \pm \sqrt{A^2-1}\right), \tag{5.7}$$

$$\alpha = \mathrm{Real}(\gamma) \times 8.686 \quad (\mathrm{dB/unit\ length}), \tag{5.8}$$

where l is the length of the feed-through structure and γ is the propagation constant.

Equation (5.9) [10–15] below is used to compute the conductor loss due to a microstrip line and (5.10) [10–15] is used to compute the attenuation due to the dielectric loss tangent of the microstrip line and the coplanar waveguide. Equation (5.11) [10–15] is used to compute the attenuation due to the metal in a coplanar waveguide. The transmission line dimensions are based on the package feed-through specified in the FEM software. The above-mentioned equations are as follows:

$$\alpha_c = \frac{8.686 R_s}{Z_0 W}, \tag{5.9}$$

Table 5.4. Theoretical calculation of attenuation for package feed-through components

Package components	Substrate	Attenuation due to metal (dB/mm)	Attenuation due to dielectric loss tangent (dB/mm)
Microstrip	LCP	0.0147	0.0121
CPWG	RO4350B	0.0067	0.0207
Via	LCP	0.011	0.0121
Bond wires	air	0.089	N/A

$$\alpha_d = 8.686 \frac{2\pi f}{c} \frac{\left[\varepsilon_r\left(\varepsilon_{re}-1\right)\right]\tan\delta}{2\sqrt{\varepsilon_{re}}\left(\varepsilon_r-1\right)} \ (\text{dB/unit length}), \tag{5.10}$$

$$\alpha_c = 4.88\times10^{-4} \frac{R_s \varepsilon_{re} Z_{ocp} P'}{\pi W_g}\left(1+\frac{S}{W_g}\right) \times \frac{\frac{1.25}{\pi}\ln\frac{4\pi S}{t}+1+\frac{1.25t}{\pi S}}{\left[2+\frac{S}{W_g}-\frac{1.25t}{\pi W_g}\left(1+\frac{\ln(4\pi S)}{t}\right)\right]^2} \ (\text{dB/unit length}), \tag{5.11}$$

where R_s is the series resistance per unit length (in ohm/m), μ_0 is the series inductance per unit length (in H/m), σ is the shunt conductance per unit length (in mho/m), α_c is the attenuation due to metal loss per unit length (in dB/unit length), Z_0 is the characteristic impedance of the microstrip, W is the width of the microstrip, α_d is the attenuation from the dielectric loss tangent, ε_r is the dielectric constant, ε_{re} is the effective dielectric constant, c is the speed of light (in m/s), h is the dielectric substrate thickness, S is the coplanar waveguide signal width, W_g is the coplanar waveguide signal-to-ground gap, t is the thickness of conductor, and Z_{ocp} is the characteristic impedance of the CPW.

Table 5.4 shows the theoretical attenuation for the package feed-through components computed from (5.9)–(5.11). Figure 5.14 shows the total attenuation of the package feed-through due to conductor loss and due to the dielectric. It can be seen that the attenuation due to conductor loss shows a logarithmic-type characteristic, the result of its proportionality to the square root of the frequency; this comes from the surface resistivity in (5.9). However, the attenuation due to the dielectric shows a linear characteristic since it is linearly proportional to the frequency, as shown in (5.10). Also, the conductor loss is higher than the dielectric loss, by at least a factor 2, up to 40 GHz. The total attenuation computed from the simulation and the total attenuation computed from the measurement data, using (5.5)–(5.8), are plotted in Fig. 5.15 along with the total theoretical attenuation. As can be seen, the theoretical

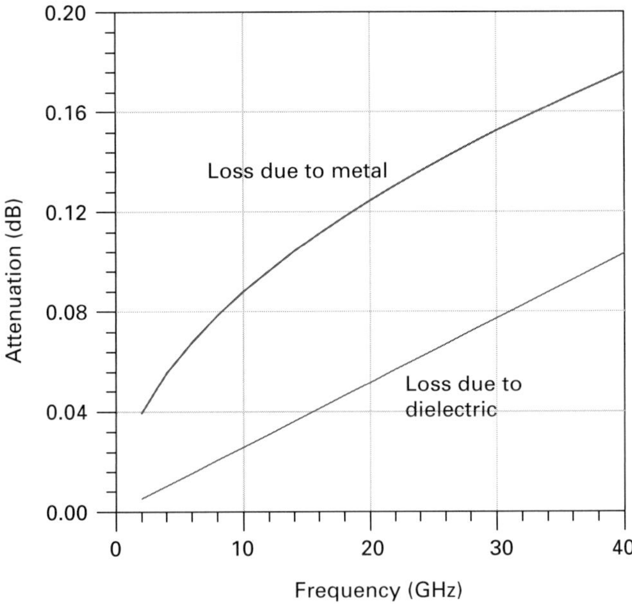

Fig. 5.14 The theoretical attenuation due to the metal and due to the dielectric loss tangent.

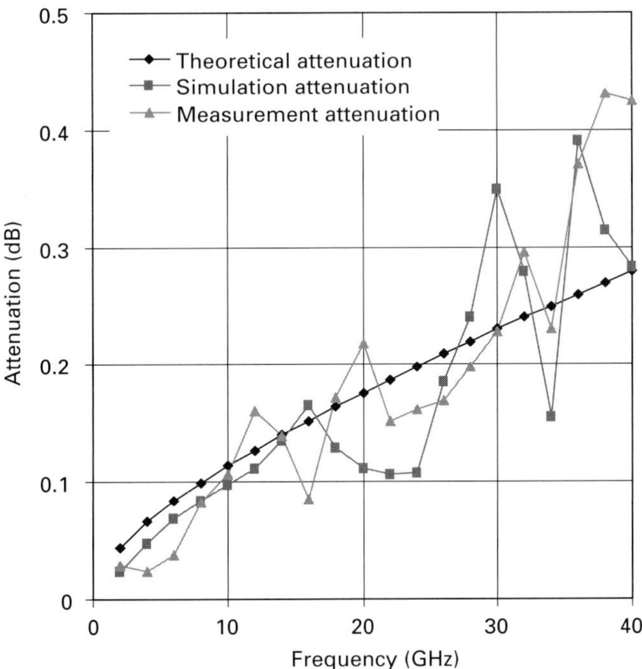

Fig. 5.15 The theoretical attenuation, simulation attenuation, and measured attenuation.

attenuation and the attenuation plots based respectively on design simulation and measurement vary by up to 0.15 dB. Although the attenuation plots computed from the simulation and from the measurements show fluctuations, their trend follows the theory fairly closely.

5.1.7 Fabrication of LCP package base

Package bases were fabricated on multilayer LCP films using a standard PCB process. Five layers of LCP film with respective thicknesses of 25.4, 50.8, 101.6, 50.8 and 25.4 μm (1 mil) were laminated to produce a 254 μm (10 mil) thick substrate. Two types of LCP with different melting temperatures, 280 °C and 315 °C, were used. These two types were staggered, with the higher melting temperature LCP at the top, bottom, and middle, to make up the five layers.

The films were pressured and preheated and then ramped up to 285 °C for lamination to be carried out in a single step, to develop a 10 mil thick sample. The bottom ground was 36 μm thick to provide mechanical support. The top metal traces were 9 μm thick. Once the multilayer LCP package base was fabricated, a package cavity was drilled using laser ablation [3], as can be seen in Fig. 5.16a. In production, LCP films can be made with holes using low-cost machining or punching techniques [16].

5.1.8 Feed-through measurement and model extraction

Prototype LCP SMT packages were mounted onto a 508 μm (20 mil) thick microwave PC board using solder. In order to characterize the package feed-through, a 50 Ω coplanar waveguide-to-microstrip adapter [17,18] was mounted inside a cavity approximately 200 μm from the LCP cavity wall to provide a micro-probe connection. The adapter consists of a 50 Ω line fabricated on a 127 μm thick alumina substrate ($\varepsilon_r = 9.6$) with a ground–signal–ground probe pattern at one end [17]. The 200 μm gap between the adapter and the cavity wall matches the condition in MMIC packaging.

This construction establishes a complete package transition from the bond wires through a vertical via to the test PC board. The input from the package feed-through was probed at a CPWG launch on the test board. The package feed-through transition was measured using a probe station, network analyzer and 150 μm pitch coplanar waveguide probes (see Fig. 5.16b). The network analyzer, was calibrated up to 50 GHz using a line–reflect–line technique. Measurements on feed-through transitions included a small grounded CPW signal launch on the PCB, a solder joint between the PCB pad and package, a vertical via transition, a short microstrip line on the package substrate, three bond wires, and a ceramic adapter (see Fig. 5.16).

Figure 5.17a shows simulation, measurement, and lumped-modeled results for the LCP package feed-through transition. The thicknesses of the copper ground and solder used in simulation were 30 μm and 36 μm, respectively. The feed-through

(a)

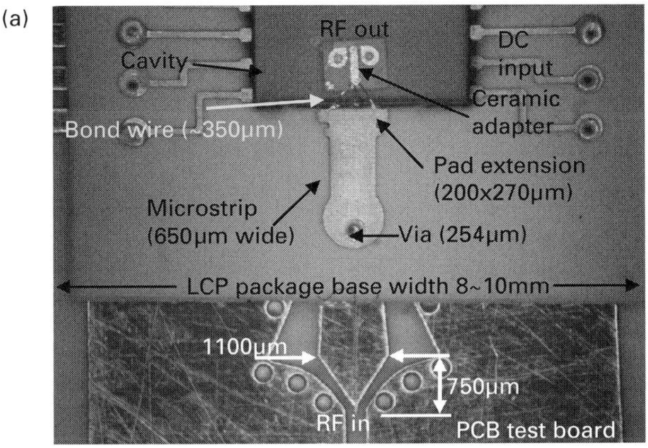

(b)

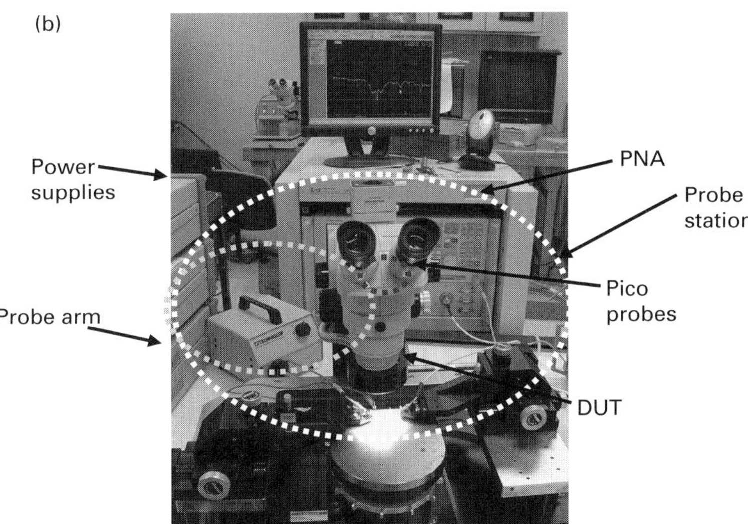

Fig. 5.16 (a) Thin-film LCP surface mount package mounted on a printed circuit board [32] (© 2008 IEEE); (b) RF/microwave probe station measurement setup.

design was optimized for Ka-band operation, but measurements from 0 to 40 GHz are shown for completeness. The measured data show a return loss of better than 20 dB and an insertion loss of 0.4 dB up to Ka-band. The simulation predicted a lower insertion loss and a better return loss than the measured values at the lower end of the Ka-band. This was due to the uneven solder thickness used in mounting the package on the PC board. We recall from the simulation in Fig. 5.12 that thicker metal degrades the electrical performance at Ka-band. In addition, the bond wire lengths were not exactly as those used in the simulation. The bond wire compensation pads were designed to compensate a specific length of bond wire inductance. Therefore, variations in wire length can cause mismatch. Overall, the experimental results demonstrate that a well-matched package feed-through can be designed

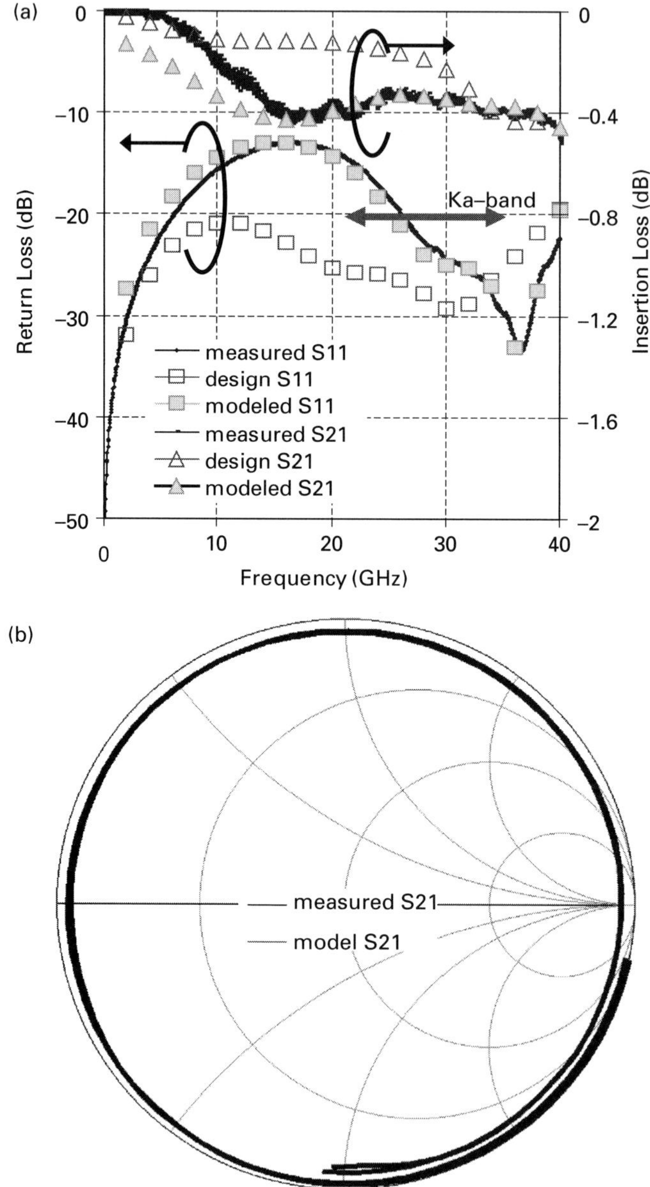

Fig. 5.17 (a) Measured, design, and modeled return loss and insertion loss for package feed-through, [32] (© 2008 IEEE); (b) the measured and modeled S21 phases of the package feed-through, on a Smith chart [32] (© 2008 IEEE).

that enables surface mount packages for millimeter-wave frequencies. Figure 5.17b shows the measured and modeled S21 phases.

Figure 5.18 shows an equivalent circuit model of a package feed-through transition; the corresponding values are shown in Table 5.5. Initially, a simple lumped

Table 5.5. Lumped element values of the package feed-through equivalent circuit model

Element	Value
*L*1	0.26 nH
*R*1, *R*2, *R*3	0.008 Ω
*C*1	0.0653 pF
*L*2	0.172 nH
*C*2	0.115 pF
*L*3	0.059 nH
*C*3	0.078 pF
*C*4	0.007 pF
*C*5	0.049 pF

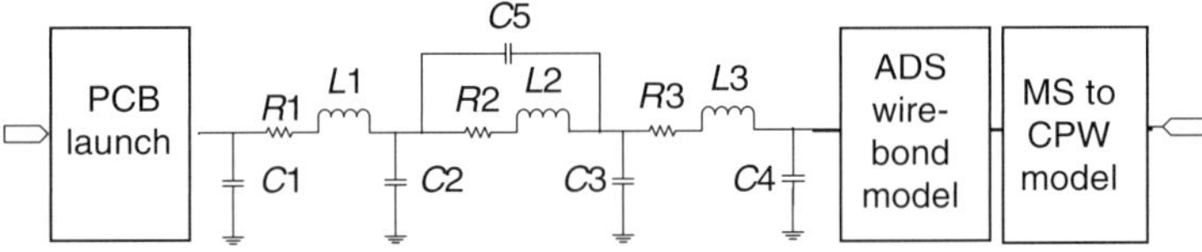

Fig. 5.18 Lumped circuit model of package feed-through transition [32] (© 2008 IEEE).

element model is generated using model extraction software. Then the circuit can be optimized using a circuit simulator to correlate the modeled S-parameters with the measurements. The resistors *R*1 and *R*3 in Fig. 5.18 represent the microstrip line conductor losses at the top and bottom of the package base; *L*1 and *L*3 model the inductance of the microstrip lines; *C*1 and *C*4 are the capacitances between the microstrip lines and the grounds; *R*2 and *L*2 model the conductor loss and the inductance of the via, respectively; *C*2 and *C*3 represent the capacitances of the via pads. The circuit also includes a PCB launch, bond wires, and a CPW-to-microstrip adapter, all modeled using circuit simulator built-in components. The modeled and measured S-parameters are well correlated, according to the dB and phase plots in Fig. 5.17.

5.1.9 Insertion loss roll-off beyond 40 GHz

Beyond 40 GHz, there is a roll-off in insertion loss. This is caused by a mismatch of the via pad capacitance to ground and the inductance of the via barrel as the frequency increases beyond 40 GHz. Such conclusion may be reached by observing the insertion loss variation with respect to changes in the lumped element values from the model. The roll-off beyond 40 GHz is compensated by lowering the via pad capacitance values *C*2 and *C*3 and the via barrel inductance *L*2. Furthermore, element *C*3, the top via pad capacitance to ground, is reduced the most, 25%, compared with the reduction in via barrel inductance, 10%, and in the bottom via pad

Table 5.6. The design dimensions and the new dimensions of the package feed-through via transition needed to reduce the insertion loss roll-off

	New dim. (μm)	Cap. (fF)	Ind. (pF)	Design dim. (μm)	Model cap. (fF)	Model ind. (pH)
Top via pad diameter	680	58.5	N/A	900	78	N/A
Bottom via diameter	800	109.5	N/A	755	115	N/A
Via pad offset	280	N/A	N/A	280	N/A	N/A
Via height	229	N/A	155	254	N/A	172
Via diameter	254	N/A	N/A	254	N/A	N/A

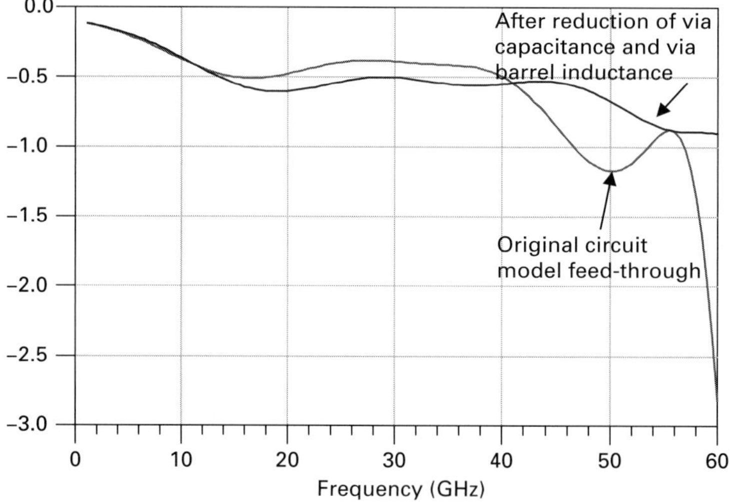

Fig. 5.19 The simulated circuit model for the feed-through before and after reductions in the via pad capacitances and via barrel inductance.

capacitance, 5%. Hence, the top via pad capacitance to ground may be the most sensitive element in a package feed-through design in improving the performance at higher frequencies. Table 5.6 shows the new dimensions of the via transition needed to achieve the new via pad capacitances and via inductance derived from the model-extraction software. The via pad offset distance and the via barrel diameter are unchanged. It may be noted that a percentage reduction in the capacitance or the inductance desired can be achieved by removing the same percentage from the diameter of via pad or the height of the via barrel, respectively. For example, if a 25% reduction in capacitance is desired for the top via pad then the diameter of the top via pad should be reduced by 25%, from 900 to 680 μm. Figure 5.19 gives for comparison a simulation of the original feed-through circuit model and a simulation for a circuit with reduced via pad capacitance and via barrel inductance. As can be seen, the insertion loss roll-off in the higher frequency range is mitigated when the via pad capacitances and the via barrel inductance are reduced.

5.1.10 Package isolation measurement

In this subsection we investigate the isolation between two feed-throughs in an LCP package. Package isolation analysis is necessary to determine the coupling between RF ports. Good isolation is desired to avoid cross-talk between the RF ports in a package. For characterization, each package feed-through is terminated with a 50 Ω resistor (Fig. 5.20a), and the two resistors are spaced 2 mm apart. Measurements are taken using the same package cavity dimensions as those used to package the amplifier in the next section. Microprobes are placed at the RF input and output on the PC board for measurement. Figure 5.20b shows the measured isolation and the simulated isolation for a package up to 60 GHZ. The simulations used FEM software. The measured data shows a better than 45 dB isolation between the two terminated feed-throughs, which were separated by 2 mm.

5.1.11 Measurement and simulation of packaged amplifier

In this section, we explain how an active device is packaged and measured to further demonstrate the performance of the developed LCP package feed-through and its circuit model. An amplifier [19] is assembled into an LCP surface mount package. This amplifier is typically used from 28–36 GHz. The MMIC and by-pass capacitors are first mounted inside a package cavity, using silver-loaded polymeric adhesive, and cured at 200 °C for 30 minutes [20]. High-temperature adhesive is used to prevent re-flowing during lid lamination. Then, bond wires are used to make the DC and RF electrical connections. Triple bond wires are used between the RF input and output of the amplifier to package the feed-throughs. Figure 5.21a shows the amplifier mounted inside the package cavity. As can be seen, there is a gap between the LCP package wall and the amplifier RF input and output. The gap is provided to match the bond wire length to the simulated values for a package feed-through. A lid is laminated onto the base to seal the package. The a package is surface mounted on a test PC board. Figure 5.21b shows a package prototype. The packaged amplifier is measured and compared with simulations of a packaged MMIC. The simulation includes the package feed-through model and the bare chip data measured at the die level using microprobes. The feed-through model was taken from Fig. 5.18 with the values listed in Table 5.5. Figure 5.21c shows a schematic diagram of the simulated components. Figure 5.21d shows the measured gain and input–output return loss of the simulated and packaged amplifiers. From 33 to 38 GHz, the measured packaged low-noise amplifier (LNA) shows up to 7 dB better input return loss, and hence the gain is up to 1 dB higher than in the simulation. Overall, the simulation S-parameters of the packaged LNA show a similar trend to the measured data for the packaged amplifier. It is reasonable to assume

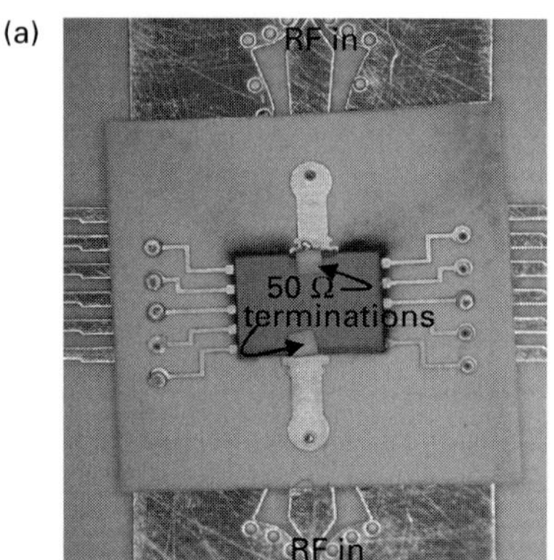

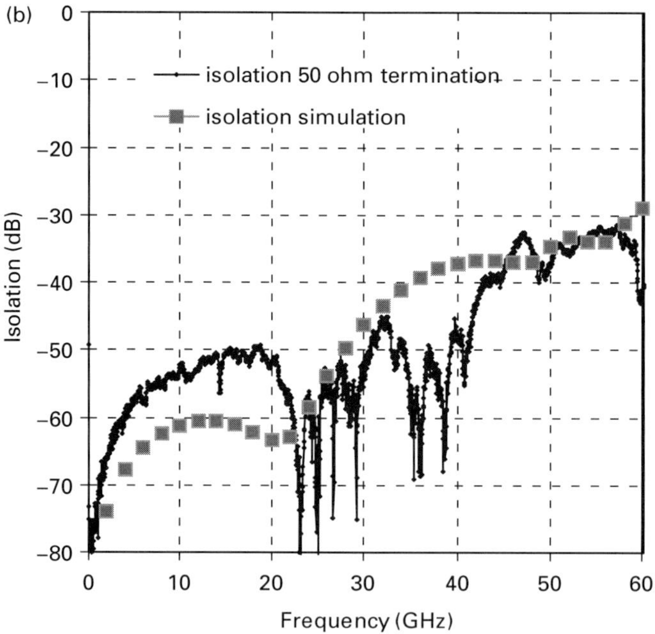

Fig. 5.20 (a) Top view of the package base demonstrating the 50 Ω termination at each feed-through for isolation measurement [32] (© 2008 IEEE); (b) the measured and designed isolation data between two terminated ports of the package feed-through [32] (© 2008 IEEE).

that the extracted package feed-through model can be simulated in conjunction with an on-wafer measurement of the amplifier to predict the characteristics of a packaged chip, because the entire package feed-through is designed to have 50 Ω characteristic impedance. It can be imagined as a 50 Ω transmission line. This work showed an excellent Ka-band package interconnect in LCP and demonstrated its performance with a packaged amplifier.

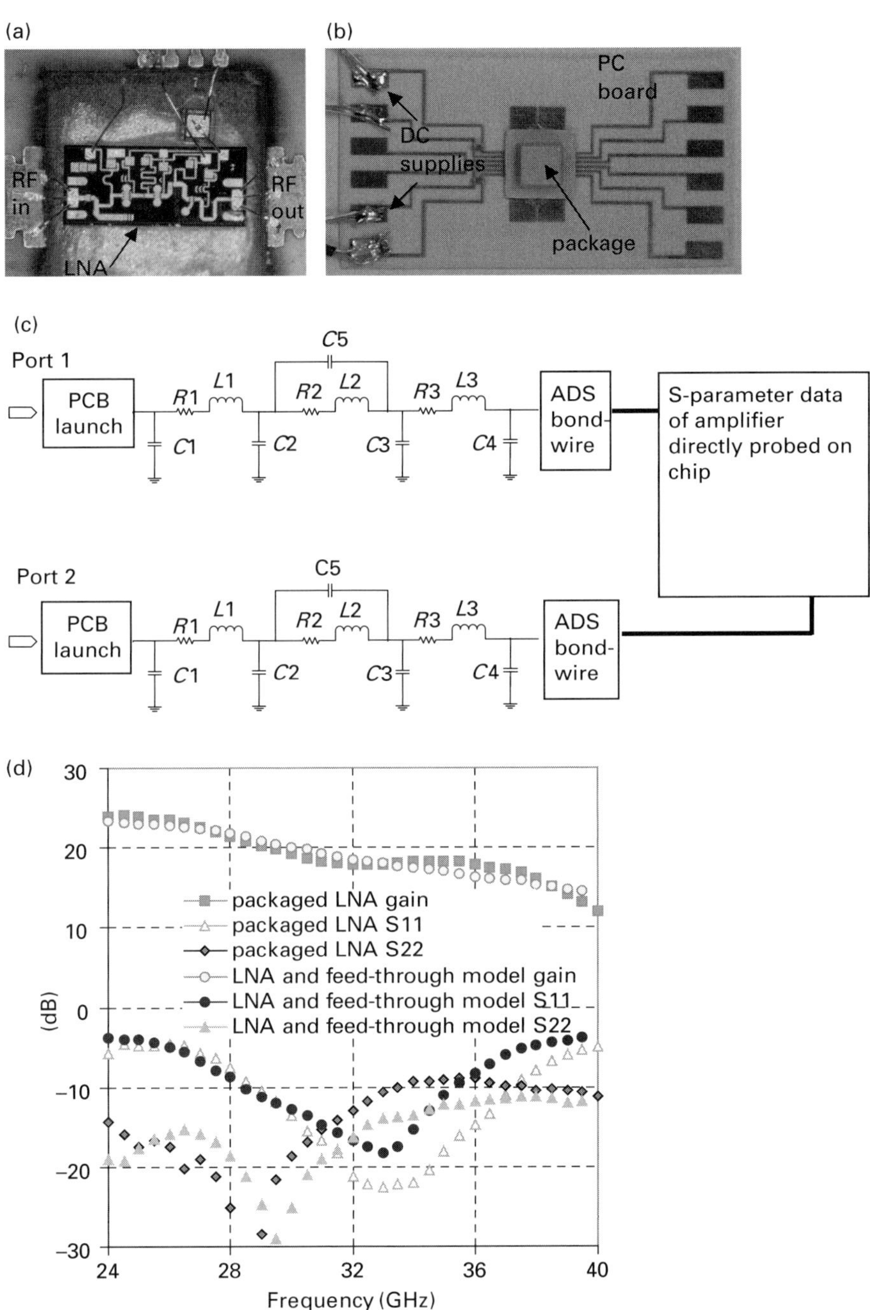

Fig. 5.21 (a) Packaged amplifier on PCB with lid [32] (© 2008 IEEE) and (b) without lid [32] (© 2008 IEEE), (c) schematic diagram of amplifier simulation and (d) comparison of gain, input return loss and output return loss between packaged amplifier and bare chip data with feed-through model simulation [32] (© 2008 IEEE).

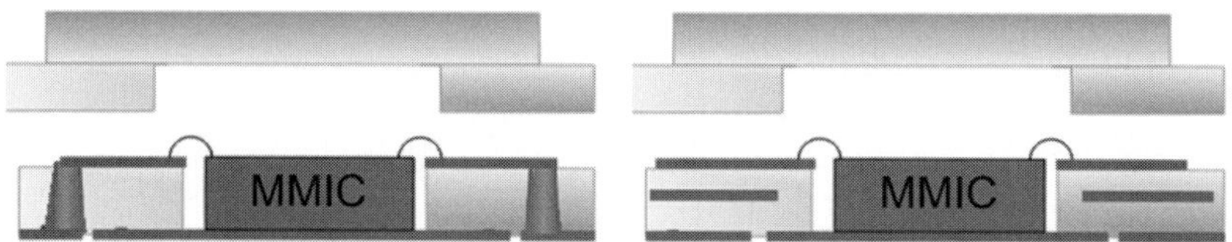

Fig. 5.22 Surface mount packages: on the left a low-pass via-based package and on the right a high-pass via-less package.

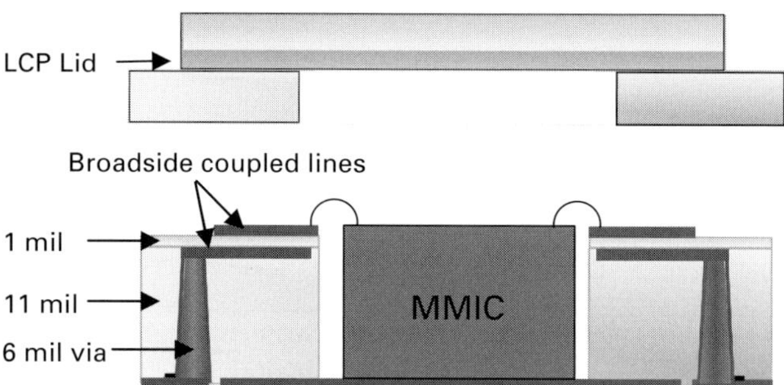

Fig. 5.23 Schematic diagram of a bandpass feed-through surface mount package [33] (© 2008 IEEE).

5.2 DC blocked coupled-line interconnect

Vias can be a source of many fabrication and reliability difficulties. Vias increase cost, create a possible path for contaminants to enter packaging, create virtual leaks in hermetic tests, and are prone to failure. A technique to eliminate vias involves a coupled line package interconnect structure. Figure 5.22 illustrates a typical low-pass SMT interconnect alongside a bandpass interconnect. Coupled-line bandpass feed-throughs are made possible by the thin-film capabilities offered by LCP. Such feed-throughs offer low-loss broadband performance, while providing inherent DC isolation and reliability. This section presents a novel bandpass package interconnect developed on multilayer LCP.

5.2.1 Coupled-line interconnect design

A bandpass interconnect structure was designed in multilayer LCP films using a compact broadside coupler. Coupled-line DC blocks were investigated for use as DC blocks first by Kajfez and Vidula [21]. The structure proposed was an edge-coupled symmetric inhomogeneous coupled line. The bandwidth and insertion loss for a given combination of even- and odd-mode impedances can be obtained. Tripathi [22] obtained ABCD parameters for asymmetric inhomogeneous lines, from which the bandwidth and insertion loss can be similarly obtained.

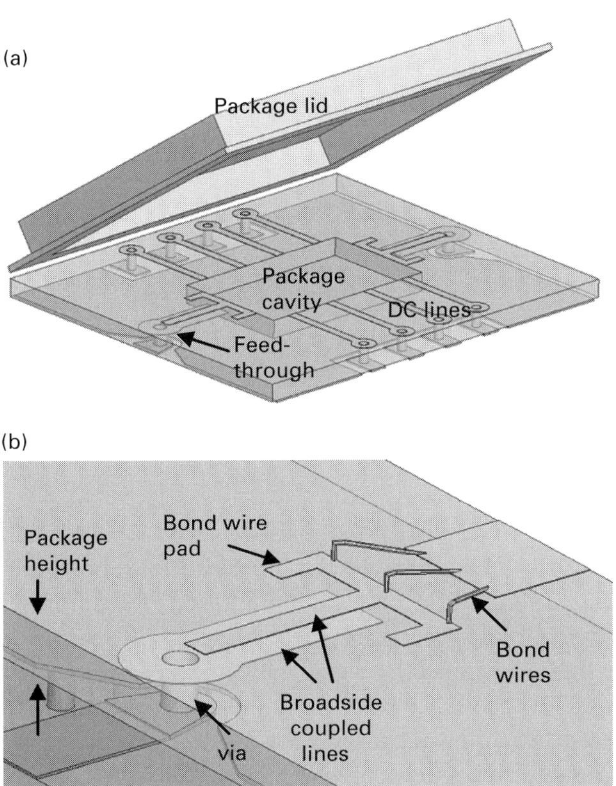

Fig. 5.24 LCP package diagrams: (a) a three-dimensional image of the surface mount package [33] (© 2008 IEEE); (b) a close-up of the structure including coupled lines, 6-mil via, and CPWG -to-microstrip transition [33] (© 2008 IEEE).

Figure 5.23 shows an LCP coupled-line interconnect package. Electrical signals enter the package through the bottom metal trace and continue to the broadside coupler through a plated via. The signals are broadside-coupled across a thin LCP dielectric. Bond wires connect the embedded MMICs to the package. Electromagnetic simulation is performed using FEM software. The resulting coupled-line package and interconnect design are shown in Fig. 5.24.

The broadside coupled strips were made ~400 μm wide. The coupled-line impedance values Z_c and Z_π were found to be 111 Ω and 6 Ω, respectively, using the symmetric approximation. The inner radius was 400 μm. The top-side coupled trace width is made less than that of the bottom, in order to effectively increase the impedance of both modes and improve the bandwidth. The top metal width represents a design trade-off between bandwidth and insertion loss. Triple bondwires were employed to reduce the package inductance. To reduce the mutual coupling, the coupler bond-pad width was increased to accommodate a wider bond spacing, as in [23]. As the spacing between the bond wires increases, the mutual inductance decreases and the bond wire resistance and inductance increase. This spacing was

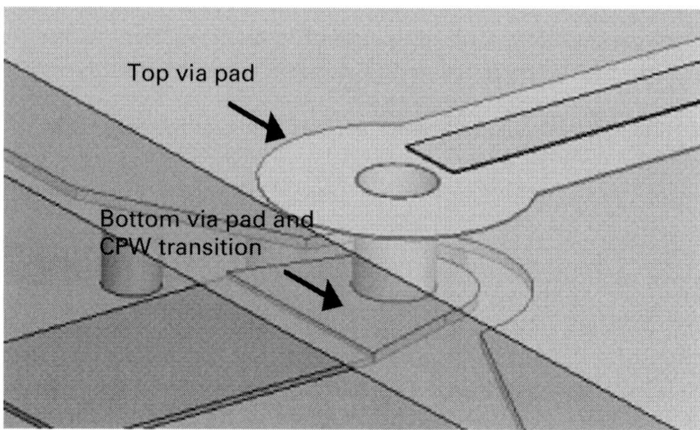

Fig. 5.25 Via transition.

optimized in simulation to be ~600 μm for bond wires of 110 μm height and 600 μm length. In addition, the package bond pads were extended to compensate the residual parasitic inductance and to approximate a lumped 50 Ω transition. The bond pads are bent, to reduce the total interconnect area. Although the performance degrades with this feature, this interconnect is acceptable for many applications.

Lastly, the design for the vias connecting coupled-line sections to the CPWG signal traces on the bottom of the package were considered, as illustrated in Fig. 5.25. The CPWG configuration was chosen to increase the top-via-pad shunt capacitance. The bottom via pad was set to the minimum value allowable (400 μm). This choice increases the top-via-pad capacitance-to-ground and minimizes the series capacitance. The top-via-pad diameters were chosen to approximate a lumped 50 Ω transmission line.

Coupled-line interconnect simulation results are presented in Figs. 5.26 and 5.27. The insertion loss is less than 0.8 dB over 20–50 GHz while the return loss is greater than 20 dB for the design with straight bond pads. The bent-bond-pad design has a return loss greater than 17 dB over a similar band.

A sensitivity analysis was performed to establish how robust this design is in relation to the fabrication and assembly uncertainties. Three specific design elements were considered: the bond wire length, the coupled-line alignment, and the coupling thickness.

Bond wire sensitivity

With automated equipment, bond wires can be reliably made with ~10% error. The results for various bond wire lengths are shown in Figs. 5.28 and 5.29. The structure performs well over 20 GHz to 50 GHz for bond wire lengths between 0.615 mm and 0.900 mm. Its performance deteriorates at the upper band edge for bond wire lengths less than 0.615 mm.

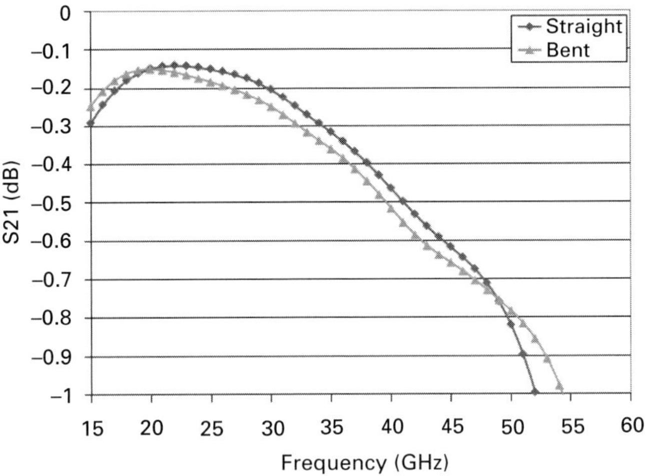

Fig. 5.26 Insertion loss for bandpass feed-through designs with bent bond pads and with straight bond pads.

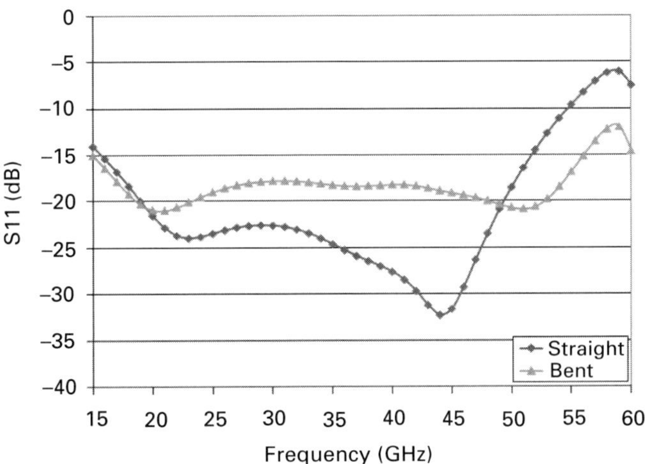

Fig. 5.27 Return loss for bandpass feed-through designs with bent bond pads and with straight bond pads.

Alignment sensitivity

During lamination, it is possible for LCP layers to shift laterally and longitudinally, as shown in Fig. 5.30. In this case, the lack of alignment of two lines forming a coupled line structure may impact performance. Offsets of 2, 4, 6, 8, and 10 mil were simulated. The results are shown in Figs. 5.31 and 5.32. For 0, 2, and 4 mil offsets in each direction, the interconnect is fully functional.

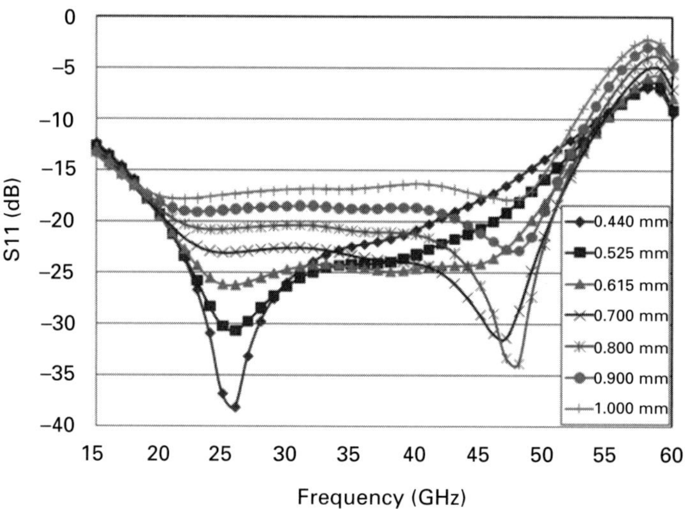

Fig. 5.28 Return loss versus bond wire length for the bandpass structure.

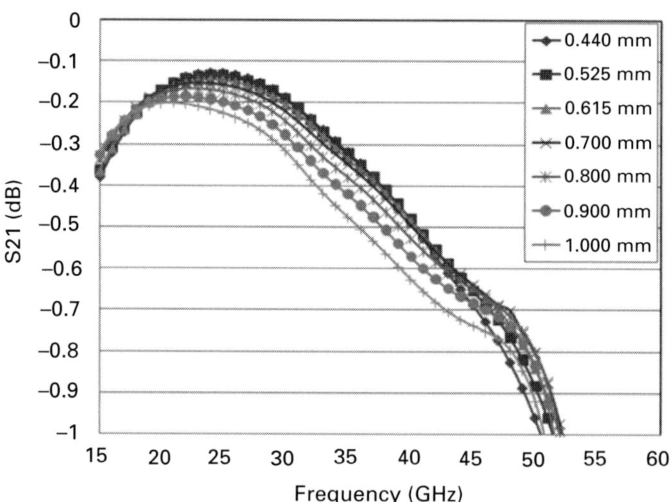

Fig. 5.29 Insertion loss versus bond wire length for the bandpass structure.

Material thickness sensitivity

The coupling thickness (Fig. 5.33) was varied between 0.6 and 1.6 mil. Coupling thicknesses less than 1 mil yielded structures that still perform quite well. As the coupling thickness increases beyond 1 mil, an in-band match is acceptable but the lower band edge increases in frequency. The coupling thickness corresponds to the odd-mode impedance created by the coupled-line capacitance. As the odd-mode impedance increases, with decreasing coupling thickness, the bandwidth decreases. However, this also changes the even- and odd-mode impedance balance,

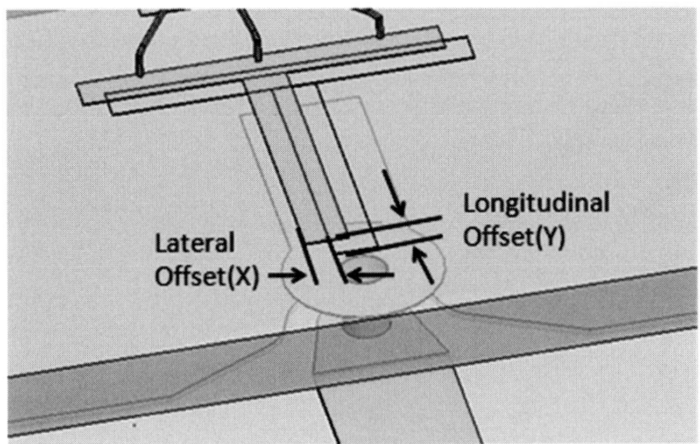

Fig. 5.30 Illustration of two-dimensional offset in the bandpass structure.

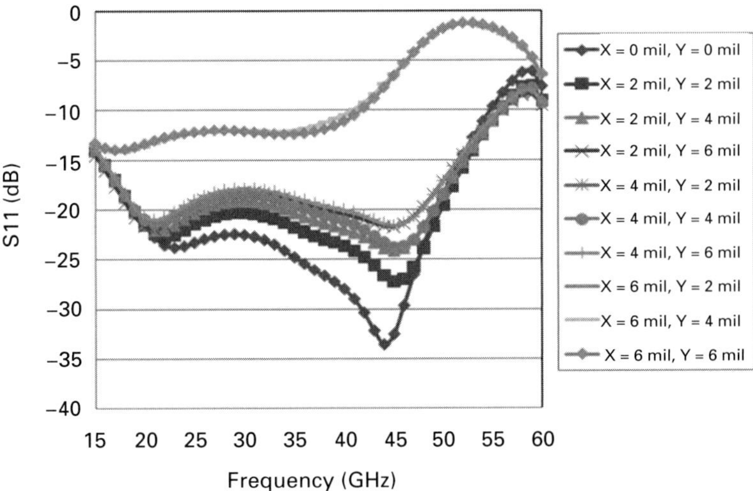

Fig. 5.31 Return loss versus offset for various combinations of lateral (X) and longitudinal (Y) offsets.

resulting in slight impedance-match degradation. The material thickness sensitivity S-parameter predictions are shown in Figs. 5.34 and 5.35.

5.2.2 Coupled-line interconnect fabrication and measurement results

Package bases were fabricated on multilayer LCP films using standard PCB processes [24]. Seven layers of LCP films with thicknesses varying from 1 mil (25.4 μm) up to 4 mil (101.6 μm) were laminated together to produce a 12 mil (304.8 μm) thick substrate, as illustrated in Fig. 5.36. In this stack, low-temperature 1 mil thick LCP bond films were inserted between the high-temperature LCP films. Initially, the bottom six layers were laminated to produce an 11 mil (279.4 μm) thick substrate.

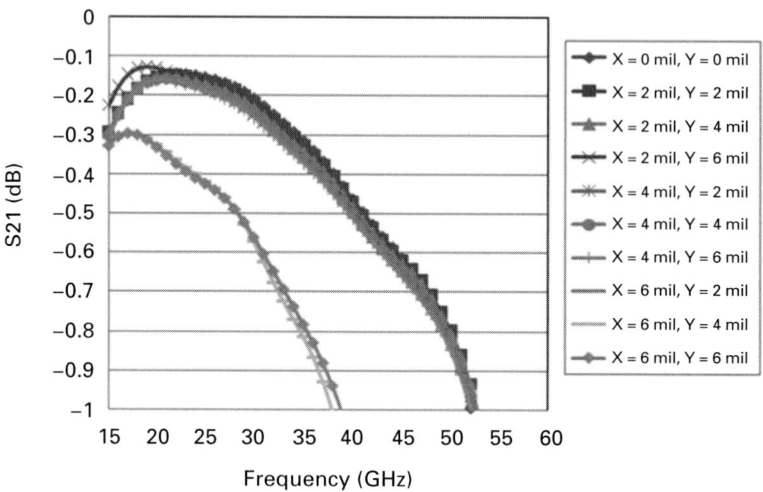

Fig. 5.32 Insertion loss versus offset for various combinations of lateral (X) and longitudinal (Y) offsets.

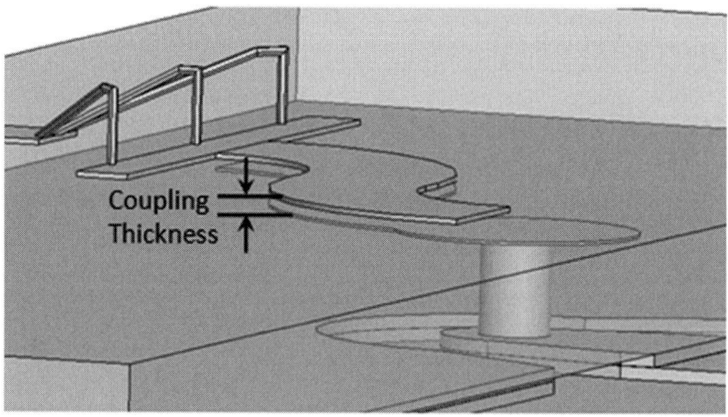

Fig. 5.33 Illustration of coupling thickness.

Secondly, internal metallization was applied. Thirdly, blind vias were drilled and plated. Fourthly, a 1 mil thick high-temperature LCP was laminated to the top of the stack to complete the package base. The bottom ground was made 36 μm thick to provide mechanical chip support. The top metal traces were ~9 μm thick. Lastly, a chip cavity was created using laser ablation (Fig. 5.36). In production, LCP films can be pierced with holes using low-cost machining or punching techniques. (See Fig. 5.24 for a complete package prototype.)

The fabricated packages were mounted on PCB with probe pads for testing. A 10 mil thick microwave circuit board was fabricated with a transition from package underside to CPWG probe pads. A 5 mil stencil was used to apply solder paste for mounting the LCP package. Finally, a microstrip-to-CPWG adapter was mounted

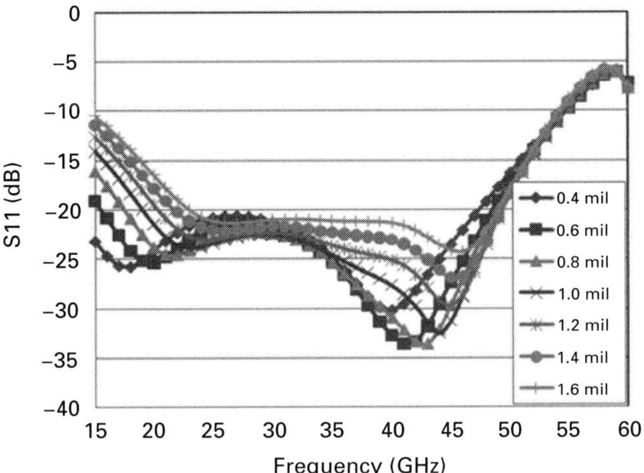

Fig. 5.34 Return loss versus coupling thickness.

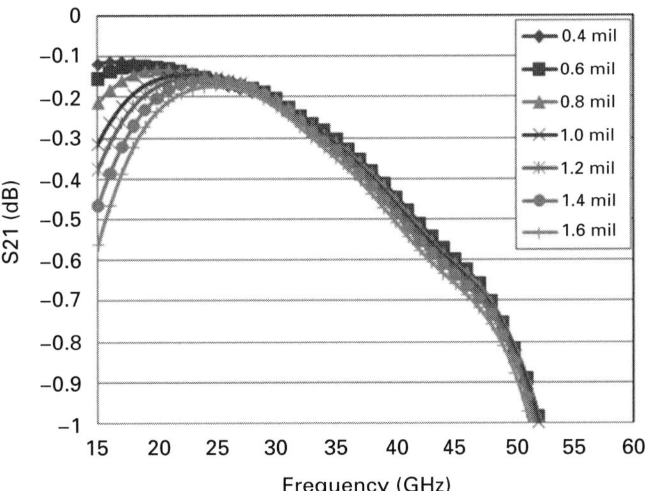

Fig. 5.35 Insertion loss versus coupling thickness.

on the PCB board to emulate the MMIC bond pads. The process is illustrated in Figs. 5.37a–d.

Sectioned views on the coupled-line interconnect were examined in order to verify fabrication. Fig. 5.38 shows where two cross-sections were made. The cross-sections were imaged using a scanning electron microscope (SEM) as shown in Fig. 5.39. The coupling cross-section enables examination of the coupled-section spacing, width, and alignment. The via cross-section enables examination of the via radius and shape and solder-paste thickness. The images show the correct fabrication of the coupler and vias.

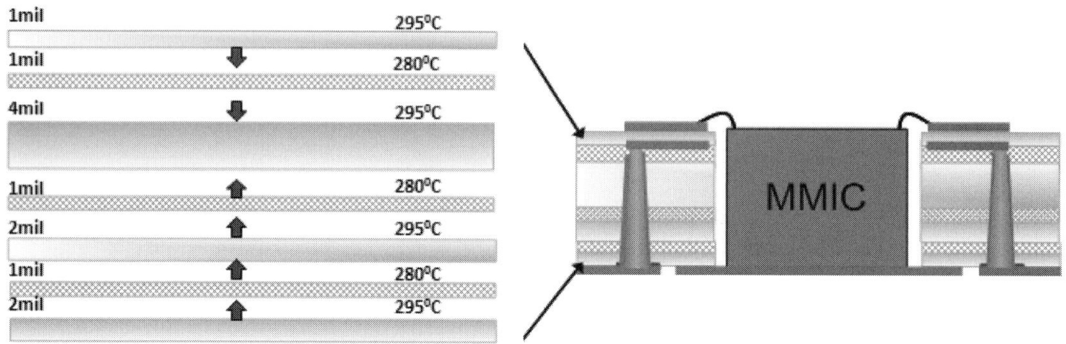

Fig. 5.36 LCP package layer stack-up.

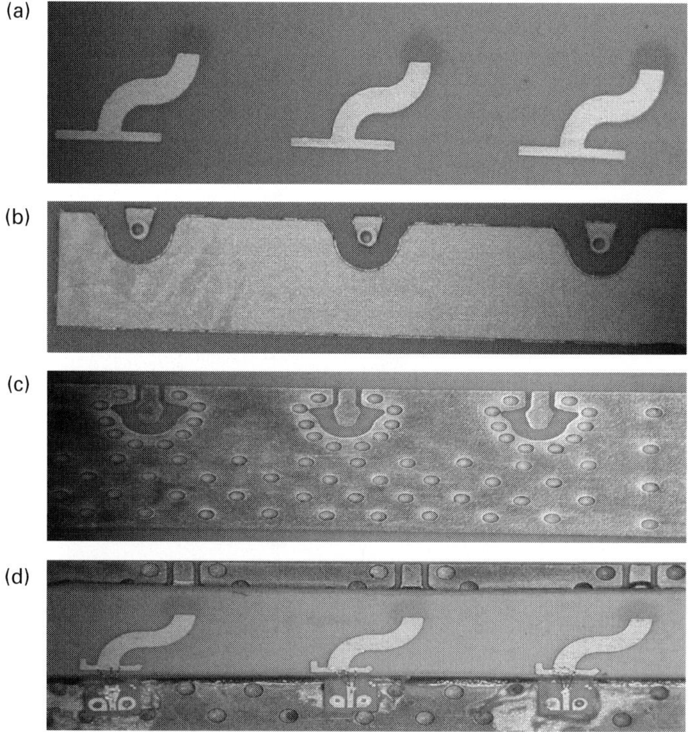

Fig. 5.37 (a) LCP package top, (b) LCP package bottom, (c) PCB top, and (d) LCP package mounted on PCB with CPWG-microstrip adaptors [33] (© 2008 IEEE).

Results for the return loss and insertion loss are shown in Figs. 5.40 and 5.41. Measurements show a better than 17 dB return loss over the 15–35 GHz frequency band. The insertion loss over this band is maintained to less than 0.6 dB. Hence, we have shown excellent performance and manufacturability for a coupled bandpass interconnect operating over 15 GHz to 35 GHz.

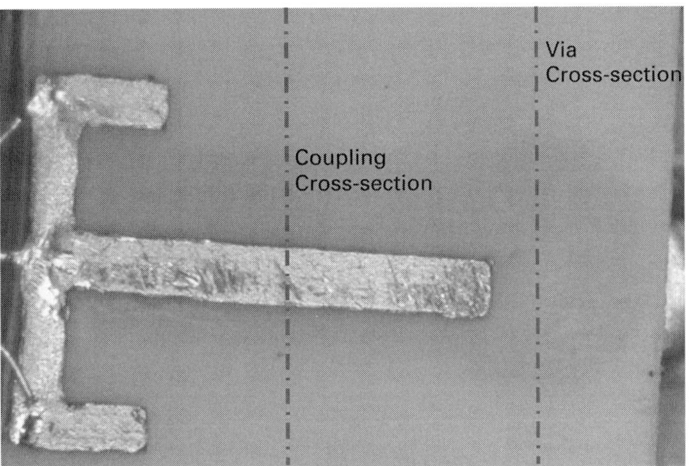

Fig. 5.38 The positions of two cross-sections made for SEM imaging.

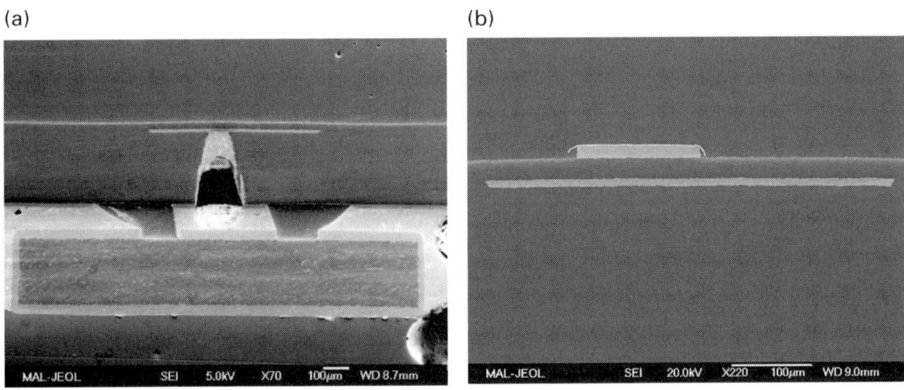

Fig. 5.39 The SEM cross-sections: (a) via cross-section and (b) coupling cross-section.

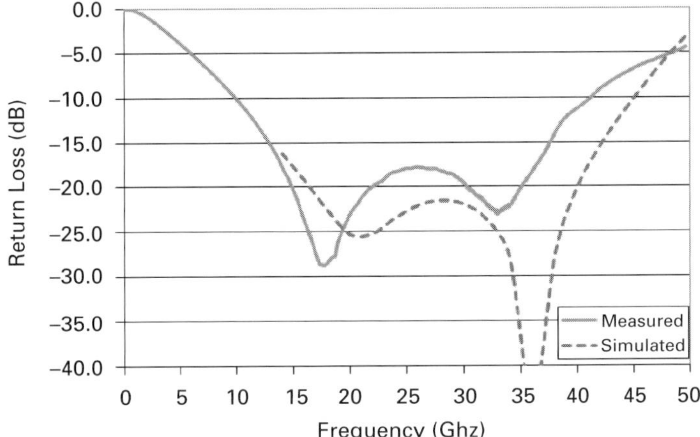

Fig. 5.40 Measured coupled-line interconnect return loss.

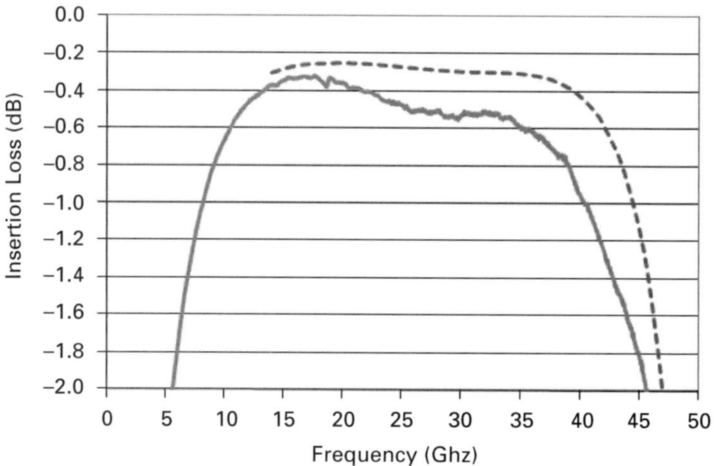

Fig. 5.41 Measured coupled-line interconnect insertion loss.

5.3 DC blocked lumped-element coupled interconnect

The bandpass feed-throughs presented in the previous section offer advantages over their low-pass counterparts, including DC blocking and reliability. However, coupled-line interconnects suffer from inherently larger sizes, as they are proportional to quarter-wavelengths. Further, coupled-line interconnects characteristically exhibit a periodic frequency response. An alternative and convenient way to achieve a DC-blocked interconnect is through a lumped-element capacitive coupled feed-through. Such a feed-through consists of a metal–insulator–metal (MIM) capacitor and a series inductor. The 1 mil LCP layer in an MCM-L package corresponds to a MIM capacitor, while the bond wires and a high-impedance line provide series inductance for frequency matching. Lumped-element interconnects have fractional bandwidth of 0.6 and require half the area of their coupled-line counterparts. In this section, the design, fabrication, and results of a lumped-element bandpass interconnect are presented.

5.3.1 Lumped-element interconnect design

A circuit diagram for the simplest possible filtered feed-through is shown in Fig. 5.42. It consists of a series *LC* circuit, a shunt capacitor to ground, and a series inductor at the output. The shunt capacitor to ground allows two poles to form, allowing for a large fractional bandwidth. Although its out-of-band rejection is poor, this circuit is inherently DC blocked and eliminates the need for vias. A lumped element LCP interconnect is shown in Fig. 5.43. A bond wire from the chip to the package feed-through serves as a series inductor. A capacitor is formed by the metallization on both sides of the 1 mil thick LCP. The capacitor's bottom-plate layer provides

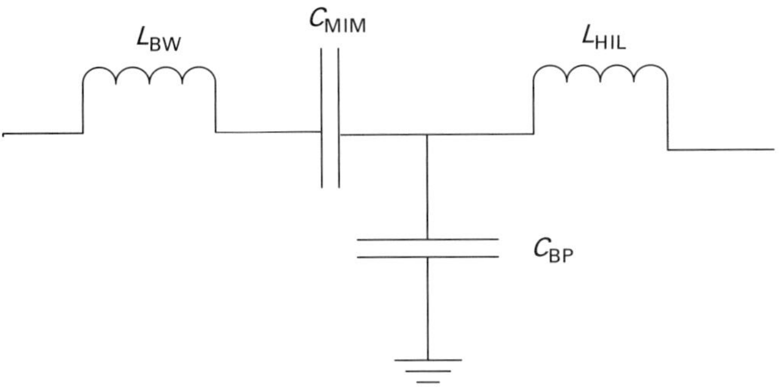

Fig. 5.42 Circuit model of lumped-element filter feed-through.

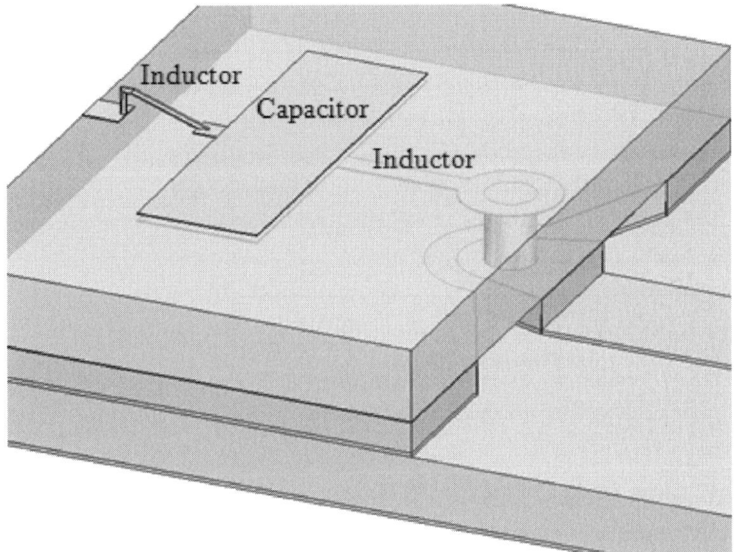

Fig. 5.43 An LCP filter bandpass feed-through.

a shunt capacitance to ground. The capacitor's bottom plate is also connected to a high-impedance line, which serves as an output series inductance. Figure 5.44 shows several options for customizing a lumped element interconnect for a wide variety of arrangements of packages by adjusting the capacitor dimensions, the high-impedance-line location, and an optional high-impedance-line meandering.

Optimizing a lumped-element interconnect in FEM simulation software gives the responses shown in Fig. 5.45. It is relevant to point out the design significance of the bottom metal layer's capacitor to ground. This capacitor requires feed-through response scaling by either changing the substrate thickness or making the top and bottom capacitor plates different sizes. The structure response of the transmitted signal is primarily due to the bond wire inductance, the high-impedance-line inductance, the series MIM capacitance, and the shunt capacitance to ground. An

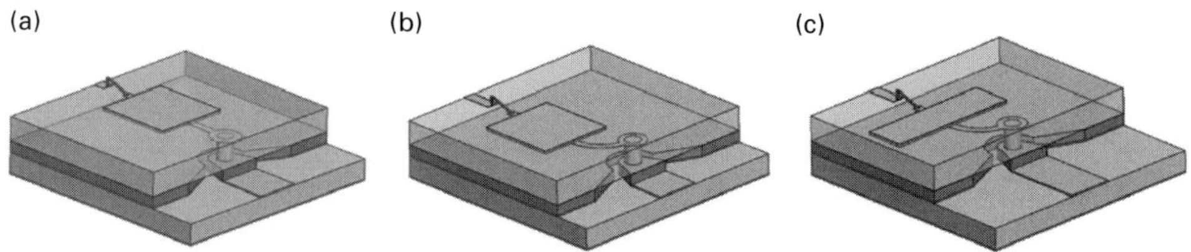

Fig. 5.44 Different possibilities for lumped element filter feed-throughs: (a) square capacitor; (b) square capacitor and curved high-impedance line; and (c) rectangular capacitor and curved high-impedance line.

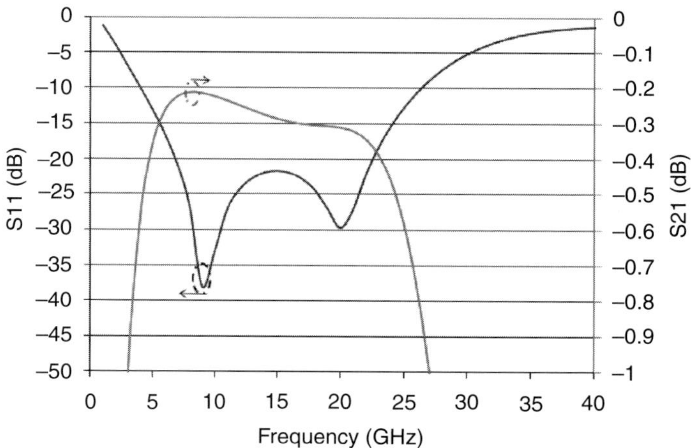

Fig. 5.45 Lumped-element filter feed-through S-parameters.

expanded circuit model, provided in Fig. 5.46, includes the bond wire resistance, the bond wire capacitance to ground, the top plate capacitance to ground, the top plate resistance, the bottom plate resistance, the high-impedance-line resistance, the high-impedance-line capacitance to ground, the via resistance, the via inductance, and the via pad capacitance to ground.

Approximate model parameters values are calculated as described below. The bond wire inductance and capacitance may be approximated by those of straight wire over a ground plane suspended in air [25]:

$$L = l \times 0.198 \ln\left[\left(1+\frac{2h}{d}\right)+2\sqrt{\frac{h}{d}\left(1+\frac{h}{d}\right)}\right](nH), \tag{5.12}$$

$$C = \frac{0.198l}{ln\left[\left(1+\frac{4h}{d}\right)+2\sqrt{\frac{h}{d}\left(1+\frac{h}{d}\right)}\right]}(pF), \tag{5.13}$$

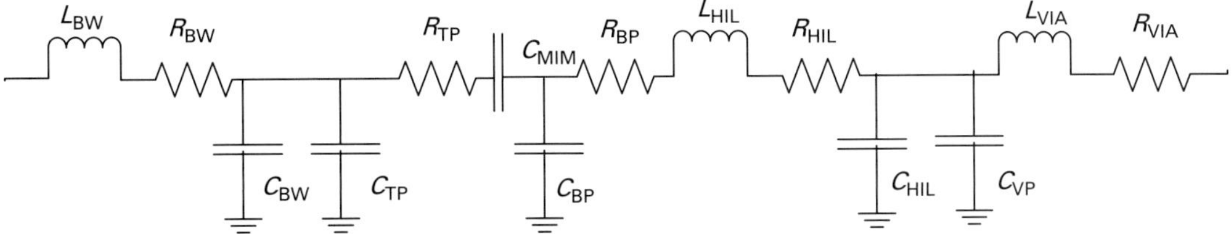

Fig. 5.46 Second-order model of lumped-element filter feed-through including the parasitics due to bond wire, LCP capacitor, high-impedance line, via pad, and via.

where h is the distance above the ground plane, d is the wire diameter, and l is the wire length (in mm). The resistance of the bond wire can be approximated by [25]:

$$R = \frac{4l}{\pi\sigma d^2}\left(0.25\frac{d}{\delta} + 0.2654\right), \tag{5.14}$$

where σ is the wire conductivity, d is the bond wire diameter (μm), l is the bond wire length (μm), and δ is the skin depth.

A high-impedance line can be approximated as a series inductor and shunt capacitance, with values derived from the lossless case given by the telegrapher's equations approximating a TEM waveguide:

$$L = l \times Z_{HIL}\ \omega/\beta, \tag{5.15}$$

$$C = \frac{l \times \omega/\beta}{Z_{HIL}}, \tag{5.16}$$

where β is a phase constant, ω is the angular frequency, and Z_{HIL} is the line impedance, which can be calculated as in [26]. The sheet resistance for a microstrip line, as well as for the LCP MIM capacitor plates, can be calculated by dividing the resistivity by the skin depth:

$$R = \sqrt{\frac{\pi f \mu}{\sigma}}\ (\Omega/\square) \tag{5.17}$$

where f is the frequency and μ_0 is the magnetic permeability of copper. The sheet resistance can then be modified according to the current distribution, as in [27]. It will be assumed, however, that all the current flows on the conductor facing the ground plane or on the inner faces for the MIM capacitor.

The via inductance was obtained through FEM simulation. An approximate resistance formula can be found in [28]. The resistance calculation is as follows:

$$R_{via} = R_{DC}\sqrt{1 + f\pi\mu_0\sigma t^2}, \tag{5.18}$$

where h is the via length, r is the via radius, R_{DC} is the via DC resistance, f is the operation frequency, σ is the metal conductivity, and t is the metal thickness. The via DC resistance R_{DC} is calculated assuming a uniform current throughout the via

Table 5.7. Model parameters

Parameter	Description	Appr. value	Tuned value
L_{BW}	Bond wire inductance	0.377 nH	0.377 nH
R_{BW}	Bond wire resistance	0.294 Ω	1.19 Ω
C_{BW}	Bond wire capacitance to ground	0.005 pf	0.01 pF
C_{TP}	Top plate capacitance to ground of LCP capacitor	0.02 pF	0.025 pF
R_{TP}	Top plate resistance of LCP capacitor	0.11 Ω	0.11 Ω
C_{LCP}	LCP capacitor value	1 pF	1 pF
C_{BP}	Bottom plate capacitance to ground of LCP capacitor	0.07 pF	0.07 pF
R_{BP}	Bottom plate resistance of LCP capacitor	0.11 Ω	0.11 Ω
L_{HIL}	Inductance of high-impedance line	0.429 nH	0.568 nH
R_{HIL}	Resistance of high-impedance line	0.23 Ω	1.359 Ω
C_{HIL}	Capacitance of high-impedance line	0.03 pF	0.06 F
C_{VP}	Via pad capacitance to ground	0.07 pF	0.07 pF
L_{Via}	Via inductance	0.12 nH	0.12 nH
R_{Via}	Via resistance	0.015 Ω	0.015Ω

cross-sectional area and then multiplying the resistivity by the via length divided by the cross-sectional area.

The via pad capacitance is found through FEM simulations by extracting the via inductance. Various LCP capacitor values were found with a two-dimensional field solver [29]. Approximate and tuned values chosen to match simulation to a lumped-element model are summarized in Table 5.7. For resistance values that vary with frequency, the model value was chosen as the resistance at 20 GHz.

The circuit model and the FEM simulation match almost exactly in-band for S21 and S11, as shown in Figs. 5.47 and 5.48, respectively. Deviations between the circuit model and the simulation at increasing frequency are expected, as the lumped element model becomes less accurate owing to transmission line effects as well as its neglect of the frequency dependence of resistance.

5.3.2 Lumped-element interconnect sensitivity analysis

The lumped-element interconnect sensitivity must be analyzed to investigate the electrical performance degradation due to fabrication and assembly tolerances. Three critical parameters were chosen for the analysis: the layer offset, the bond wire length, and the coupling thickness. The changes in these parameters can be directly related to the various parameters used to describe the second-order lumped-element model. The results show that performance is maintained well over variations in the above parameters.

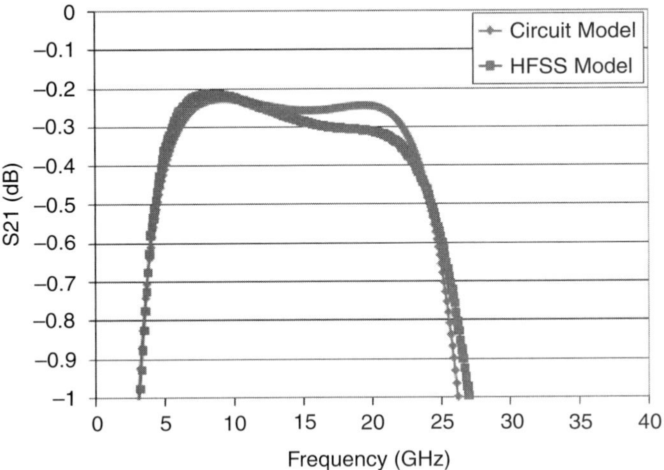

Fig. 5.47 The insertion loss for the second-order circuit and for the FEM model.

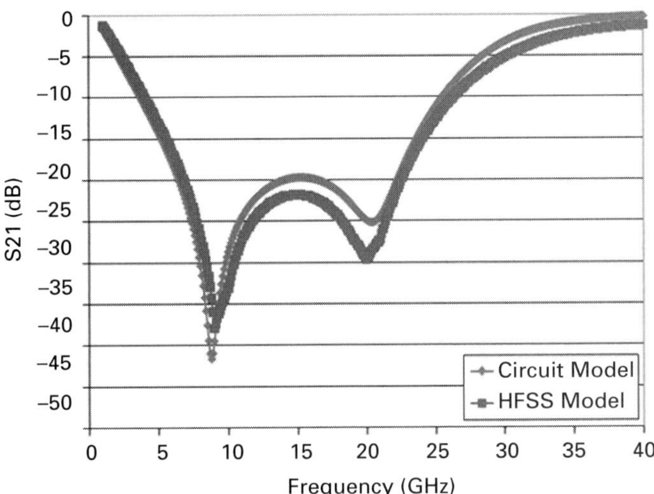

Fig. 5.48 The return loss for the second-order circuit and for the FEM model.

Lateral-offset sensitivity

The first parameter to be analyzed is the lateral offset between layers. Figure 5.49 shows how this offset is defined. Increasing the offset decreases the LCP MIM capacitance while increasing the top plate capacitance to ground. Figure 5.50 shows the return loss S11 for offset values ranging from 0 to 12 mil, or 0 to ~300 μm. The bandwidth response narrows with increasing offset. The response for a 4 mil offset is still within reason, and offsets can typically be held to 4 mil or less. Figure 5.51 illustrates the insertion loss S21 versus the offset. The response is maintained well for offsets 4 mil and under. An insertion loss of 0.3 dB or less

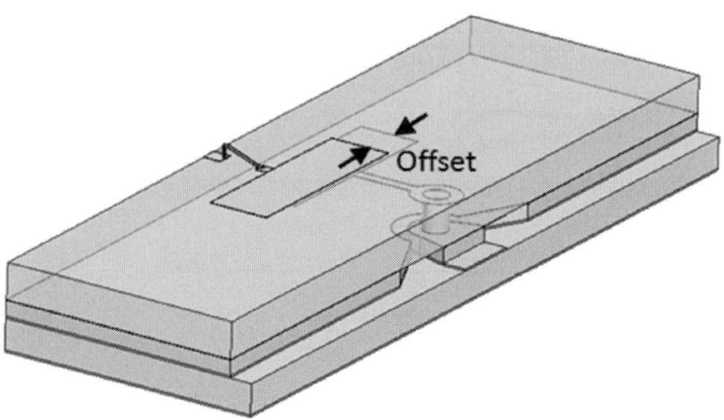

Fig. 5.49 The top and bottom metal offset as used in sensitivity analysis.

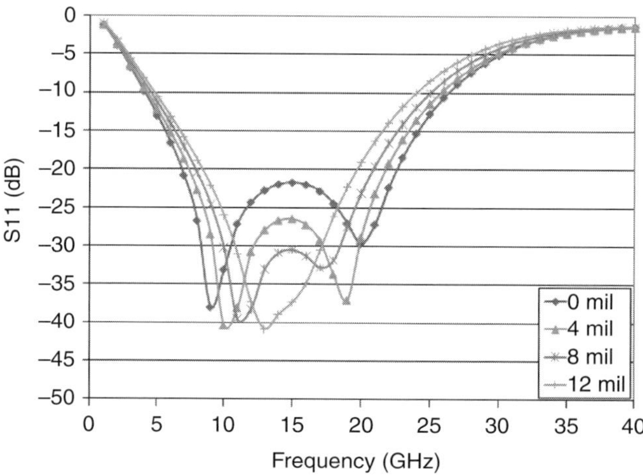

Fig. 5.50 Return loss for different values of the offset between top and bottom.

is seen from 5 to 20 GHz. An insertion loss of 0.4 dB or less is maintained from 4 to 24 GHz.

Coupling-layer-thickness sensitivity

The coupling layer thickness was also investigated because of its potential sensitivity impact on the LCP capacitance. Figure 5.52 illustrates the coupling thickness under consideration. The coupling layer thickness was varied from 0.5 mil to 1.5 mil. The SEM measurements on the coupling thickness (see Fig. 5.39) are well within this range. The primary effect of changing the material thickness is to modify the LCP capacitor while producing a negligible effect in the top plate capacitance to ground. Simulation results are shown in Figs. 5.53 and 5.54. The bandwidth increases for coupling thicknesses less than 1 mil. The return loss does increase in-band up to −15 dB, while S21 degrades to 0.37 dB in-band for the smallest coupling thickness. As the coupling thickness increases, the feed-through bandwidth starts to narrow.

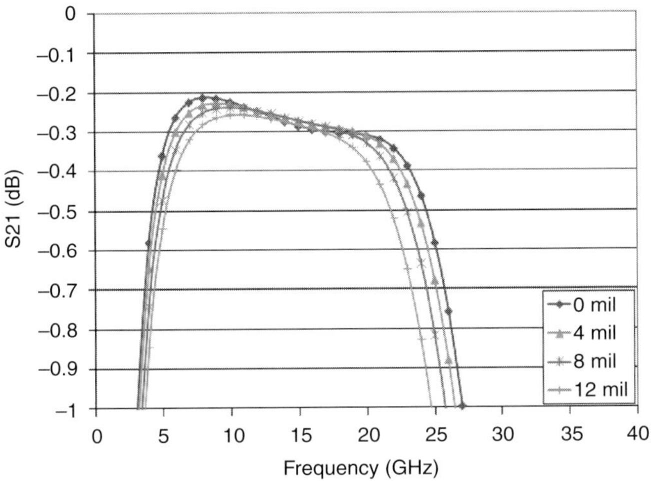

Fig. 5.51 Insertion loss for different values of the offset between top and bottom.

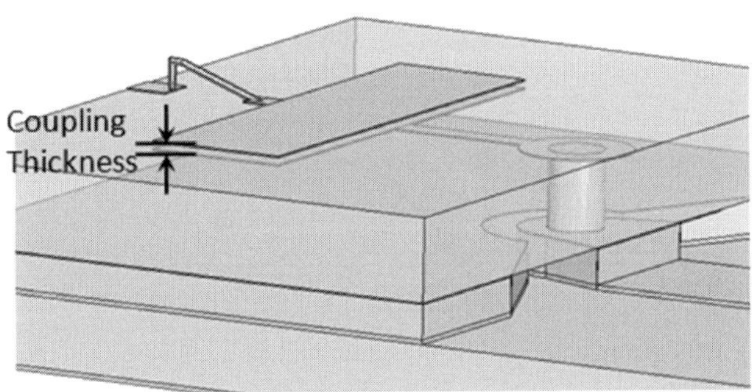

Fig. 5.52 Illustration of coupling layer thickness.

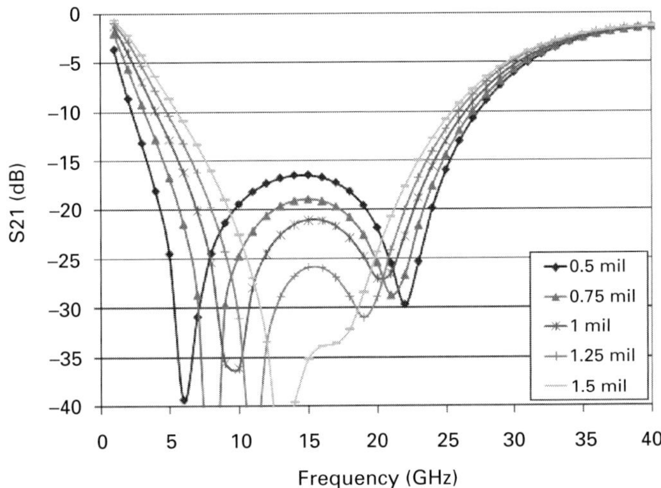

Fig. 5.53 Return loss for various values of the thickness of the dielectric layer between the top and bottom metal plates of the capacitor.

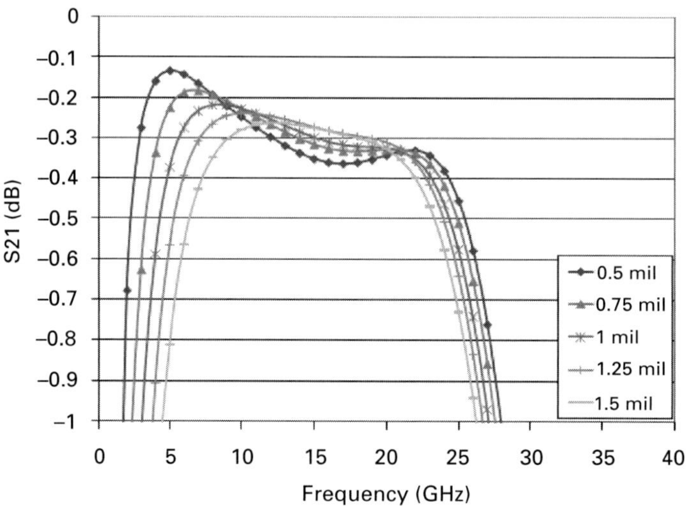

Fig. 5.54 Insertion loss for various values of the thickness of the dielectric layer between the top and bottom metal plates of the capacitor.

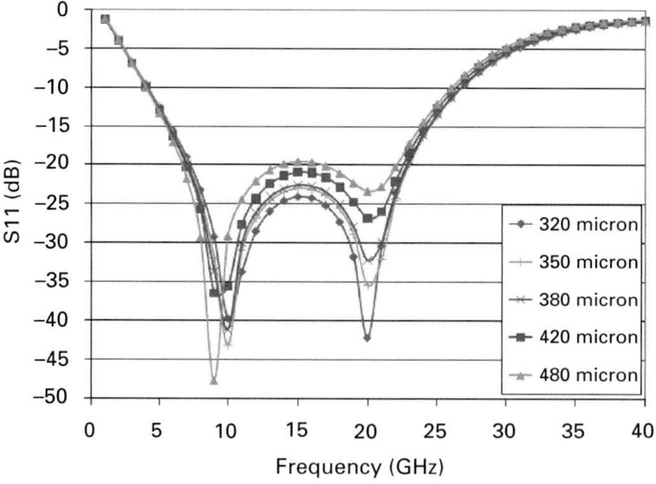

Fig. 5.55 Return loss for various values of the bond wire length.

Bond wire sensitivity

Changes in the bond wire length affect the inductance, resistance, and capacitance to ground. Its nominal value is 380 μm. From Figs. 5.55 and 5.56, it is seen that decreasing the bond wire length improves the return loss and the in-band insertion loss. Clearly, for processes capable of producing reliable bond wire lengths of this range, filter feed-through becomes increasingly attractive. For bond wire lengths within 20% of nominal, minimal degradation occurs.

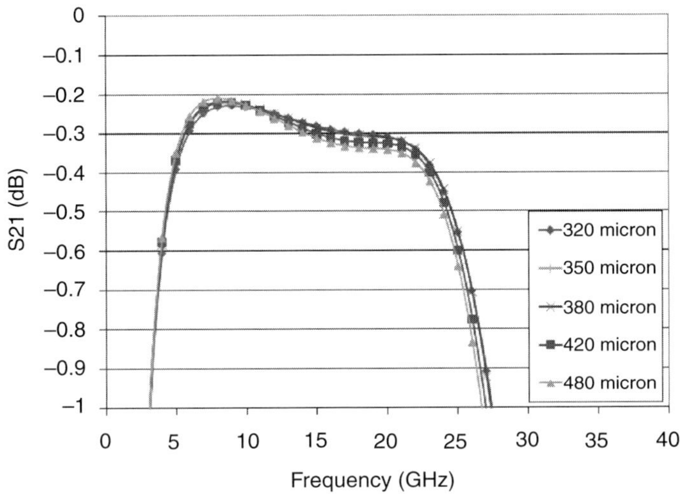

Fig. 5.56 Insertion loss for various values of the bond wire thickness.

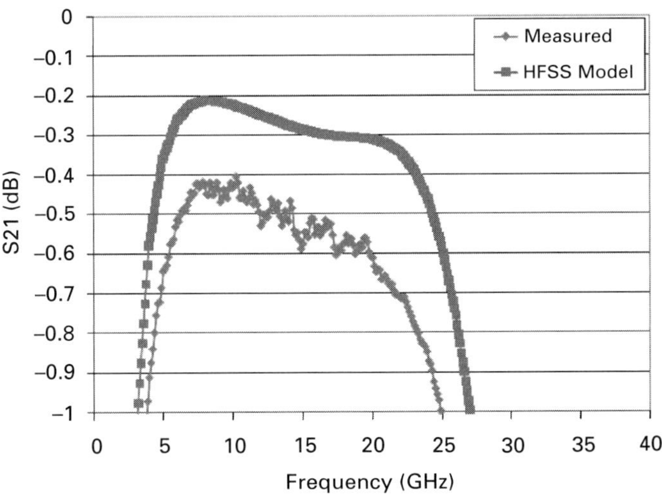

Fig. 5.57 Insertion loss of filter feed-through.

5.3.3 Lumped-element interconnect measurement results

A lumped-element feed-through as designed above was fabricated and measured. The results are provided in Figs. 5.57 and 5.58. There appears to be an excess loss of 0.3 dB when the two insertion loss curves of Fig. 5.57 are compared. This can be attributed to via process variations. The return loss in Fig. 5.58 shows very good agreement between the simulation and the measurement. Hence, we have demonstrated a via-less lumped-element bandpass interconnect with excellent performance from 5 GHz to 20 GHz.

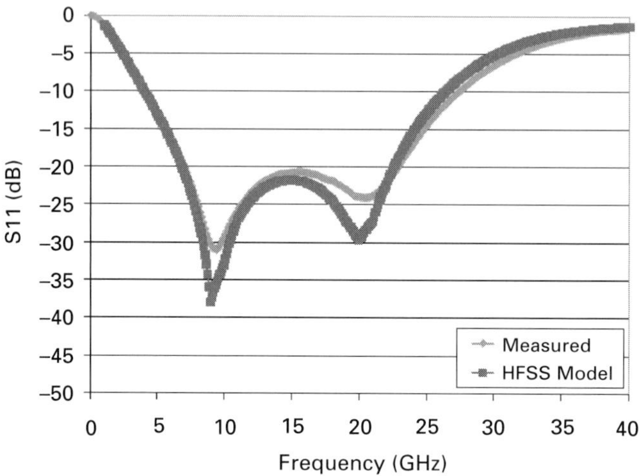

Fig. 5.58 Return loss of filter feed-through.

5.4 Ka-band down-converter multi-chip module (MCM) using LCP SMT packages

This section takes the 40 GHz package feed-through introduced in the previous section and incorporates the design into multi-chip Ka-band receiver down-converter modules using thin-film LCP surface-mount packages. The module down-converts a Ka-band input signal to a 1.2 GHz intermediate frequency (IF) output. A down-converter is an important subsystem component in a receiver; it tunes an RF signal to the IF range so that an analog-to-digital converter (ADC) is able to digitize an incoming message within its bandwidth and data processing can take place.

Section 5.4.1 gives a schematic-level design, the components used, and the proposed package cross-section. Section 5.4.2 presents the design, layout and prototypes of the down-converter LCP module. In section 5.4.3 we describe the assembly of the module and demonstrate an LNA measurement in the module, its conversion gain, and a loss analysis.

5.4.1 Schematic diagram, chip components, and cross-section of module

Figure 5.59 shows a schematic diagram of the down-converter module that was developed. The components inside the broken-and-dotted rectangle are included in the LCP module. There are two inputs and two outputs. The inputs to the package are an RF signal from the antenna and a local oscillator (LO) to drive a mixer. Both the RF signal and the LO are at Ka-band. The outputs are two IF signals generated by the image rejection mixer. The key components are an LNA, to amplify the RF signal and set a noise standard, a buffer driver amplifier, to boost the LO signal,

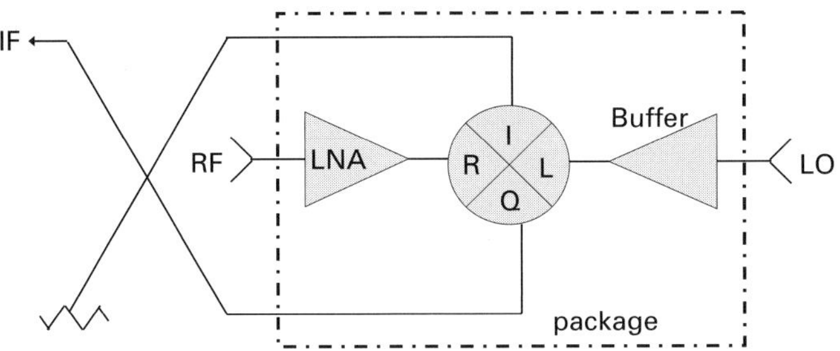

Fig. 5.59 Schematic diagram of a down-converter [34, 35] (© 2007 IEEE).

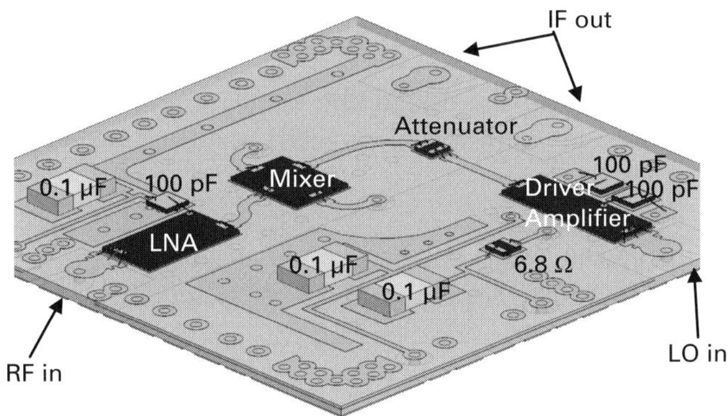

Fig. 5.60 Proposed image of the LCP down-converter module [34] (© 2007 IEEE).

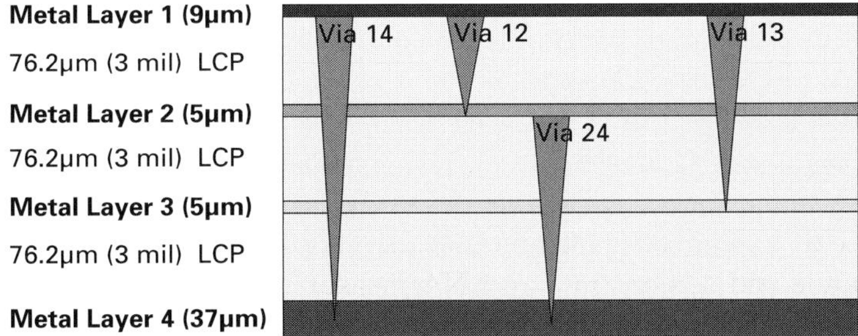

Fig. 5.61 Multilayer LCP package cross-section.

and a mixer, which receives the LO signal and mixes down the Ka-band signal to the IF outputs. In addition to the three main components there are seven bypass capacitors and a resistor, and an attenuator housed inside the package. Figure 5.60 shows the proposed multi-chip down-converter module package base design.

A cross-section of the multilayer LCP package is shown in Fig. 5.61. There are four metal layers. From top to bottom, they are referred to as metal layers 1, 2, 3, and 4. There are three dielectric layers between the metal layers. Metal layer 1 is 9 μm thick, metal layers 2 and 3 are 5 μm thick, and metal layer 4 is 37 μm (1 oz of copper) thick for rigidity of the package base. All visible copper traces and planes are gold plated to prevent oxidation. Each dielectric layer is 3 mil thick. Via 14 connects metal layers 1 and 4; via 12 connects metal layers 1 and 2; via 24 connects metal layers 2 and 4; and via 13 connects metal layers 1 and 3. Via 14 and via 13 are 8 mil in diameter, via 12 has 6 and 8 mil diameters, and via 24 has a 12 mil diameter. All vias were assigned dimensions that can be mechanically machined, for cost efficiency. Any smaller via may require laser drilling through the dielectric material, which adds extra cost and lead time.

5.4.2 Design and simulation of signal traces

In this subsection we consider the design and simulation of a microstrip line to bond wire transition used at microstrip line to chip connections and an IF stripline used inside the package base.

Package base microstrip line to chip pad bond wire transition

The RF input of the module ranges from 35.2 to 39.4 GHz, and the LO input ranges from 34 to 38.2 GHz. The package feed-through design is the same as that described in section 5.1. On the RF path a signal travels through an LNA, then back to the package trace with 3 mil dielectric thickness to ground using bond wires, and finally connects to a mixer using bond wires. There are two chip pad to package trace bond wire transitions that need to be designed for Ka-band operation. Note that the first RF bond wire transition to the LNA is already incorporated in the feed-through design from the previous section. Along the LO path there are four chip pad to package trace bond wire transitions. They are the buffer amplifier output to the package trace, the package trace to the 2 dB attenuator chip pad, the attenuator pad to the package trace, and finally the package trace to the mixer LO input pad. In all simulations it is assumed that the chip pads are tied to a 50 Ω characteristic impedance within the MMIC. Hence, the bond wire inductance is tuned out by varying the package trace width where the bond wires make connections. Figure 5.62 illustrates the microstrip-line-to-bond wire transition structure, and Fig. 5.63 shows the simulation results. The simulation indicates a 17 dB return loss and a 0.1 dB insertion loss up to Ka-band.

Stripline design for IF output from mixer

A signal leaving the mixer requires routing to metal layer 3 in order to cross the LO trace on metal layer 1 without interaction, or cross-talk. The IF is designed to go through an additional via transition because of its lower operation frequency compared with the Ka-band LO. Metal layer 2 serves as a ground layer for the LO trace and the IF outputs. A stripline is used in metal layer 3 to route the IF output from the mixer; it is placed in between two 3 mil thick LCP layers. The equation used to determine the stripline width, modified from [30], is as follows:

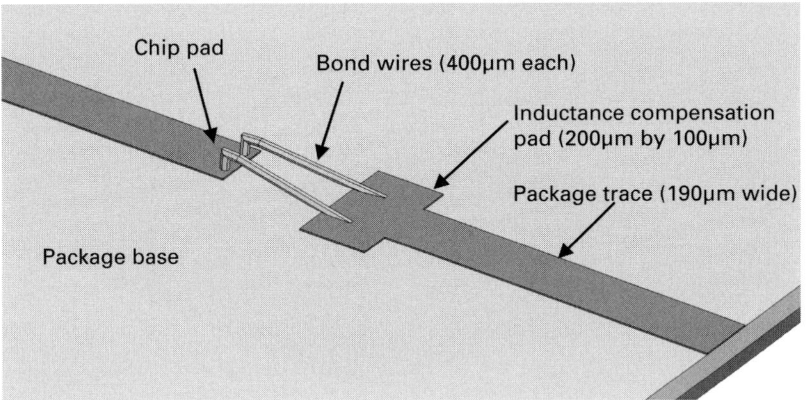

Fig. 5.62 Simulation structure of a package base microstrip connection to a chip pad.

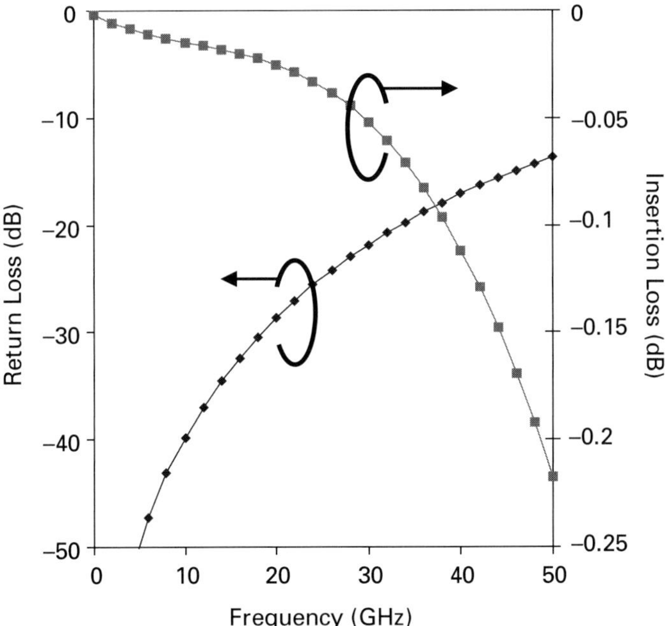

Fig. 5.63 Simulation predictions for the package base microstrip to chip pad connection.

$$W = \frac{1.9(2h+t)}{0.8e^{Z_0\sqrt{\varepsilon_r}/60}} - \frac{t}{0.8}, \tag{5.19}$$

where t is the signal conductor thickness of the stripline, h is the distance between the signal trace and ground planes, W is the signal trace width, and ε_r is the dielectric constant of LCP, which equals 2.9. Inserting 50 Ω as the characteristic impedance yields 3.31 mil for the stripline width. Figure 5.64 shows a simulation in FEM software of a stripline. In addition to the stripline, a microstrip line operating at

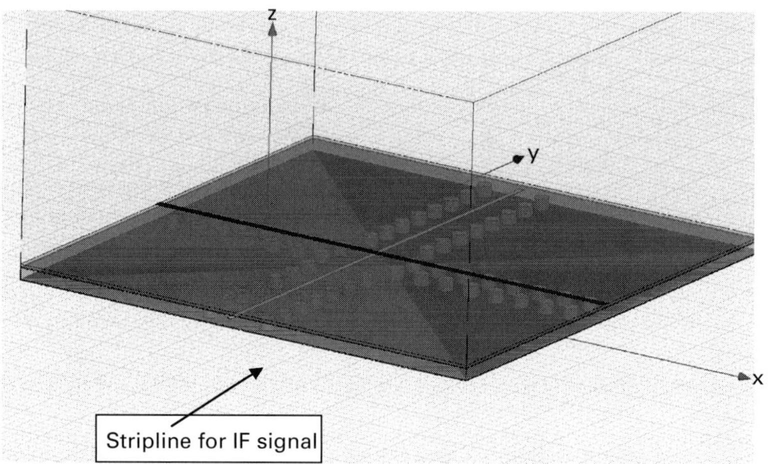

Fig. 5.64 The PCB stripline structure simulated with an FEM solver.

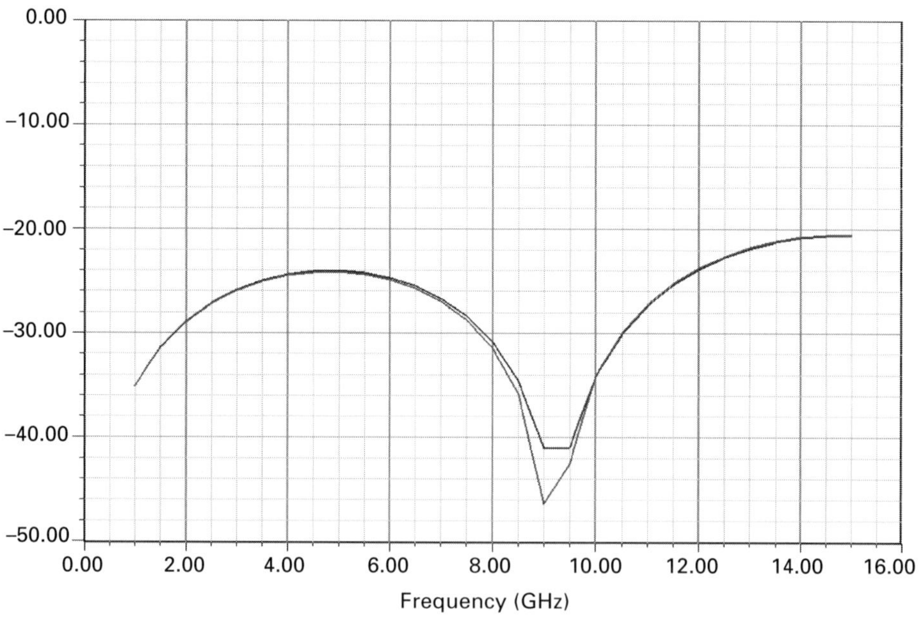

Fig. 5.65 An FEM simulation result for a stripline design.

Ka-band to feed the mixer LO input is laid above the stripline to investigate whether there is cross-talk. To minimize cross-talk, a ground plane is inserted between the stripline and the microstrip line. The simulation prediction in Fig. 5.65 is up to Ku-band. This is sufficient since the stripline will be used by the IF signal leaving the mixer at 1.2 GHz. The return loss remains greater than 20 dB up to 15 GHz.

(a)

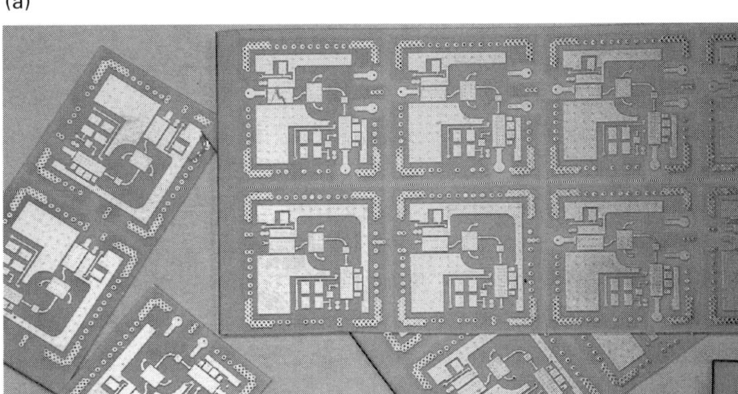

(b)

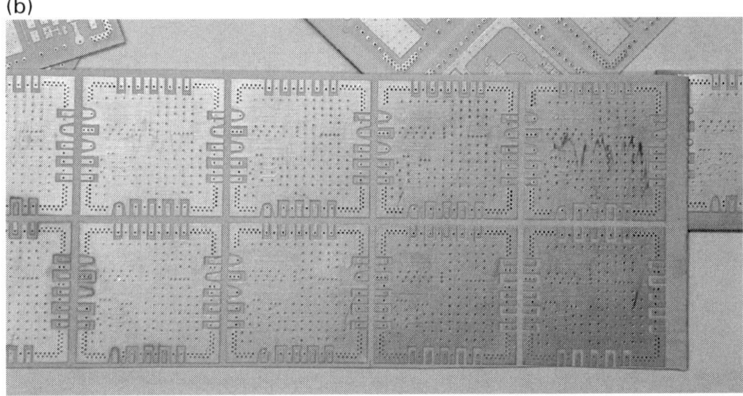

Fig. 5.66 (a), (b) Multilayer LCP down-converter module package base prototypes.

Down-converter package prototypes

Figures 5.66a, b show the top and back sides of a receiver down-converter module base in panel form. A panel is cut into individual package bases for assembly and testing. Once the assembly of the down-converter components is complete and the lid sealed, the module is mounted on a PC board, as shown in Fig. 5.67, with pads designed to match the back side pattern of a package base. The RF and LO inputs on the PC board are laid out using the feed-through design from section 5.1.

5.4.3 Assembly and measurements of Ka-band down-converter module

Figure 5.68a shows a cross-section of an LCP down-converter module during a silver-loaded polymeric adhesive curing process at 200 °C lasting 30 minutes [20]. The curing temperature may vary with respect to the curing time. Initially, a package base cut from a panel shown in Fig. 5.66 is cleaned with isopropanol using a cotton swab. Then the silver-loaded adhesive is deposited with an x-acto knife onto the contact pads where the MMICs and bypass components will be

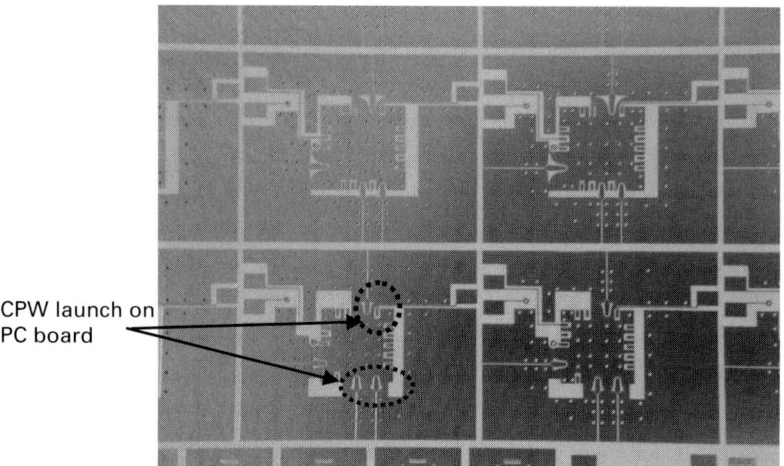

Fig. 5.67 A PCB test board for measurement on a down-converter module.

mounted. The coating is done under microscope and the adhesive thickness is made approximately 50 μm (referenced with 1 mil diameter bond wire) using the knife. The chip components are placed on the adhesive coated pads using a tweezer. The hand holding the tweezer is pressed firmly against a chuck where a module package base sits. The other hand should support the hand holding the tweezer by pressing against the direction in which the chip needs to travel to reach a contact pad. The above two techniques allow chip mounting with accuracy and significantly reduce the handshake seen under the microscope. High-cure-temperature silver adhesive is used to prevent re-flow during LCP lid lamination: a drop of adhesive is placed on a ground plane. The module is finally cured at 200 °C for 30 minutes.

After curing, the silver-adhesive quality is examined by using a probe or any sharp-tipped object. If the probe tip is not able to pierce through the adhesive, then the whole module can be considered cured. Next, bond wire connections are made throughout the module. Triple wires are used between the package feed-through and the chip. Double wires are used everywhere else for internal RF path-matching and robust DC connections. Once all the wire connections are made, the module is sealed with an LCP lid using the technique described in chapter 3. Finally, the sealed module is mounted onto a PC board for testing (see Fig. 5.68b). Eutectic tin–lead solder is used in this process and is heated to above 183 °C. The adhesive used inside the module is able to withstand 183 °C [31] since it was cured at 200 °C, and this prevents the re-flow of any component inside the module. Figure 5.69 gives a top view of the assembled LCP down-converter module before the lid was put in place.

(a)

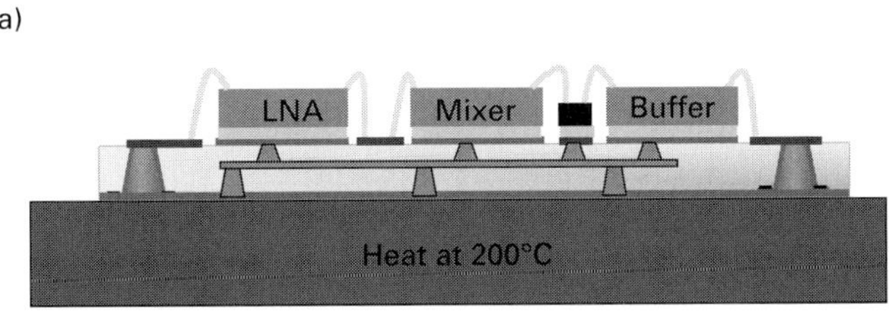

(b)

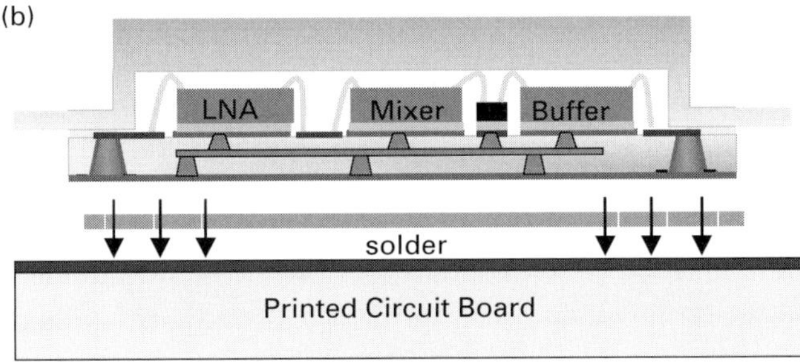

Fig. 5.68 Cross-sections of (a) the LCP downconverter module during adhesive curing with silver epoxy and (b) the module while mounted on a PC board for measurement.

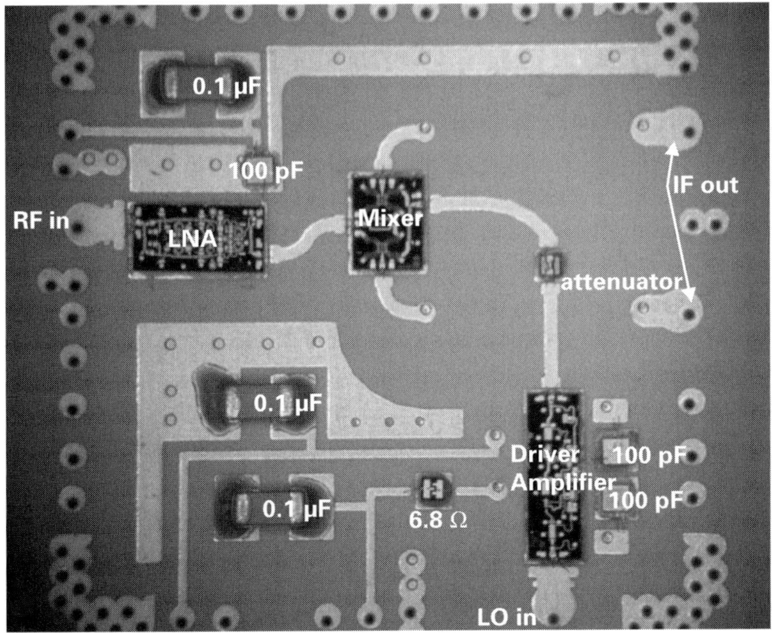

Fig. 5.69 Top view of the assembled LCP down-converter module.

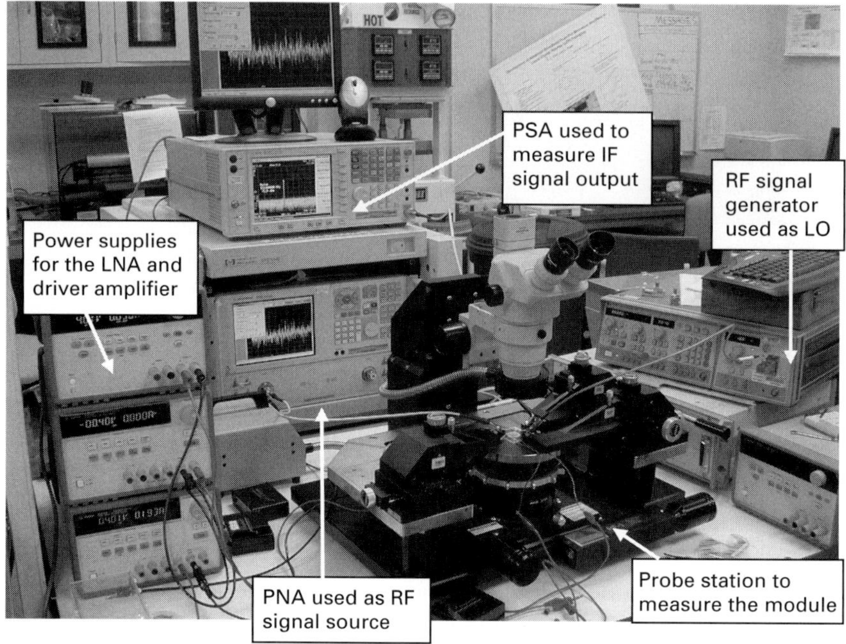

Fig. 5.70 Measurement setup for a down-converter.

Measurement setup

The module package was mounted on a 20 mil thick PC board using eutectic tin–lead solder. Coplanar waveguide (CPW) launches on the test PCB were probed for measurement (see Fig. 5.67). The RF input power at the receiver was varied from −39 dBm to −12 dBm, the RF frequency was varied from 37 to 39.4 GHz, the LO frequency was varied from 35.8 to 38.2 GHz, and the IF output power was measured at 1.2 GHz. The RF signal was generated using a network analyzer, the LO was generated by a sweep oscillator, and the IF output was measured by a spectrum analyzer. The LO input power entering the package was 2 dBm, and the driver amplifier provided 24 dB gain, with 17 dBm P_{1dB}, to supply the mixer with greater than 17 dBm power as required for the mixer to function fully. The 2 dB attenuator mounted in between the driver amplifier and mixer serves as isolation between the two components and attenuates the extra power provided by the driver amplifier. Figure 5.70 illustrates the measurement setup.

Initial LNA gain measurement

Prior to measuring the conversion gain of the LCP Ka-band down-converter module, the gain of the LNA is measured to observe its characteristic. Knowing its characteristic may help to explain the conversion gain observed in the next section of the circuit. Figure 5.71 shows a top view of the LNA used in the down-converter module between the RF input and the mixer. The gain measurement is shown in

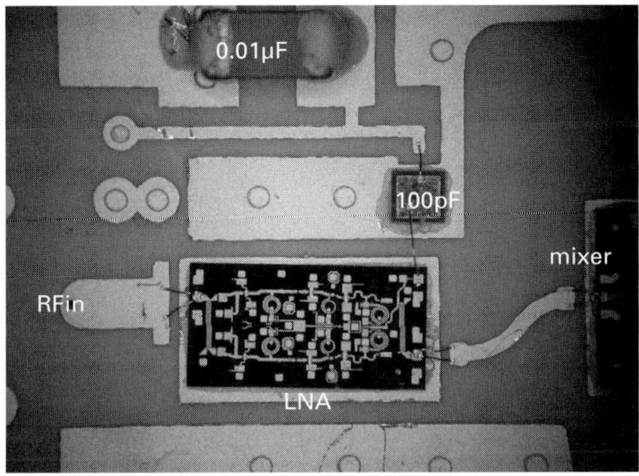

Fig. 5.71 The LNA used in the down-converter module.

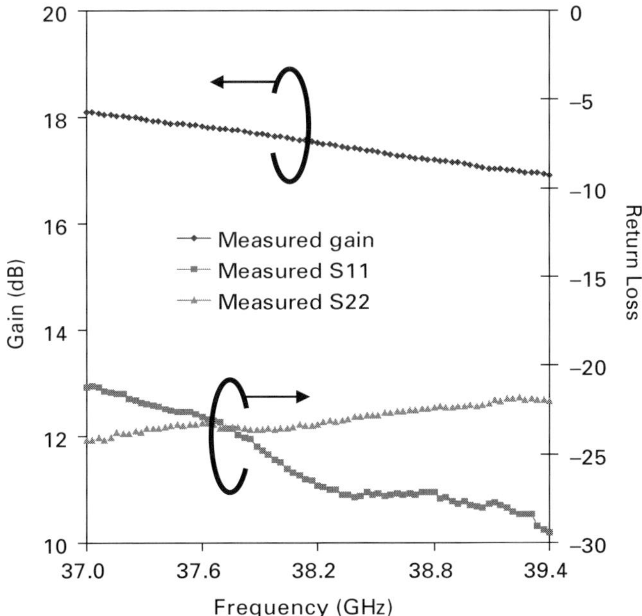

Fig. 5.72 Gain and input and output return loss measurements for the LNA used in the module.

Fig. 5.72. At 37 GHz RF input frequency, the gain was measured to be ~18 dB. The gain decreases to ~16.9 dB at 39.4 GHz.

Conversion gain measurement

Figure 5.73 shows the assembled receiver down-converter module in an LCP package, with and without LCP lid sealing, mounted on the PCB for measurement. Figure 5.74 shows the conversion gain of the packaged down-converter. The

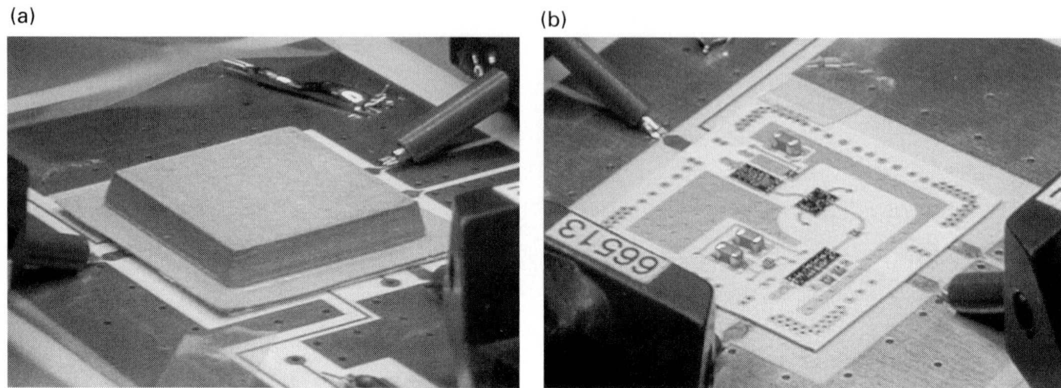

Fig. 5.73 Images of (a) sealed and (b) unsealed down-converter modules, mounted for measurement [34] (© 2007 IEEE).

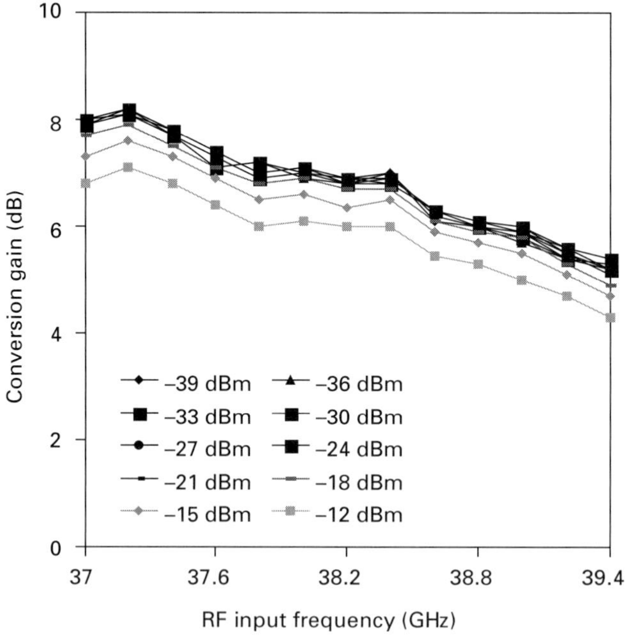

Fig. 5.74 Conversion gain measurement of Ka-band down-converter module [35].

horizontal axis shows the input RF signal frequency, and the vertical axis gives the conversion gain achieved at 1.2 GHz on a dB scale. The legend shows the various input RF power values. As can be seen, at the high end of the Ka-band the conversion gain is maintained at between 6 and 8 dB. Conversion gain above 37 GHz drops off due to the gain roll-off of the LNA, seen in Fig. 5.72, and is not caused by package feed-through performance degradation. Also, notice that the conversion gain for RF input powers of −18, −15, and −12 dBm are compressed, compared

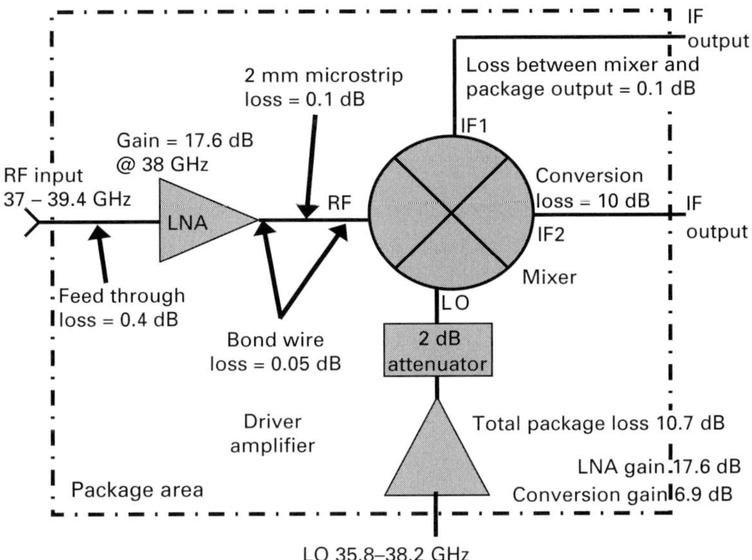

Fig. 5.75 Conversion gain and loss distribution of down-converter module at 38 GHz RF input frequency.

with the data for the RF input power for –39 to –21 dBm. This is due to the P_{1dB} of the LNA, which is rated at 6 dBm.

Loss analysis of down-converter module

Here we investigate the loss that accumulates throughout the module. Figure 5.75 gives a schematic diagram of the down-converter module with the appropriate loss values labeled at each subsection of the circuitry. For example, the module feed-through sees 0.4 dB as a signal enters the package. While an RF signal achieves 17.6 dB gain at 38 GHz, two bond wire transitions are seen in between the LNA output and input of the mixer. A bond wire transition consists of a double wire connection between the chip pad and the microstrip line on the package's top surface. The design was discussed at the beginning of section 5.4.2. Also, the microstrip line connecting the two chips accrues a 0.1 dB loss through 2 mm of microstrip length. Overall, the RF input to the down-converter module sees a 17 dB gain before reaching the mixer.

The mixer accumulates a 10 dB conversion loss when mixing a Ka-band millimeter-wave signal down to an 1.2 GHz IF output. From the mixer output, the IF signal sees an estimated 0.1 dB loss through the stripline, package feed-through, RF probe, and RF cable before reaching the spectrum analyzer (see the PSA in Fig. 5.70). Overall, the module achieves a 6.9 dB conversion gain at 38 GHz RF input frequency. Table 5.8 shows the gain and loss distribution seen by a 38 GHz RF input signal throughout the package until the signal leaves the package after being down-converted. In this table it is assumed that a power level of –18 dBm is

Table 5.8. Loss distribution throughout the package at 38 GHz

Description	Loss (dB)	Gain (dB)	Power (dBm)
Receiver package RF input			−18
Package feed-through	0.4		−18.4
LNA gain		17.6	−0.8
Bond wires connecting LNA and microstrip on package	0.05		−0.85
Microstrip between LNA and mixer (~2 mm)	0.1		−0.95
Bond wires connecting microstrip to mixer	0.05		−1
Mixer conversion loss	10		−11
Loss from mixer wirebonds, stripline, package feedthrough, probe and cable connecting to PSA at IF frequency output	0.1		−11.1
Conversion gain		6.9	

presented at the package RF input. From the table it can be seen that the package feed-through loss is approximately 0.4 dB, as seen above.

5.5 Chapter summary

In this chapter, the extensive design, characterization, and analysis of various LCP SMT package feed-through structures was carried out. The chapter demonstrated a variety of high-performance millimeter-wave MCM interconnects and packages using LCP packaging technology. Package feed-through structures were optimized for minimal loss up to Ka-band. Conductor and dielectric attenuation in package feed-throughs were computed to validate the optimized design structure in FEM simulation by comparing the theory with measured attenuation values. The characterization showed an insertion loss of 0.4 dB and a return loss of better than 20 dB in the Ka-band. A circuit model was derived and validated by comparing measurements on a packaged amplifier to simulations based on on-wafer bare-die measurements in conjunction with the package feed-through circuit model.

Section 5.2 introduced an alternative feed design: a bandpass feed-through that is aimed at ultimately removing vias from package feed-throughs in order to eliminate potential moisture-leak paths into the package cavity and to overcome the Z-expansion disadvantage of LCP due to its temperature cycle. In section 5.3 we improved the bandpass concept by reducing the feed size. This was achieved by utilizing a lumped-element filter concept. The design, measurement, model, and sensitivity analysis were explored.

The last section expanded on a single-chip package to a receiver down-converter using LCP. The feed design developed in section 5.1 was incorporated in a multi-chip down-converter module using LCP packages. The module was designed to down-convert a Ka-band signal to 1.2 GHz IF output to allow digital processing. The conversion gain was measured and loss analysis was given to investigate the loss accumulated at each microwave transition, such as the bond wires connecting a chip pad and a package microstrip line, throughout the RF and IF signal paths.

References

[1] http://www.ceramics.nist.gov/srd/summary/scdaos.htm

[2] http://www.rogerscorp.com/documents/730/acm/ULTRALAM-3000-LCP-laminate-data-sheet-ULTRALAM-3850.aspx

[3] M. Li, K. Hix, L. Dosser, K. Hartke, J. Blackshire, "Micromachining of liquid crystal polymer film with frequency converted diode-pumped Nd:YVO4 laser," in *Proc. SPIE Conf.* vol. **4977**, pp. 207–218, July 2003.

[4] http://www.ansoft.com

[5] http://eesof.tm.agilent.com/products/ads_main.html

[6] H. R. Kaupp, "Characteristics of microstrip transmission lines," *IEEE Transactions on Electronic Computers*, vol. **EC-16**, no. 2, pp. 185–193, April 1967.

[7] E. Bogatin, "Design rules for microstrip capacitance," *IEEE Transactions on Components, Hybrids and Manufacturing Technology*, vol. **10**, no. 3, pp. 253–259, September 1988.

[8] http://www.technick.net/public/code/cp_dpage.php?aiocp_dp=util_pcb_imp_microstrip

[9] K. Kiziloglu, N. Dagli, G.L. Matthaei, S.I. Long, "Experimental analysis of transmission line parameters in high-speed GaAs digital circuit interconnects," *IEEE Transactions on Microwave Theory and Techniques*, vol. **39**, no. 8, pp. 1361–1367, August 1991.

[10] K.C. Gupta, R. Garg, I.J. Bahl, *Microstrip Lines and Slotlines*, Artech House. 1979.

[11] L. Chai, A. Shaikh, V. Stygar, "LTCC for wireless and photonic packaging applications," in *Proc. IEEE 4th Int. Symp. on Electronic Materials and Packaging*, Taiwan, 2002, pp. 381–385.

[12] D. M. Pozar, *Microwave Engineering,* John Wiley & Sons, 1998.

[13] T. Edwards, *Foundations for Microstrip Circuit Design,* John Wiley & Sons, 1982.

[14] K. Chang, *Microwave Ring Circuits and Antennas,* John Wiley & Sons, 1996.

[15] R. Hoffmann, *Handbook of Microwave Integrated Circuits,* Artech House, 1987.

[16] Private communications with Nippon Steel Chemicals.

[17] A.-V. Pham, J. Laskar, J. Schappacher, "Development of on-wafer microstrip characterization techniques," in *Proc. 47th IEEE ARFTG Conf. Dig.*, June 1996, pp. 85–94.

[18] A. Fraser, J. Schappacher, "Test adapter substrates ease the task of measuring PHEMT FETs," *Microwave Journal*, vol. **38**, pp. 120–122, March 1995.

[19] http://www.triquint.com/docs/t/TGA4507

[20] http://www.diemat.com/docs/products/thermals/DM5030P.php

[21] D. Kajfez, B. S. Vidula, "Design equations for symmetric microstrip DC blocks," *IEEE Transactions on Microwave Theory and Technique*, vol. **MTT-20**, August 1972, pp. 555–556.

[22] V. K. Tripathi, "Asymmetric coupled transmission lines in an inhomogeneous medium," *IEEE Transactions on Microwave Theory and Techniques*, vol. **MTT-23**, September 1975, pp. 734–739.

[23] K. Aihara, A.-V. Pham, "Development of thin-film liquid crystal polymer surface mount packages for Ka-band applications," in *Proc. IEEE MTT-S Int. Microwave Symp. Dig.*, San Francisco, June 2006, pp. 956–959.
[24] http://www.microconnex.com
[25] I. Bahl, *Lumped Elements for RF and Microwave*, Artech House, 1999.
[26] A. Wheeler, "Properties of a strip on a dielectric sheet on a plane", *IEEE Transactions*, vol. **MTT-25**, pp. 631–647, August 1977.
[27] R. E. Collin, *Foundations for Microwave Engineering*, IEEE, 2001.
[28] M. E. Goldfarb, R. A. Pucel, "Modeling via hole grounds in microstrips," *IEEE Microwave Guided Wave Letters*, vol. **1**, June 1991.
[29] A. Djordjevic, M. Bazdar, T. Sarkar, R. Harrington, *LINPAR for Windows – Matrix Parameters for Multiconductor Transmission Lines,* software and user's manual, version 2.0, Artech House, 1999.
[30] R. E. Canright Jr., "A simple formula for dual stripline characteristic impedance," in *Proc. IEEE Southeastcon*. vol. **3,** April 1990, pp. 903–905.
[31] http://www.chemguide.co.uk/physical/phaseeqia/snpb.html
[32] K. Aihara, M. J. Chen, A.-V. Pham, "Development of thin-film liquid crystal polymer surface mount packages for Ka-band applications," *IEEE Transactions on Microwave Theory and Techniques*, vol. **56**, no. 9, pp. 2111–2117, September 2008.
[33] M. P. Mcgrath, K. Aihara, A. Pham, S.R. Nelson, "Development of LCP surface mount package with a bandpass feedthrough at K-band," in *Proc. IEEE MTT-S Int. Microwave Symp. Dig.,* Atlanta, June 2008, pp. 93–96.
[34] K. Aihara, A. Pham, D. Zeeb, T. Flack, E. Stoneham, "Development of multi-layer liquid crystal polymer Ka-band receiver modules," in *Proc. IEEE Asia-Pacific Microwave Conf.*, December 2007.
[35] K. Aihara, Z. Zhang, A. Pham, D. Zeeb, T. Flack, E. Stoneham, "Development of multi-layer liquid crystal polymer Ka-band receiver modules," *Microwave and Optical Technology Letters*, vol. **51**, no. 2, pp. 364–367.

Appendix Conductor and dielectric attenuation

The details of computing the theoretical, simulation, and measurement attenuation due to the conductor and the dielectric are presented in this section. Equations (A.1)–(A.7) below are used to compute the attenuation from the simulation and measurement S-parameter data. Equations (A.1) and (A.2) are used to convert the S-parameters to the ABCD matrix. From the ABCD matrix, elements A and D are related to the propagation constant γ by (A.3). Having solved for γ, one obtains (A.5) and (A.6). Finally, taking the real part of γ in A.7 gives the total attenuation seen from a set of S-parameter data. The equations are as follows:

$$A = \frac{(1+\mathrm{S11})(1-\mathrm{S22})+\mathrm{S12}\times\mathrm{S21}}{2\times\mathrm{S21}}, \tag{A.1}$$

$$D = \frac{(1-\mathrm{S11})(1+\mathrm{S22})+\mathrm{S12}\times\mathrm{S21}}{2\times\mathrm{S21}}, \tag{A.2}$$

$$A = D = \cosh \gamma l, \tag{A.3}$$

$$\gamma = \alpha + j\beta, \tag{A.4}$$

$$\gamma = \frac{1}{l}\cosh^{-1}(A), \tag{A.5}$$

$$\gamma = \frac{1}{l}\ln\left(A \pm \sqrt{A^2 - 1}\right), \tag{A.6}$$

$$\alpha = \text{Real}(\gamma) \times 8.686\,(\text{dB/unit length}), \tag{A.7}$$

where l is the length of feed-through structure simulated, γ is the propagation constant, α is the attenuation constant, and β is the phase constant.

Equations (A.8)–(A.10) below are used to compute the attenuation due to the conductor on a microstrip line. Equation (A.8) computes the surface resistivity of the conductor. Equation (A.9) computes the characteristic impedance of the microstrip line. The attenuation due to the conductor can be computed using (A.10) values obtained from the computed surface resistivity and characteristic impedance. Equation (A.11) is used to compute the attenuation due to the dielectric material of a microstrip line and coplanar waveguide. Equations (A.12)–(A.23) are used to compute the attenuation due to conductor loss of a coplanar waveguide. The equations are as follows:

$$R = \sqrt{\frac{\pi f \mu_0}{\sigma}} \tag{A.8}$$

$$Z_0 = \frac{377}{2\pi\sqrt{(\varepsilon_r + 1)/2}}\left[\ln\frac{8h}{W} + 0.125\left(\frac{W}{2h}\right)^2 - 0.5\left(\frac{\varepsilon_r - 1}{\varepsilon_r + 1}\right)\left(\ln\frac{\pi}{2} + \frac{1}{\varepsilon_r}\ln\frac{4}{\pi}\right)\right], \tag{A.9}$$

$$\alpha_c = \frac{8.686 R_s}{Z_0 W}, \tag{A.10}$$

$$\alpha_d = 8.686 \times \frac{2\pi f}{c}\frac{\left[\varepsilon_r(\varepsilon_{re} - 1)\right]\tan\delta}{2\sqrt{\varepsilon_{re}}\,(\varepsilon_r - 1)}\ (\text{dB/unit length}), \tag{A.11}$$

$$k = \frac{S}{S + 2W_s}, \tag{A.12}$$

$$\Delta = \frac{1.25t}{\pi}\left(1 + \ln\frac{4\pi W_g}{t}\right), \tag{A.13}$$

$$k_e = k + \frac{(1 - k^2)\Delta}{2W_g} \tag{A.14}$$

$$k' = \sqrt{1-k^2} \tag{A.15}$$

$$\frac{K(k)}{K'(k)} = \begin{cases} \dfrac{1}{\pi}\ln 2\left(\dfrac{1+\sqrt{k}}{1-\sqrt{k}}\right) & \text{for } 0.707 \le k \le 1, \\ \pi\left[\ln 2\left(\dfrac{1+\sqrt{k'}}{1-\sqrt{k'}}\right)\right]^{-1} & \text{for } 0 \le k \le 0.707, \end{cases} \tag{A.16}$$

$$k_e' = \sqrt{1-k_e^2}, \tag{A.17}$$

$$\frac{K(k_e)}{K'(k_e)} = \begin{cases} \dfrac{1}{\pi}\ln 2\left(\dfrac{1+\sqrt{k_e}}{1-\sqrt{k_e}}\right) & \text{for } 0.707 \le k_e \le 1, \\ \pi\left[\ln 2\left(\dfrac{1+\sqrt{k_e'}}{1-\sqrt{k_e'}}\right)\right]^{-1} & \text{for } 0 \le k_e \le 0.707, \end{cases} \tag{A.18}$$

$$\varepsilon_{re}^t = \varepsilon_{re} - \frac{0.7(\varepsilon_{re}-1)t}{W_g}\left(\frac{K(k)}{K'(k)} + 0.7\frac{t}{W_g}\right)^{-1}, \tag{A.19}$$

$$Z_{ocp} = \frac{30\pi}{\sqrt{\varepsilon_{re}^t}}\left(\frac{K(k_e)}{K'(k_e)}\right)^{-1}, \tag{A.20}$$

$$P = \begin{cases} \dfrac{k}{\left(1-\sqrt{1-k^2}\right)(1-k^2)^{3/4}} & \text{for } 0 \le k \le 0.707, \\ \dfrac{1}{(1-k)\sqrt{k}}\left(\dfrac{K(k_e)}{K'(k_e)}\right)^{-2} & \text{for } 0.707 \le k \le 1, \end{cases} \tag{A.21}$$

$$P' = P\left(\frac{K(k)}{K'(k)}\right)^2 \tag{A.22}$$

$$\alpha_c = 4.88\times 10^{-4}\,\frac{R_s \varepsilon_{re} Z_{ocp} P'}{\pi W_g}\left(1+\frac{S}{W_g}\right) \times \frac{\dfrac{1.25}{\pi}\ln\dfrac{4\pi S}{t} + 1 + \dfrac{1.25t}{\pi S}}{\left[2+\dfrac{S}{W_g} - \dfrac{1.25t}{\pi W_g}\left(1+\ln\dfrac{(4\pi S)}{t}\right)\right]^2} \ (\text{dB/unit length}) \tag{A.23}$$

where R_s is the series resistance per unit length (in ohm/m), μ_o is the series inductance per unit length (in H/m), and σ is the shunt conductance per unit length (in mho/m). Furthermore, α_c is the shunt capacitance per unit length (in F/m), Z_0 is the characteristic impedance of the microstrip, W is the width of the microstrip, α_d is the attenuation from the dielectric loss tangent, ε_{re} is the effective dielectric constant, ε_r is the substrate dielectric constant, c is the speed of light (in m/s), h is the substrate thickness, S is the coplanar waveguide signal width, W_g is the coplanar waveguide signal-to-ground gap, t is the thickess of conductor, and Z_{ocp} is the characteristic impedance of the CPW.

6 LCP for passive components

Hai Ta, Morgan J. Chen, Kunia Aihara, Andy C. Chen, Jia-Chi Samuel Chieh, Anh-Vu H. Pham

It has become increasingly apparent that LCP provides an ideal form, fit, and function for many broadband passive components. Since LCP is available in thicknesses less than 1 mil with low dielectric constant, this enables easy design of varying controlled impedances. Further, LCP's property of being its own adhesive layer provides for a high layer count in a multilayer stack while simultaneously maintaining high-frequency performance that otherwise would be detuned by poor electrical ply layers. This chapter provides design and development examples of broadband passives that benefit from LCP. In section 6.1 we describe a broadband Marchand balun implemented on multilayer LCP covering 4–20 GHz and in section 6.2 a broadband Wilkinson power divider–combiner operating over 2–18 GHz. Section 6.3 presents a novel hybrid coupler using multilayer LCP to achieve a broadband design within a compact area.

6.1 Broadband LCP Marchand balun

A balun converts differential "balanced" signals into single-ended "unbalanced" signals, and vice versa. Marchand baluns are found in numerous microwave circuit designs owing to their characteristically wide bandwidth, low imbalance, and symmetric balanced ports. To achieve a wide bandwidth ratio, a Marchand balun is realized with multilayered broadside coupled microstrip lines implemented on LCP. A novel twin-thickness thin-film [1] structure has been devised specifically to reduce balun conduction loss without sacrificing operation bandwidth.

This section begins with a background on Marchand baluns in section 6.1.1. The origins of, and methods to minimize, balun imbalance are considered here. Next, fabrication procedures and balun topology are discussed in section 6.1.2. Then, in section 6.1.3, design and analysis will be presented for a balun implemented with multilayered broadside-coupled microstrip lines enabled by LCP. In section 6.1.4, measurement results will be reported that show that the balun achieves a 5 : 1 bandwidth and covers a 4–20 GHz frequency range when implemented on LCP. Section 6.1.5 presents a technique that improves the bandwidth of the balun discussed in sections 6.1.3 and 6.1.4. A summary will be given in section 6.1.6.

Table 6.1. Marchand balun topologies and measured bandwidth

Balun topology	Coupling mechanism	Maximum in-band insertion loss and bandwidth	Bandwidth ratio	Reference
Coaxial	—	0.5 dB	13 : 1	[3]
Coupled microstrip	Edge coupling	0.5 dB 5–11 GHz	2.2 : 1	[4]
Coupled microstrip	Edge coupling	0.5 dB 15–38 GHz	2.7 : 1	[5]
CPW	Edge coupling	40–90 GHz	2.3 : 1	[6]
Multilayer microstrip	Broadside coupling	0.7 dB 6–21GHz	3.5 : 1	[7]
Multilayer	Broadside coupling	8.2–30 GHz	3.7 : 1	[11]
Multilayer microstrip	Broadside coupling	0.5 dB 4–17 GHz 0.7 dB 4–20 GHz	5 : 1	[1]

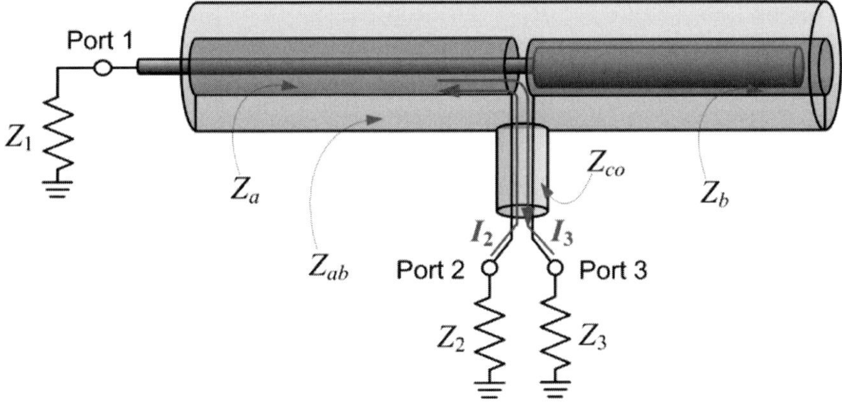

Fig. 6.1 Coaxial Marchand balun. For an explanation of the symbols, see (6.3).

6.1.1 Balun background

Here, we explain how the Marchand balun [2, 3] has been converted into a planar structure for simple integration. The performances of various planar baluns are listed in Table 6.1. These configurations include edge-coupled microstrip lines [4, 5], edge-coupled co-planar waveguide (CPW) lines [6], and multilayered broadside-coupled microstrip lines [7–11]. Single-layer baluns built upon an edge-coupled microstrip or a CPW are convenient to fabricate but have bandwidth ratios lower than 2.7:1 [4], as reported to date. When large bandwidths are required, multilayered planar baluns may be needed.

As shown in Fig. 6.1, balanced signals coming from balanced ports (ports 2 and 3) are ideally equal in amplitude and 180° out of phase. For a real balun, performance deviates from ideal conditions for numerous reasons, and balanced signals will not be perfectly differential. This deviation is known as balun imbalance and is quantified as follows:

$$\text{amplitude imbalance} = \text{dB}(|S21|) - \text{dB}(|S31|), \tag{6.1}$$

$$\text{phase imbalance} = 180° - |\angle(S21) - \angle(S31)|. \tag{6.2}$$

A folded Marchand balun was first reported by Roberts [12] using coaxial transmission lines (Fig. 6.2a). Its equivalent circuit is shown in Fig. 6.2b. A planar printed folded balun circuit (Fig. 6.2c) was later implemented by Bawer and Wolfe [9] using two metal layers across a dielectric board. The impedance Z looking outward from the transmission line impedance Z_a (Fig 6.2d) is the sum of the open stub impedance Z_b and the short stub impedance Z'_{ab} in parallel with R:

$$Z = -jZ_b \cot\theta_b + \frac{jRZ'_{ab}\tan\theta_{ab}}{R + jZ'_{ab}\tan\theta_{ab}}, \tag{6.3}$$

where Z'_{ab} is the characteristic impedance of the balanced lines residing in the lower metal layer (Fig. 6.2c), θ_{ab} is the electrical length of the balanced line, and R is the balanced load impedance.

6.1.2 Fabrication of and lamination process for the LCP balun

Integration and fabrication techniques for Marchand baluns in multilayered LCP laminates are discussed in this subsection. The baluns are constructed using broadside-coupled microstrip lines, with signal lines implemented across 1 mil thick LCP. The great advantages of this implementation are that no air bridges or extra dielectric materials other than LCP are required for its construction and that the balun can be fabricated with standard PCB equipment and technology. LCP's low dielectric constant and low loss tangent enable low-loss passive circuits to be implemented at microwave frequencies.

A dielectric thickness combination of $H_1 = 9$ mil and $H_2 = 1$ mil was selected for LCP balun implementation. This combination gives a bandwidth ratio (BWR) of approximately 6, which is sufficient to cover the 4–20 GHz bandwidth. A bottom 9 mil thick dielectric is formed with an LCP combination of 2, 4, 2, 1 mil thick laminates (Fig. 6.3). The low-melting-temperature layers are stacked in between high melting layers for adhesiveless lamination.

Copper traces are formed and etched on the high-melting-temperature LCP layers. A stack of these five LCP layers is then laminated together by allowing the low-melting-temperature layers to melt for bonding purposes. The final stack-up is completed by laser drilling and plating processes to form blind and through vias; LCP lamination is then performed as described in Chapter 3. The LCP laminates used for this balun have heat distortion temperatures 15°C lower than their melting temperatures. Stack alignment was performed by means of tooling marks

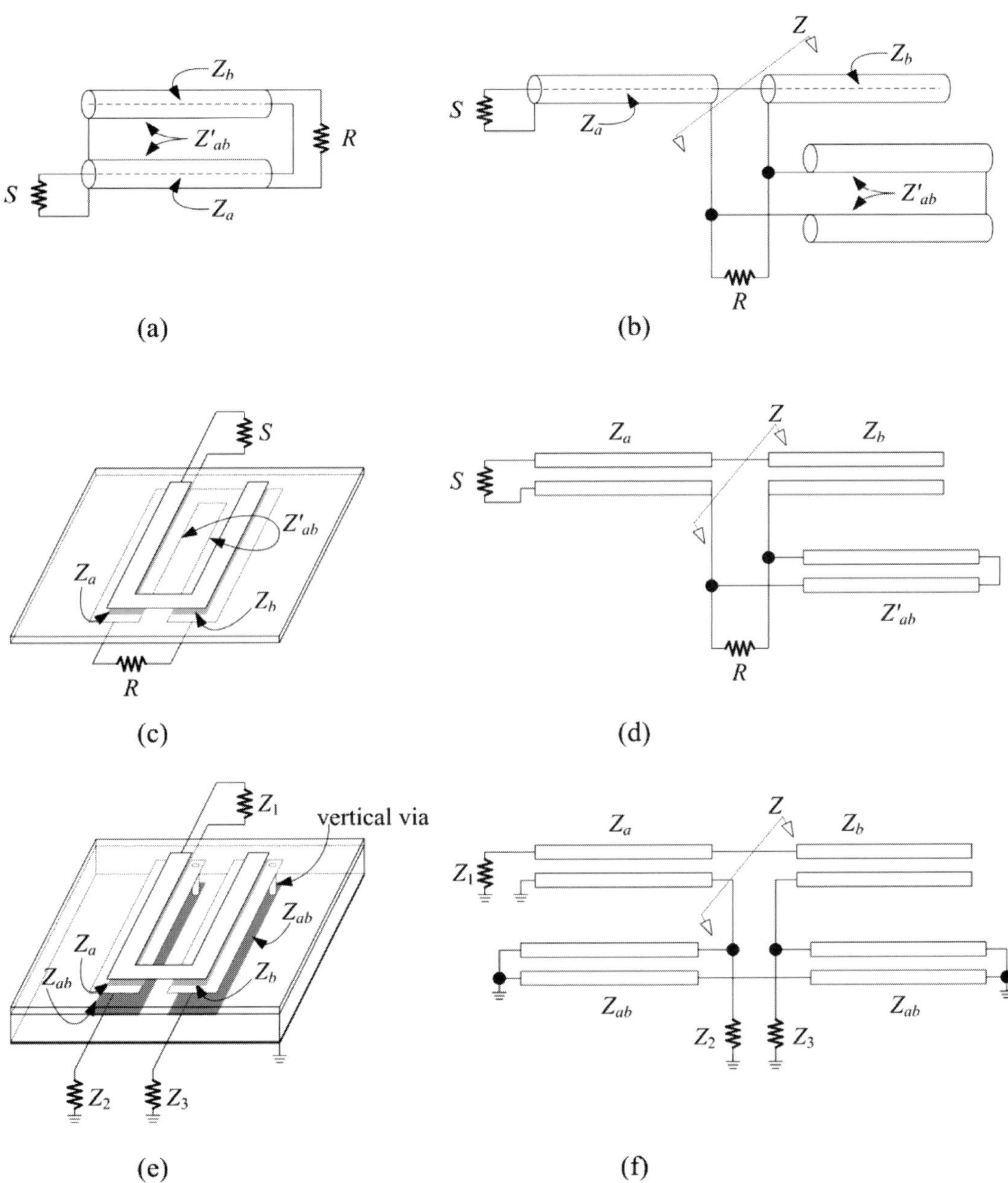

Fig. 6.2 Evolution of a multilayered microstrip Marchand balun. (a) Folded Marchand balun in coaxial implementation by Roberts [12], and (b) its equivalent circuit; (c) a printed circuit board version of a folded balun by Bawer and Wolfe [9], and (d) its equivalent circuit; (e) a multilayer balun derived from (c) with a background plane, and (f) its simplified equivalent circuit. The balanced port termination loads are designated as Z_2 and Z_3, where $Z_2 + Z_3 = R$ [20] (© 2007 IEEE).

(a)

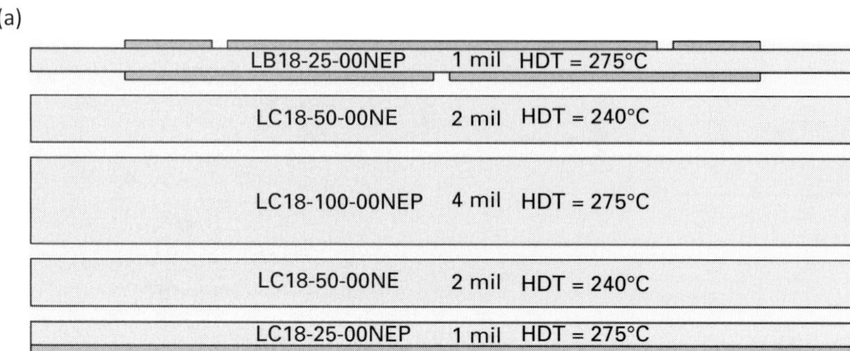

(b)

Fig. 6.3 Side view of the LCP balun (a) before and (b) after the lamination process [20] (© 2007 IEEE).

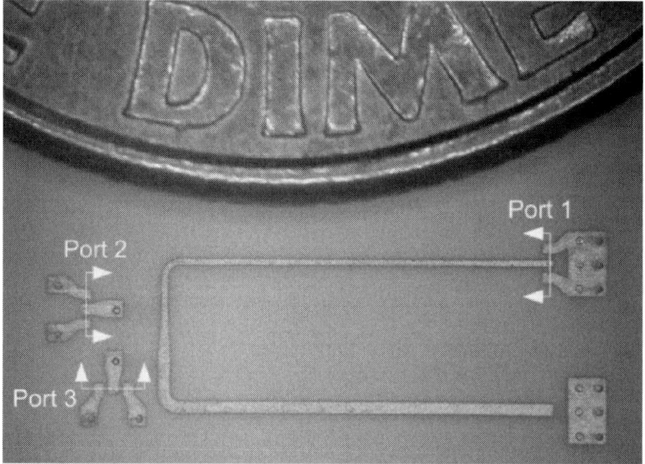

Fig. 6.4 Photograph of a fabricated LCP balun showing the port assignment and ground–signal–ground (GSG) probe launches [20, 21] (© 2005 IEEE).

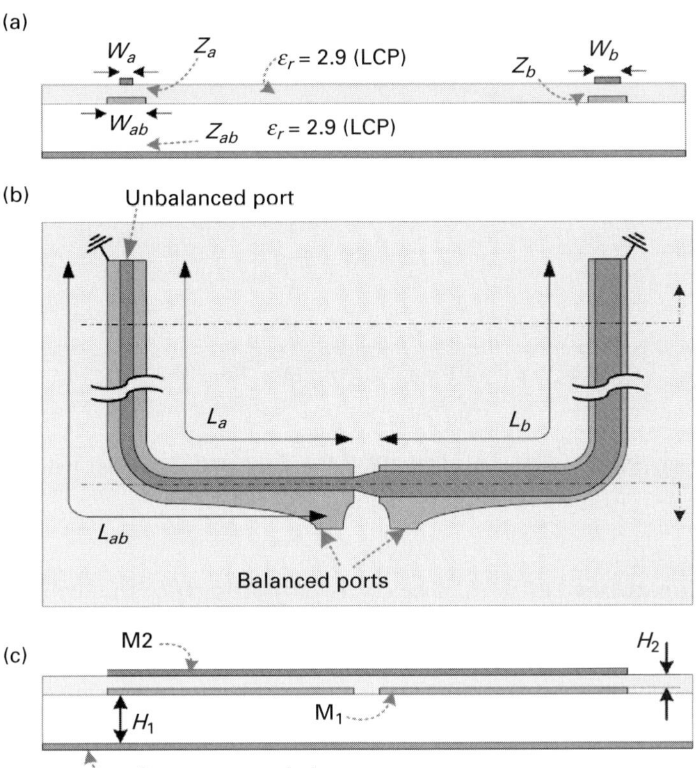

Fig. 6.5 A multilayer LCP balun using broadside coupled microstrip lines (shown without the GSG probe launches) [1, 20, 22] (© 2005 IEEE). (a) Side view showing the unbalanced ports; (b) top view; (c) side view showing the balanced ports.

along the circuit board edges prior to lamination. Figure 6.4 shows an LCP balun after fabrication.

6.1.3 Design and simulation of an LCP balun

Microwave baluns typically have narrower bandwidths and higher imbalances than transformers, owing to the nature of transmission lines. Furthermore, impedance-matching and return-loss considerations add design complexity. A cross-sectional view of the multilayered LCP balun construction is shown in Fig. 6.5. In order to obtain the widest possible operational bandwidth, the characteristic impedance Z_b needs to be kept as small as possible while the characteristic impedance Z_{ab} needs to be as large as possible. LCP laminate 1 mil (25 μm) thick is used for the top dielectric layer H_2, to facilitate achievement of the necessary characteristic impedances. Different combinations of 1, 2, and 4 mil thick LCP laminates can be used to stack up the bottom substrate layer H_1. The achievable bandwidth ratio (BWR) for this

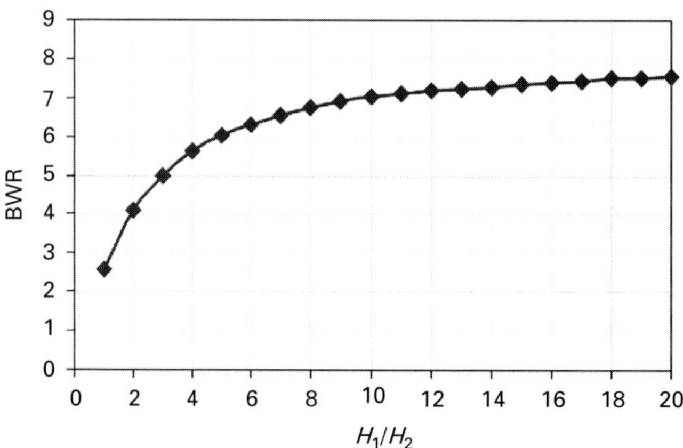

Fig. 6.6 Theoretical achievable bandwidth ratio (BWR) versus H_1/H_2 for planar baluns implemented across multiple LCP layers [20] (© 2007 IEEE).

type of balun, defined as in (6.3) below, is proportional to the dielectric thickness ratio H_1/H_2. Figure 6.6 plots BWR versus H_1/H_2 using the model shown in Fig. 6.2f. The BWR is defined as follows:

$$\text{BWR} = \frac{f_H}{f_L}, \tag{6.3}$$

where f_H is high-side passband cutoff frequency when the insertion loss equals 1 dB and f_L is the low-side cutoff frequency. The large bottom dielectric thicknesses give large BWRs. This improvement in the BWR levels off for high H_1/H_2 ratios, where one needs to take the higher-order microstrip propagation modes into consideration (see Fig. 6.6).

Even-mode balun matching network

A Marchand balun has a relatively high common-mode reflection coefficient at its balanced ports. Common-mode signals are completely rejected and reflected back from a lossless balun since only differential signals are converted to single-ended signals and allowed transmission. The isolation between the two balanced ports (ports 2 and 3 in Figs. 6.1 and 6.4) is also low. As in the case of a Wilkinson power divider, a resistive matching network can be added to improve the match. An even-mode matching network can be easily implemented with microstrip lines and integrated into an LCP balun. Agilent's advanced design system (ADS) software may be used to model and simulate the complete equivalent circuit, as shown in Fig. 6.7, using ideal transmission lines and components. The parameters used in the equivalent circuit (Fig 6.7) are listed in Table 6.2.

The ADS simulation results are shown in Fig. 6.8. These synthesized balun circuits cover a 4–20 GHz bandwidth where the insertion loss is less than 1 dB. The unbalanced port return loss (|S11|) can be synthesized to be greater than 10 dB for

Table 6.2. Circuit values used in the simulation of the balun and matching network

Parameter	Value (Ω)	Parameter	Value (Ω)
S	50	Z_T	58
R	100	R_T	75
Z_b	31.25	Z_{ce}	52
Z_{ab}	78.87	Z_{co}	35
Z_m	92	Z_r	46

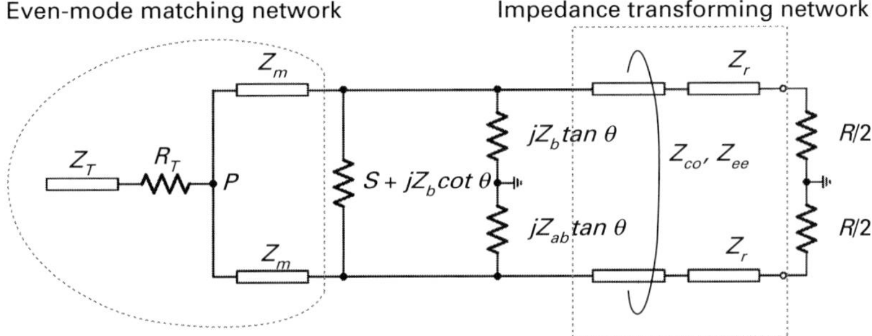

Fig. 6.7 Proposed Marchand balun circuit with a broadband even-mode matching network [23] (© 2009 IEEE). The values of the symbols are given in Table 6.2.

4–20 GHz. The balanced-port return losses (|S22|, |S33|) and isolation (|S32|) can be increased to 20 dB for 8–16 GHz (Fig. 6.8a) without the open-ended stub Z_T, and to 6–18 GHz (Fig. 6.8b) with the open-ended stub.

Figures 6.9 and 6.10 show a Sonnet simulation model used for simulating a Marchand balun with an even-mode matching network (EMMN) and the corresponding simulated results.

6.1.4 Electrical measurements on an LCP balun

Figure 6.11 shows S-parameter measurement results for a pair of back-to-back connected LCP baluns with $H_1 = 9$ mil and $H_2 = 1$ mil. This plot demonstrates that a wider bandwidth can be achieved by assigning high H_1/H_2 ratios. The back-to-back configuration is much more convenient to measure. However, now the imbalance between the balanced ports cannot be accurately redressed.

Measured Marchand balun with even-mode matching

A Marchand balun with EMMNs was fabricated in a multilayer laminated LCP substrate with the layer build-up shown in Fig. 6.3. An 0201 SMT 75 Ω resistor was used for the resistor R_T. A photograph of a fabricated balun with an EMMN is presented in Fig. 6.12. The measured three-port S-parameters are shown in Figs. 6.13

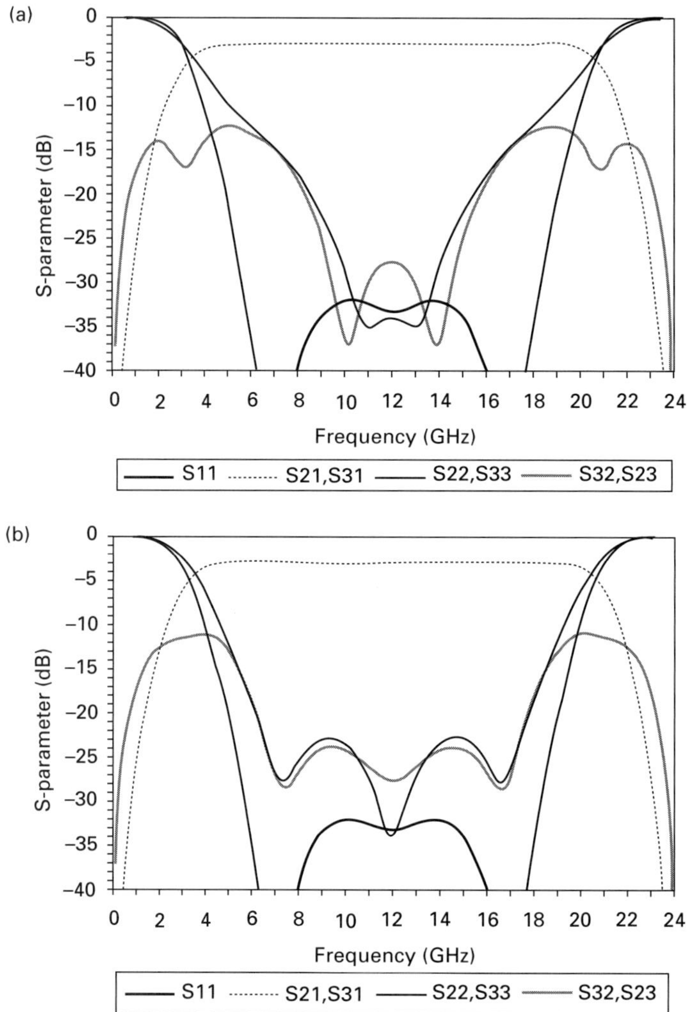

Fig. 6.8 Simulated S-parameters of Marchand baluns with even-mode matching networks: (a) the results when the matching network does not include the open-ended stub and (b) the results when the open-ended stub is included [23] (© 2009 IEEE).

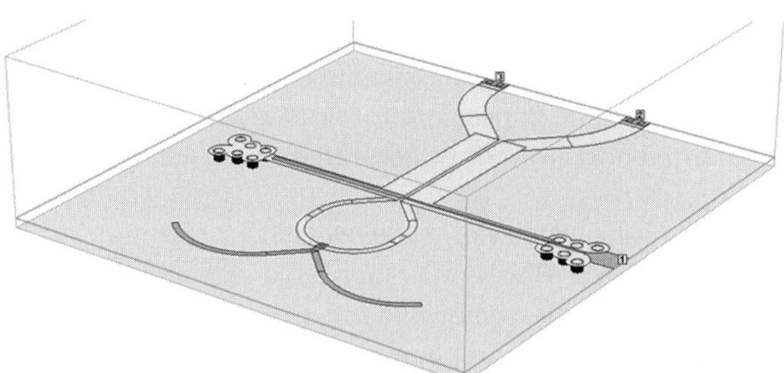

Fig. 6.9 Sonnet simulation model of the Marchand balun with an EMMN.

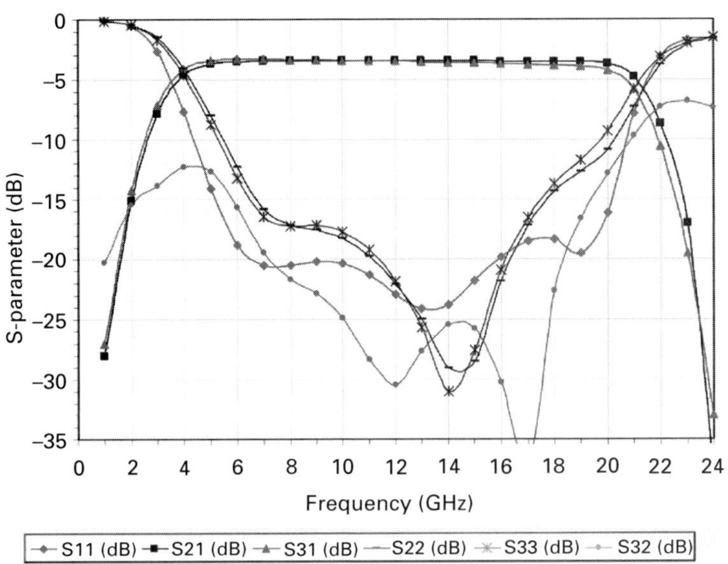

Fig. 6.10 Sonnet simulation results for the Marchand balun with an EMMN [23] (© 2009 IEEE).

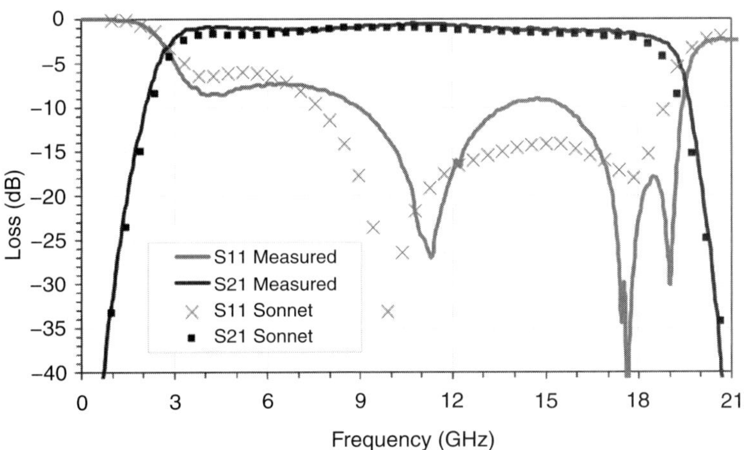

Fig. 6.11 Measured S-parameter results on a pair of back-to-back LCP baluns with $H_1 = 9$ mil, $H_2 = 1$ mil [19, 22] (© 2005 IEEE).

through 6.16. Figure 6.17 provides amplitude and phase imbalance plots. Figure 6.18 gives the measured group delay.

In Fig. 6.13, the measured transmission response S21 covers a −1 dB bandwidth of 4 to 20 GHz, which is almost the same as that predicted by the ideal transmission line model. The measured and Sonnet-simulated S21 responses are lower than that predicted from an ideal model, owing to conduction loss. The parameters S21 and S31 are identical in an ideal model because the model is perfectly symmetrical, whereas the Sonnet and measured results have an imbalance due to coupled-line

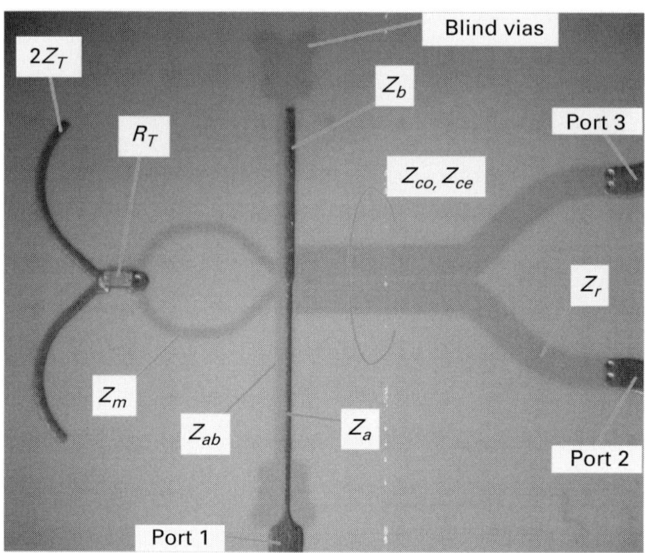

Fig. 6.12 Photograph of a fabricated Marchand balun with an EMMN [23] (© 2009 IEEE).

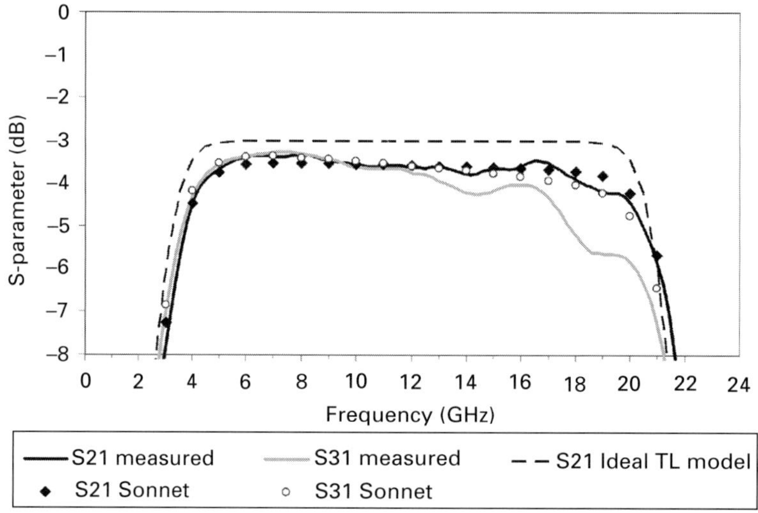

Fig. 6.13 Measured insertion losses S21, S31 on an LCP multilayer balun with even-mode matching [23] (© 2009 IEEE).

multi-mode propagation. Furthermore, the microstrip to unbalanced port transition (shown in Fig. 6.9) produces additional imbalance especially at high frequencies. As observed from the Sonnet simulation, this imbalance is caused by EM-field coupling between the microstrip and the middle metal layer when the signal enters the unbalanced port. This coupling can be reduced by adding extra blind vias to terminate this field. The transition requires careful alignment between the two

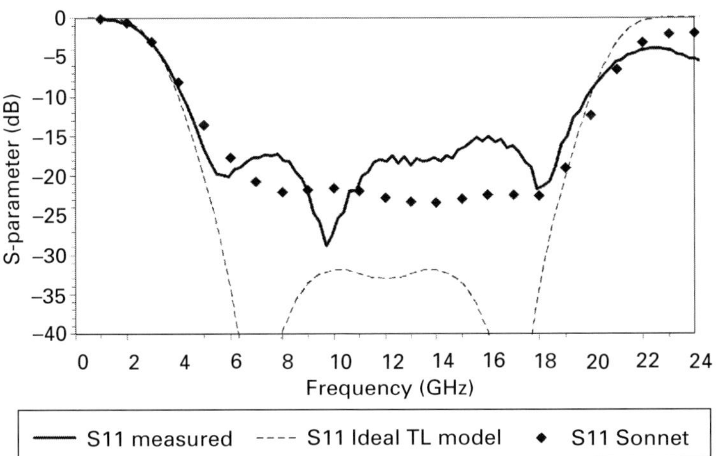

Fig. 6.14 Measured unbalanced-port return loss S11 on an LCP multilayer balun with even-mode matching [23] (© 2009 IEEE).

metal layers during fabrication, and this is arguably the most critical feature for minimizing imbalance.

In Fig. 6.14, the measurement shows a greater than 15 dB unbalanced-port return loss between 5 GHz and 19 GHz. The Sonnet simulation shows about 20 dB return loss within the same bandwidth. Both results are less than that predicted by an ideal transmission line model, which assumes that Z_a is equal to S. This assumption does not take multi-mode propagation effects into account either.

In Figs. 6.15 and 6.16, simulated results from an ideal balun model (Fig. 6.7) with and without the open-circuit stub Z_T are shown. The compensated results, with the open-circuit stub Z_T used, show a bandwidth increased to span over 6 to 18 GHz; the bandwidth is defined here as the frequency range over which the balanced-port return loss is greater than 20 dB ($-20 \log_{10}$ (|S22| > 20 dB). In the measured results, the isolation S32 between the balanced ports is higher than 15 dB from 4 to 19 GHz. The measured return loss at the balanced port, S22, is greater than 15 dB over 6 to 17 GHz. The discrepancies between the measured data and the Sonnet simulation are largely due to layer misalignments and to the via-plating quality, which is expected to improve as LCP fabrication technology matures and becomes more widely commercialized. The results for S33 and S23 are almost identical to those for S22 and S32 and are omitted from the plots.

Figure 6.17 shows the amplitude and phase imbalances calculated from the measured three-port S-parameters as respectively $20\log_{10}|S21/S31|$ and $180° - |\angle (S21) - \angle (S31)|$. The measured phase imbalance is less than 6° from 4 to 20 GHz, and the amplitude imbalance is less than 1 dB from 4 to 18 GHz. The much higher amplitude imbalance from 18 to 20 GHz also shows up in Fig. 6.13.

The measured group delay was calculated from the S21 data shown in Fig. 6.18, which, as can be seen, matches very well the predictions from an ideal transmission

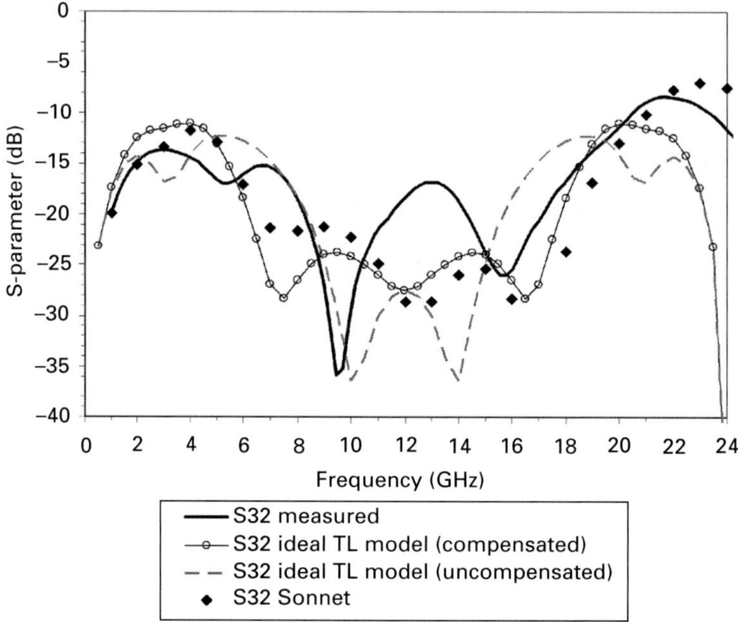

Fig. 6.15 Measured isolation between balanced ports S32 on an LCP multilayer balun with even-mode matching [23] (© 2009 IEEE).

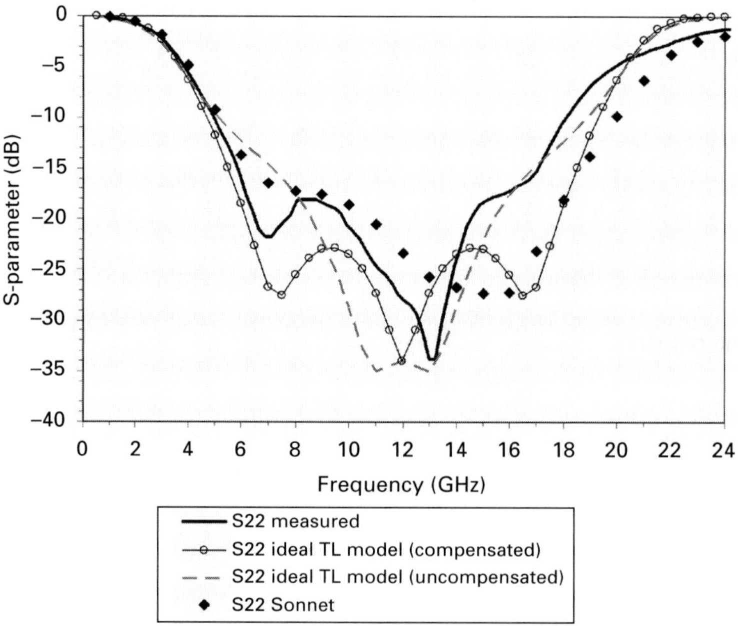

Fig. 6.16 Measured balanced-port return loss (S22) on an LCP multilayer balun with even-mode matching [23] (© 2009 IEEE).

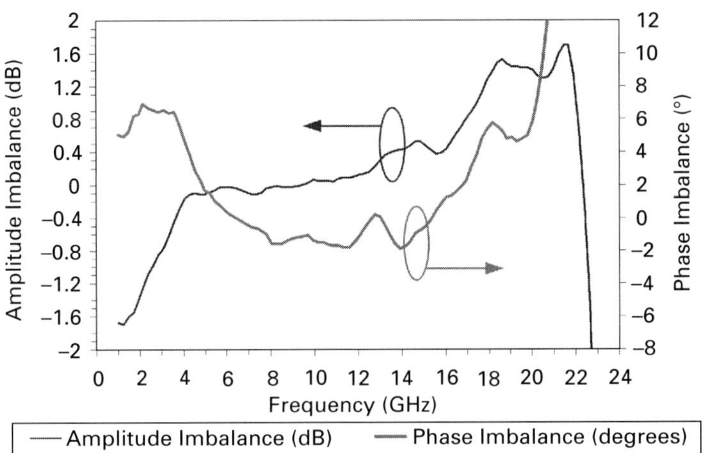

Fig. 6.17 Measured amplitude and phase imbalances on an LCP multilayer balun with even-mode matching [23] (© 2009 IEEE).

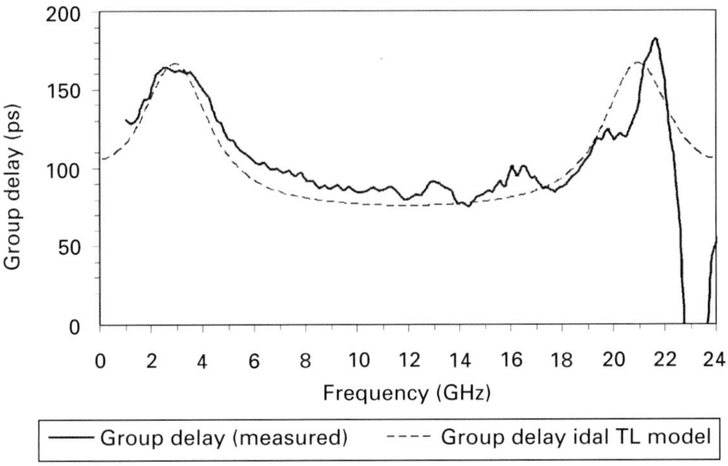

Fig. 6.18 Measured group delay on an LCP multilayer balun with even-mode matching [23] (© 2009 IEEE).

line model. The group delay results show only a very small variation across the passband, which means that the balun has low dispersion and will not greatly distort signals passing through it.

6.1.5 Defected ground structure balun

As discussed in section 6.1.1, the bandwidth of a Marchand balun depends mainly on how large a characteristic impedance of the two short-circuited transmission lines, $Z_{ab,}$ can be achieved. The larger the value of Z_{ab}, the wider the Marchand

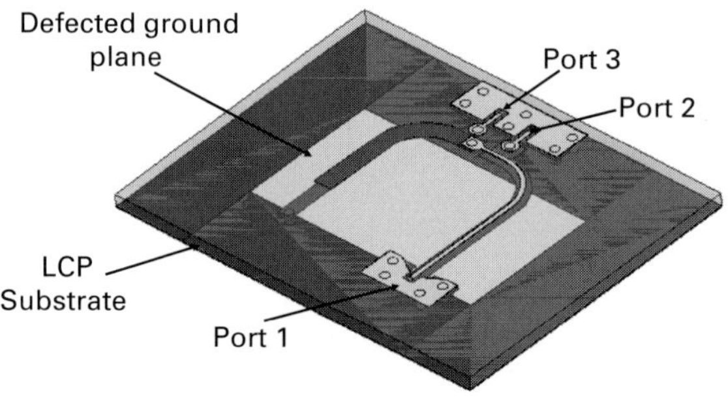

Fig. 6.19 Three-dimensional structure of the balun [24] (© 2010 IEEE).

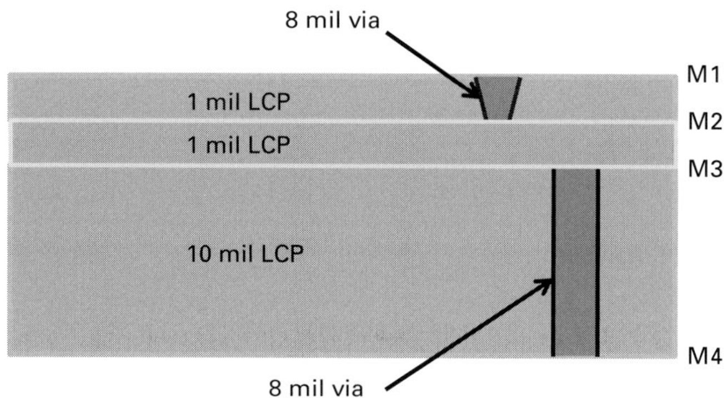

Fig. 6.20 Cross-section of the balun [24] (© 2010 IEEE).

balun bandwidth. One technique that can improve the characteristic impedance of a microstrip transmission line is the use of a defected ground structure (DGS). As reported in [13], a DGS can be used to create high-characteristic-impedance microstrip transmission lines in a branch-line hybrid coupler design. A DGS helps to increase the inductance per unit length while decreasing the capacitance per unit length of the DGS transmission-line section. This means that the characteristic impedance of the DGS transmission-line section increases; effectively, the characteristic impedance of the whole transmission line increases. The three-dimensional structure and the cross-section of the balun with a DGS are shown in Figs. 6.19 and 6.20, respectively.

The balun is designed on a multilayer LCP substrate that has four metal layers M1–M4 and three LCP layers as shown in Fig. 6.20. The topology of the balun is the same as that described in section 6.1.3 except for a rectangular-shaped defected ground structure on the bottom ground. This defected ground structure helps to increase the characteristic impedance of the two short-circuited transmission lines in the balun. This leads to the bandwidth improvement. Figure 6.21 shows the fabricated prototype of the balun. The total size of the balun is 4.4 mm × 5.6 mm.

(a)

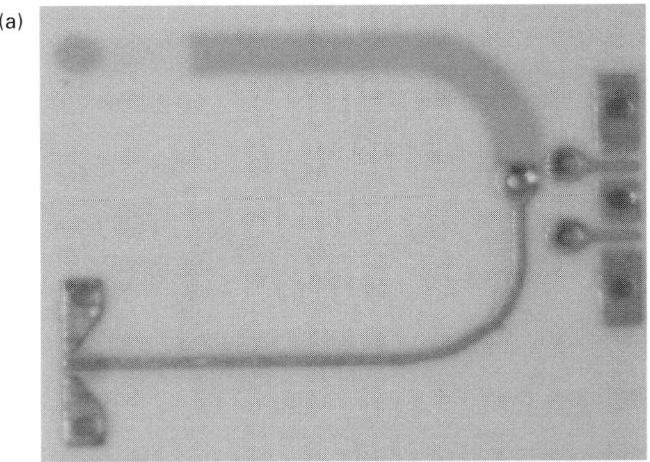

(b)

Fig. 6.21 Fabricated balun: (a) top view; (b) bottom view [24] (© 2010 IEEE).

A Cascade Microtech RF probe station and Agilent E8364 two-port network analyzer were used to measure the electrical performance. The probes were calibrated by using an on-wafer through–reflect–line on a GGB CS-5 impedance standard substrate. Since the network analyzer has only two ports, the unused port of the balun is terminated with a 50 Ω load. For example, when S21 is measured, port 3 is then terminated with a 50 Ω load while ports 1 and 2 are connected to the network analyzer. Figure 6.22 shows the measurement results.

The measured S-parameters S11, S21 and S31 of the balun are shown in Fig. 6.22a. As can be seen, the balun has a measured return loss better than 10 dB from 1.9 to 21.5 GHz. The measured average insertion loss is approximately 0.5 dB (in addition to 3 dB) over 1.9 to 18 GHz and 1 dB from 18.5 to 20 GHz (Fig. 6.22b). The balun achieves a less than 1.4 dB measured amplitude imbalance from 1.9 to 21.5 GHz and a phase imbalance of 5° from 1 to 19 GHz (Fig. 6.22c, d).

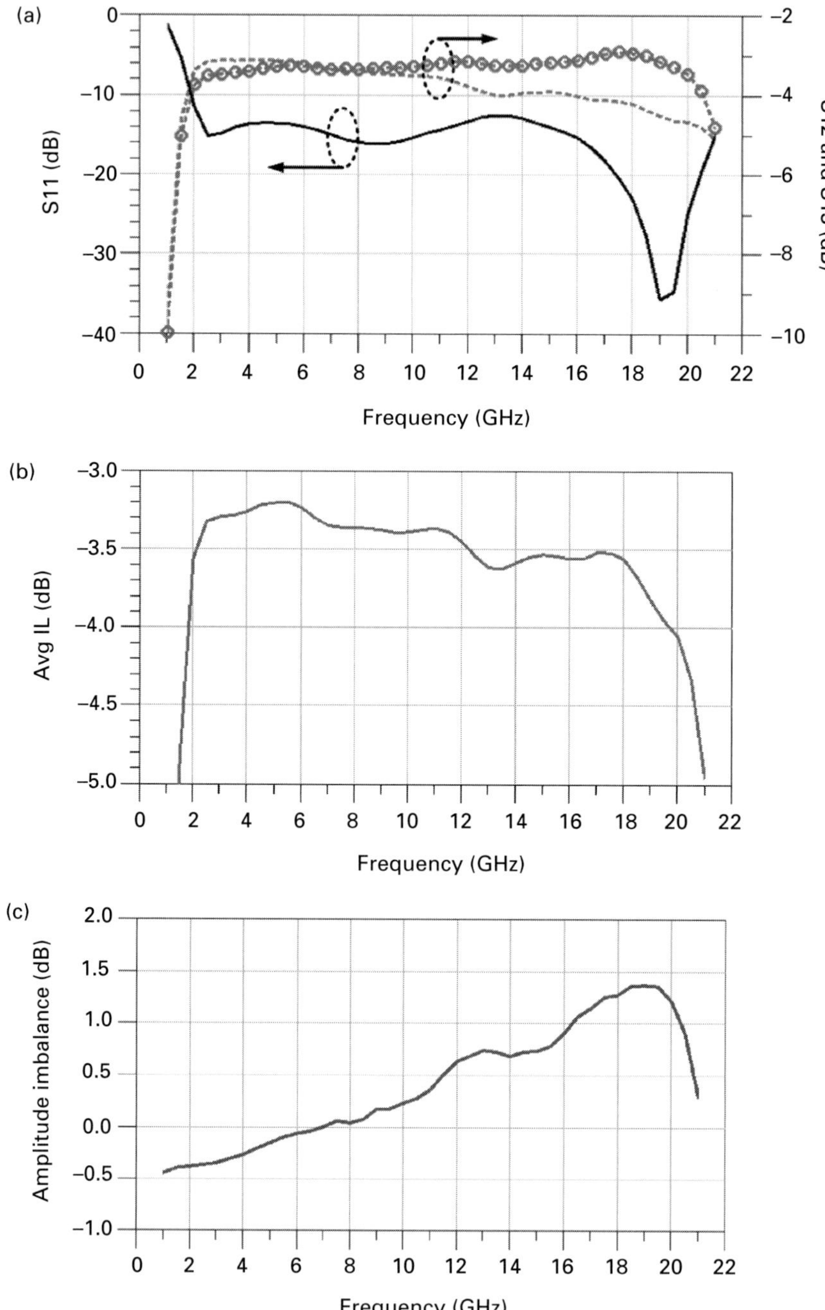

Fig. 6.22 Measurement results for the balun: (a) insertion loss and return loss; (b) average insertion loss; (c) amplitude imbalance; (d) phase imbalance [24] (© 2010 IEEE).

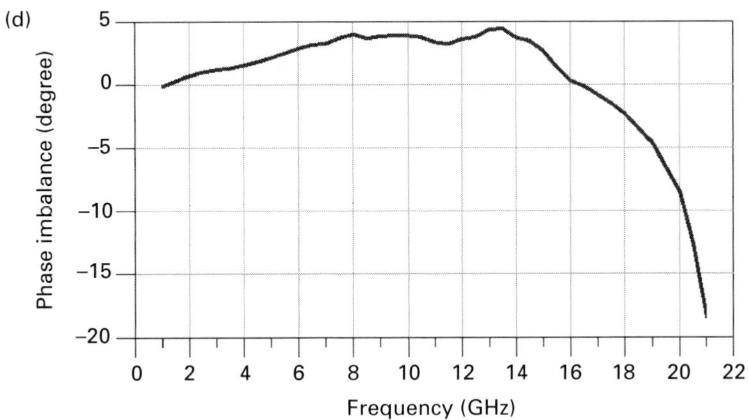

Fig. 6.22 (*cont.*)

6.1.6 LCP balun summary

A Marchand balun has been demonstrated to achieve a broadband performance enabled by LCP as a convenient thin multilayer platform. An even-mode matching network was devised in order to improve the balanced-port return loss and isolation. Measurements on an even-mode matched balun show less than 1 dB insertion loss from 4 to 20 GHz. The return losses at the balanced ports are measured to be greater than 15 dB from 6 to 17 GHz, and the isolation between them is higher than 15 dB from 4 to 19 GHz. The bandwidth of this balun can be extended further by using a defected ground structure. With the defected ground structure embedded, the Marchand balun on a multilayer LCP substrate can achieve a 10 : 1 bandwidth ratio from 2 to 20 GHz.

6.2 Wilkinson power combiner section

Power combining and dividing techniques are readily found in much of today's cutting-edge technology. This feature is critical in the corporate-feed networks commonly found in beam-forming antenna arrays. Often the feed network is the most space-hungry component in a phased-array antenna, and the miniaturization of this basic building block can yield significant space savings. The Wilkinson power divider was conceived by Ernest Wilkinson [14] and later broadband-characterized by Seymour Cohn [15].

Recently, researchers have fabricated Wilkinson power dividers on LCP with integrated resistors operating in the V-band and the W-band [16], with bandwidth ratios 1.6 : 1 and 1.5 : 1 respectively. Their designs are compact and broadband because the fractional bandwidth is higher at higher operating frequencies. However, many applications, such as UWB (3.1–10.6 GHz), require a wide bandwidth at lower operating frequencies. Wide-bandwidth Wilkinson power combiner–dividers can

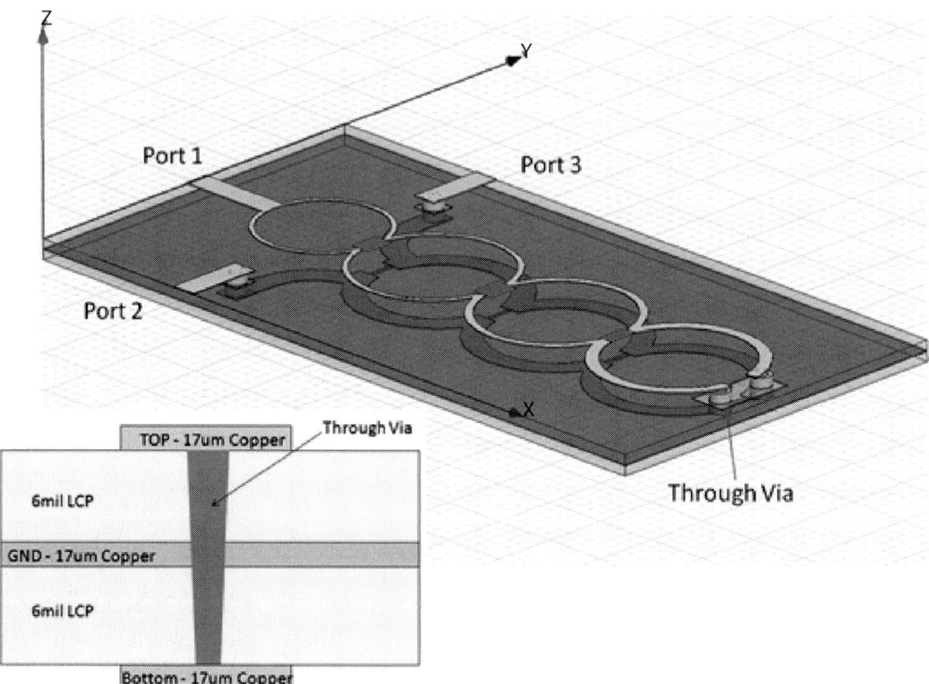

Fig. 6.23 Diagram of a folded Wilkinson power combiner on a multilayer LCP substrate [25].

be realized by cascading multiple sections, but this can lead to large sizes. A folded Wilkinson power divider, proposed in [17], was implemented on a GaAs substrate with bandwidth ratio 3 : 1, from 15 to 45 GHz.

Here we present a folded compact Wilkinson divider (Fig. 6.23) that achieves a measured voltage standing wave ratio (VSWR) of better than 1.6 : 1 with bandwidth ratio of 9 : 1 and size reduction of 43%. This is the most compact Wilkinson divider and also has the highest bandwidth ratio to date. This Wilkinson power divider is implemented on a multilayer LCP substrate and is conceptually applicable for low-frequency arrays. Section 6.2.1 describes the background and design for a broadband Wilkinson divider. Section 6.2.2 presents and discusses measurements, and lastly, a summary is provided in section 6.2.3.

6.2.1 Background and design

In the 2–18 GHz frequency band, a quarter-wavelength transmission line at 10 GHz center frequency is 30 mm long. Both folding and circular [18] section techniques can be used to miniaturize the Wilkinson divider. LCP acts as an enabling technology for this broadband application by offering an adhesiveless multilayer build composed of thin microwave substrates. As shown in Fig. 6.23, four circular quarter-wavelength transmission line sections are printed on top of a multilayer LCP substrate. Another three sections are printed on the bottom side. A metal sheet is

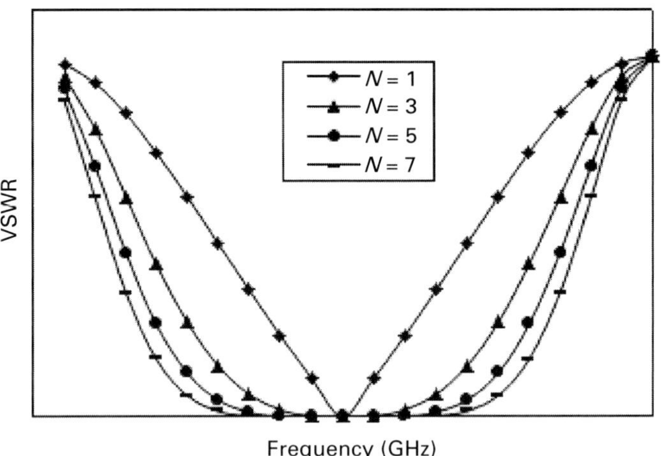

Fig. 6.24 Voltage standing-wave ratio (VSWR) for differing numbers of cascading sections centered at 10 GHz [26] (© 2009 IEEE).

inserted into the center of the LCP board to provide a ground plane for the top and bottom circular sections. Each LCP layer is 6 mil thick. Ansys HFSS software is employed to design and optimize the vias used to connect the sections of the folded structure. The vias are 8 mil in diameter with 15 mil diameter annular pads.

In Cohn's original paper [15], the bandwidth is enhanced by cascading quarter-wavelength transmission lines optimized using Chebyshev polynomials. The Chebyshev multisection-matching transformer optimizes the bandwidth at the cost of ripples in the passband. In [15], a binomial multisection-matching transformer was used to ensure a maximally flat passband and also to simplify calculations. As shown in Fig. 6.24, the use of cascading quarter-wavelength sections increases the bandwidth dramatically. A seven-stage Wilkinson divider was chosen in the present case.

The analysis is performed using both and odd-mode and even-mode half circuits. For the odd mode, the half circuit is essentially a multisection quarter-wavelength impedance transformer. In this case, a binomial matching transformer is used, defined by the function [19]

$$\Gamma(\theta) = A\left(1 + e^{-2j\theta}\right)^N, \tag{6.4}$$

where Γ is the reflection coefficient, N is the number of cascaded sections, and

$$\theta = \frac{\pi f}{2 f_0}, \tag{6.5}$$

$$A = 2^{-N} \frac{Z_L - Z_0}{Z_L + Z_0}. \tag{6.6}$$

Using a binomial expansion in (6.4) and equating it to the reflection coefficients derived from the small-reflection approximations, one obtains the following results for the impedance of each quarter-wavelength section:

Table 6.3. Power combiner parameters

Stage N	Z_N (Ω)	R_N (Ω)
1	50.26	440
2	52.21	615
3	58.50	450
4	70.80	320
5	85.46	220
6	95.76	130
7	99.46	150

$$Z = \exp\left(\ln Z_n + 2 - NC_n^N \ln \frac{Z_L}{Z_0}\right), \qquad n = 0, 1, \ldots, N, \tag{6.7}$$

where the binomial coefficient is defined by

$$C_n^N = \frac{N!}{(N-n)!n!}, \qquad n = 0, 1, \ldots, N. \tag{6.8}$$

The resistor values affect both the isolation between the output ports and the output return loss. Network theory can be applied to solve for the appropriate resistor values. For $N = 2$ an analytical approach can be taken but for greater values of N an iterative approach is necessary. For the work described above, the optimization function in Agilent's Advanced Design System (ADS) was used to solve for resistor values that optimize the output return loss. Alternatively, optimization for the isolation between the output ports could have been applied. Table 6.3 shows the calculated values for each cascading section and the optimized values for the resistors used.

6.2.2 Measurement

Resistor test structures were manufactured for characterization along with the Wilkinson divider to ensure a broadband response. The measured response of a 100 Ω test resistor is shown in Fig. 6.25 and confirms a broadband response. No more than a 1% deviation from the design value is observed throughout the frequency band.

A planar version of the Wilkinson power divider was designed and fabricated for comparison. Figure 6.26 demonstrates prototypes of both Wilkinson power dividers. Figures 6.26a, b respectively show the top and bottom sides of a folded multi-layered power combiner. Figure 6.26c shows an equivalent planar implementation. The planar Wilkinson power divider is 16.3 mm long while the folded circuit is only 9.3 mm long, resulting in a 43% size reduction.

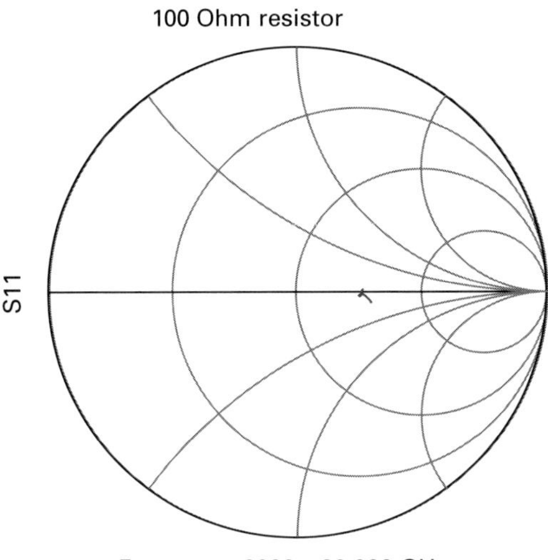

Fig. 6.25 Measured frequency response of thin-film 100 Ω resistor [26] (© 2009 IEEE). The plot shows the nearly perfect 100 Ω design up to 20 GHz. In the ideal case the small trace at center right would be a point.

Measurements were made using an Agilent E8361A network analyzer in conjunction with a cascade RF probe station. A short–open–load–through (SOLT) calibration was used to calibrate the reference plane to the probe tips. Two-port measurements were taken, with the third unused port terminated in a broadband 50 Ω load. The measurements were performed on polystyrene foam in order to approximate the dielectric constant of air. Figure 6.27 shows the physical test setup used to make measurements.

The measured and simulation results are shown in Fig. 6.28. Figures 6.28a, b show a better than 1.6 : 1 measured VSWR for both input and output ports from 2 to 18 GHz. The measured insertion loss, shown in Fig. 6.28c, never exceeds 1.6 dB excess loss. The measured isolation, shown in Fig. 6.28d, is better than 12 dB. The simulation results for the insertion loss, input and output return losses, and isolation were obtained using HFSS software and are plotted with the measurements. The measured input and output return losses closely match the simulation data. The insertion loss is also close to the simulation data. The isolation, however, suffers some deviation from the simulation owing to resistor variability when the feature sizes approach the minimum line widths.

6.2.3 LCP Wilkinson summary

We have presented the design and development of a broadband miniaturized multilayered Wilkinson power divider on LCP with integrated thin-film resistors. The design achieves a 9 : 1 bandwidth ratio, a 1.6 : 1 measured VSWR, less than 1.6 dB excess insertion loss from 2 to 18 GHz, and 12 dB isolation across the full band.

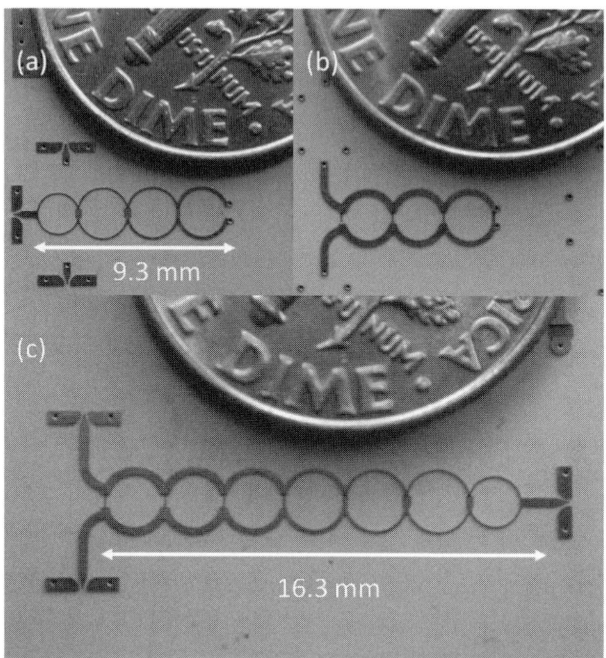

Fig. 6.26 (a) Top side and (b) bottom side photographs of folded multilayer Wilkinson power combiner. (c) Photograph of equivalent planar implementation [25].

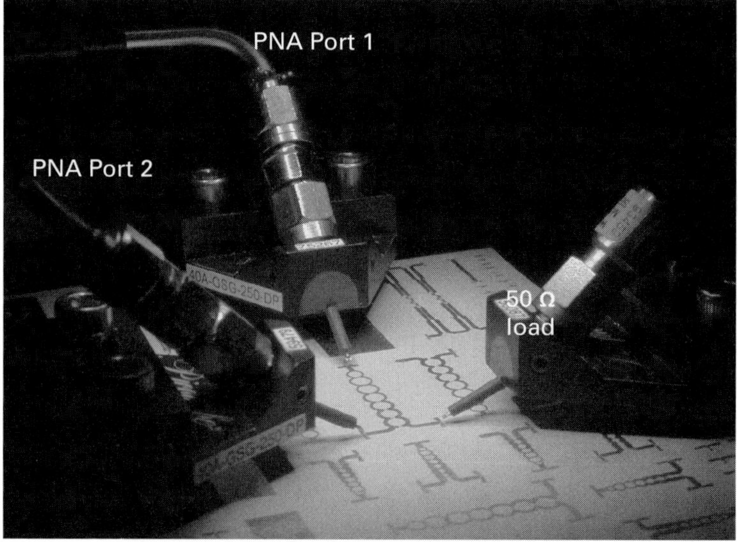

Fig. 6.27 Photograph of physical test setup used to make measurements, showing the performance network analyzers (PNAs) and the load [26] (© 2009 IEEE).

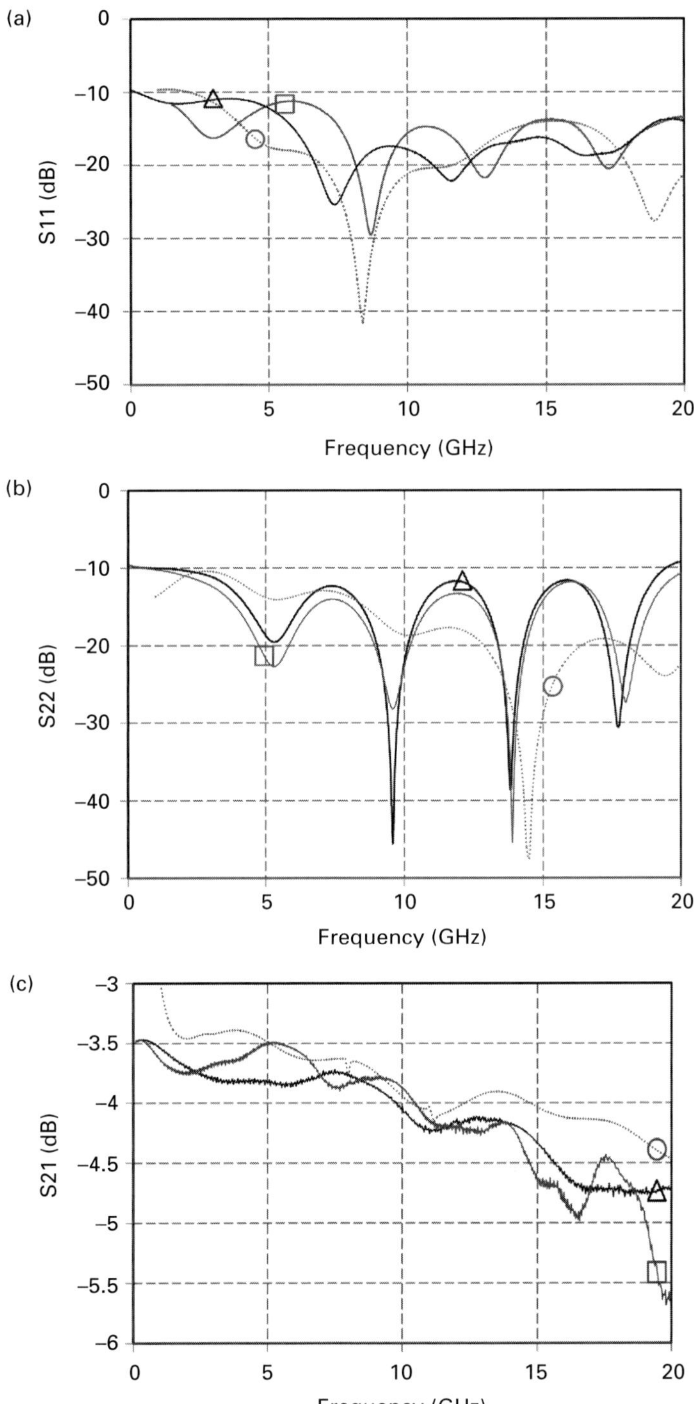

Fig. 6.28 Simulation and measurement results for (a) S11, (b) S22, (c) S21, and (d) S23 [25]. Triangles, multilayer simulated; circles, planar Wilkinson measured; squares, multilayer Wilkinson measured.

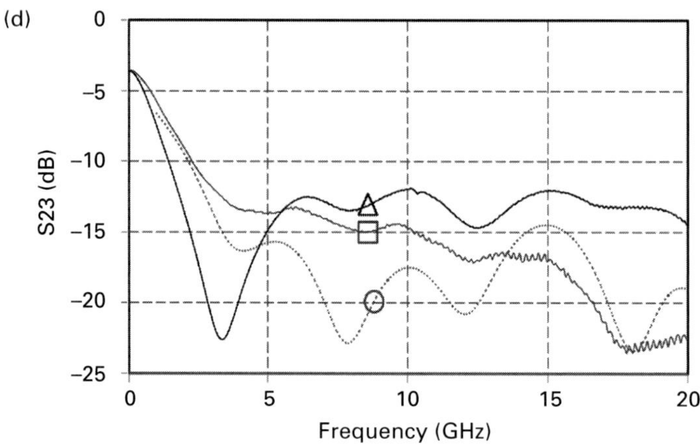

Fig. 6.28 (*cont.*)

This Wilkinson design achieves the highest bandwidth ratio, in proportion to its size, reported in the literature to date. Using multilayered LCP films, a 43% size reduction is achieved providing the enabling technology for the miniaturized feed networks that are critical for beam-forming arrays.

6.3 Folded broadband LCP hybrid coupler

This section presents the design and fabrication of a compact wide-bandwidth folded hybrid coupler on a multilayer LCP substrate. A wide-bandwidth 3 dB coupler is typically designed by cascading multiple coupled-line sections, each having the appropriate even- and odd-mode impedances, and is a quarter-wave long at the center frequency. A major challenge with this design technique is that the coupling of the center section is required to be much tighter than that of the others. For planar microstrip lines small gaps are required to achieve tight coupling, and these cannot be fabricated with standard printed circuit board processes. There are discontinuities between adjacent sections since each section has different even- and odd-mode impedances. These discontinuities cause additional reactance and decrease coupler directivity.

There have been proposals for wide-bandwidth 3 dB couplers with tandem connections and non-uniform structures [27, 28]. In [27], Uysal and Aghvami realized a 2–18 GHz 3 dB coupler that avoids abrupt changes. They used non-uniform structures and provided crossovers for tandem connections using bond wires. This coupler had large dimensions (~25.4 mm by 25.4 mm), and the bond wires had a significant effect on the coupler's performance. In [28], Salem *et al.* developed a 2–18 GHz 3 dB coupler using the same method as in [27]. Offset parallel coupled-strip transmission lines were used to realize the tandem connection. The coupler achieved good amplitude balance, but the design resulted in large dimensions, ~6.3 mm by 44.5 mm, because nine coupled-line sections were used.

Table 6.4. Normalized even-mode impedances

Z_{e1}, Z_{e5}	1.09039
Z_{e2}, Z_{e4}	1.23760
Z_{e3}	2.09021

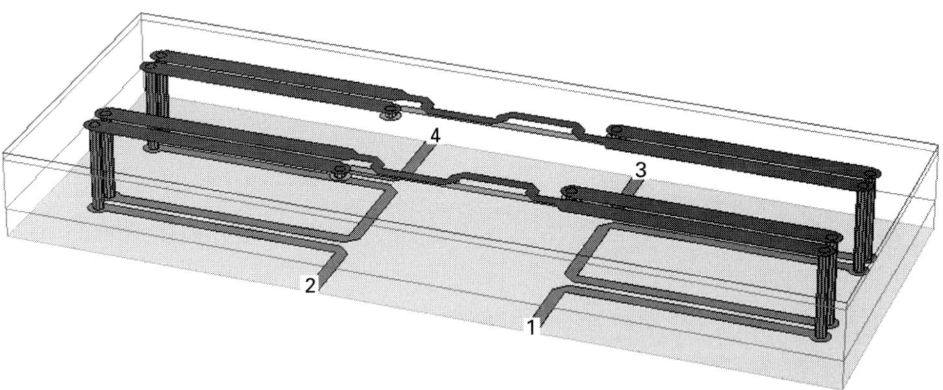

Fig. 6.29 Three-dimensional view of the multilayer broadband hybrid coupler [32] (© 2010 IEEE).

By using tandem connections of two five-section 8.34 dB couplers, a wide-bandwidth 3 dB coupler was constructed that was folded in a multilayer LCP substrate. The coupler was designed with a 9 : 1 bandwidth ratio over 2–18 GHz. The folded coupler measured 6.3 mm by 14.4 mm, which corresponds to a 42% size reduction.

6.3.1 Design of the hybrid coupler

The three-dimensional structure of the coupler is shown in Fig. 6.29. The coupler was constructed from two five-section 8.34 dB couplers tandem-connected together. The even- and odd-mode impedances of the *i*th section in each five-section 8.34 dB coupler are related by

$$Z_{0oi} = \frac{Z_0^2}{Z_{0ei}},$$

where Z_0 is the terminating port's impedance. From [29], the normalized even-mode impedance values for a five-section 8.34 dB coupler with a 9 : 1 bandwidth ratio can be determined. The values selected for this design are shown in Table 6.4; note that $Z_{ei} = Z_{0ei}/Z_0$.

To achieve for the center section (corresponding to section 3 of 5, see Fig. 6. 31) a normalized even-mode impedance value of 2.09, a coupled transmission line requires a less than ~9 μm gap, which is not compatible with PCB processes; the

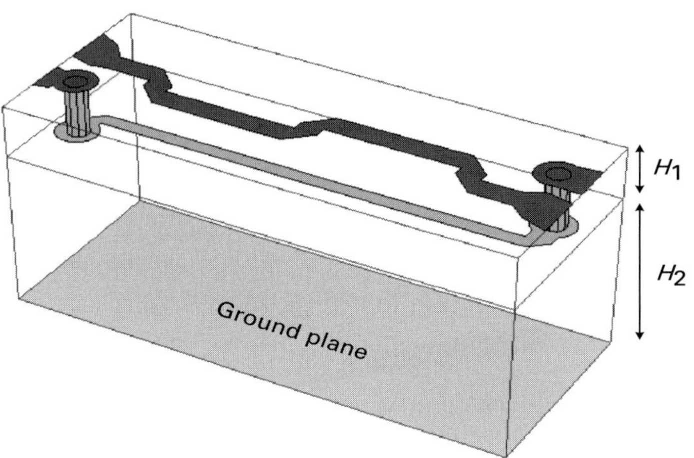

Fig. 6.30 Center section of the folded hybrid coupler [32] (© 2010 IEEE).

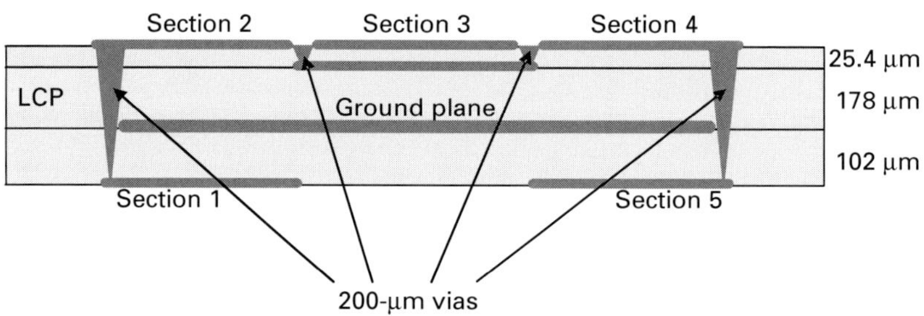

Fig. 6.31 Cross-section of a multilayer LCP board [32] (© 2010 IEEE).

minimum line width and spacing are typically 75 μm for PCB fabrication. In this design, a novel asymmetric broadside-coupled-microstrip-line with meandered upper-layer conductor is proposed, to realize the tightly coupled center section needed. Figure 6.30 illustrates this technique. The center section can easily achieve the high normalized even-mode impedance value of 2.09 while providing crossover for a streamlined tandem connection with the proposed structure. After all sections have been designed with predetermined coupling factors, they are connected together, forming a complete 8.34 dB coupler. Two 8.34 dB couplers are then connected in tandem to form a 3 dB coupler.

The cross-section of the multilayer LCP substrate used in this design is shown in Fig. 6.31. Coupled-line sections 1 and 5 are formed on the bottom metal layer, and sections 2, 3, and 4 are constructed on the top metal layer. Sections 1 and 5 are respectively connected to sections 2 and 4 through 200 μm diameter vias. A common ground is embedded in the stack between the dielectric layers. Prototypes for a folded coupler are provided in Fig. 6.32.

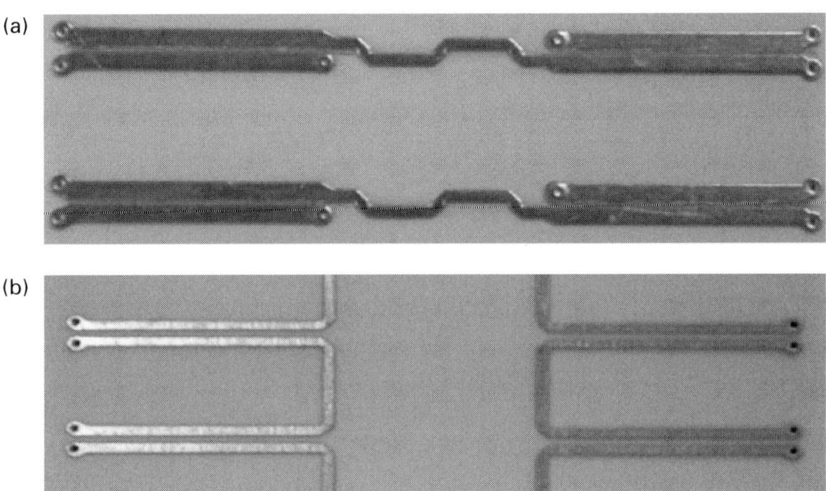

Fig. 6.32 Prototype for multilayer compact hybrid coupler on LCP. (a) Top view, (b) folded sections on the bottom metal layer [32] (© 2010 IEEE).

6.3.2 Measurement results

Sonnet EM software [30] was used to simulate and optimize the S-parameters of the coupler. A cascade Microtech RF probe station with an Agilent E8364 two-port network analyzer was used to measure its electrical performance. The probes were calibrated on-wafer using a through–reflect–line method on a GGB CS-5 substrate [31]. A two-port network analyzer was used for measurements while the two unused ports of the coupler were each terminated with a broadband 50 Ω resistor. For example, when port 1 and port 2 are connected to the network analyzer for S21 measurement, ports 3 and 4 are connected to 50 Ω terminations. Measurement and simulation results are shown in Fig. 6.33.

As can be seen in Fig. 6.33, the measurements on the coupler correlate well with the simulation results. The coupler achieves better than 20 dB return loss up to 8 GHz, better than 15 dB up to 14 GHz, and a minimum of about 10 dB around 18 GHz. The isolation is better than 16 dB up to 17 GHz. The measured phase difference over 2 to 17 GHz between the coupled and through port is $\pm 7^{\circ}$ around 90°. The measured coupled power deviates a little from the simulation. Theoretically, there should be 0.8 dB flatness, i.e. a coupling factor around 3 ± 0.8 dB [29]. The coupling factor begins to deviate from the theory owing to fabrication tolerances, losses, and discontinuities between adjacent sections. For example, each 25 μm of coupled-line misalignment in the center section decreases the coupling factor by 2.5% and changes the phase difference between the coupled and through ports of the center section by 2.3%. The folded coupler is 6.3 mm × 14.4 mm and is 42% smaller than the unfolded hybrid coupler. Table 6.5 enables a comparison of this design with published wide-bandwidth couplers on the same operating-frequency range.

Table 6.5. Various hybrid coupler designs, for comparison

References	Frequency range (GHz)	Isolation (dB)	Return loss (dB)	Phase imbalance (degree)	Dimensions (mm × mm)
[27]	2–18	15	15	±12	25.4 × 25.4
[28] (simulation)	2–18	14	19	±4	6.3 × 44.5
Present design [31]	2–18	16.8	12.84	±7	6.3 × 14.4

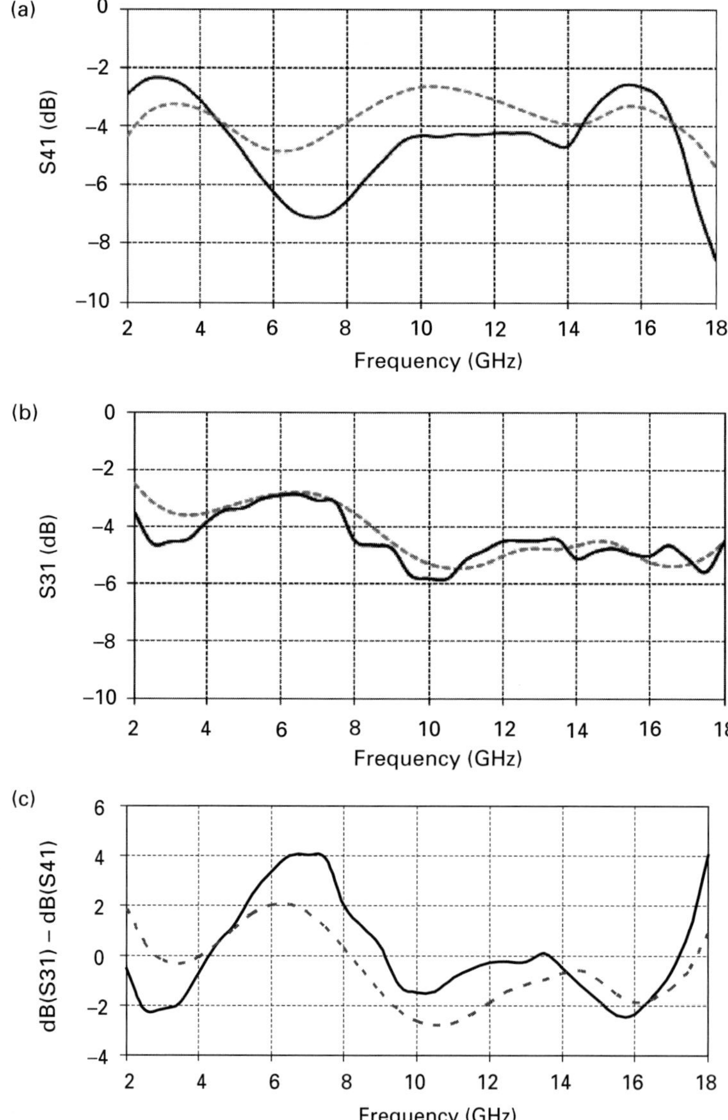

Fig. 6.33 Measurement results for the hybrid coupler: solid lines, measurements; broken lines, simulations. (a) Coupled power S41 (b) through power S31 (c) amplitude imbalance (d) isolation (e) return loss (f) phase imbalance [32] (© 2010 IEEE).

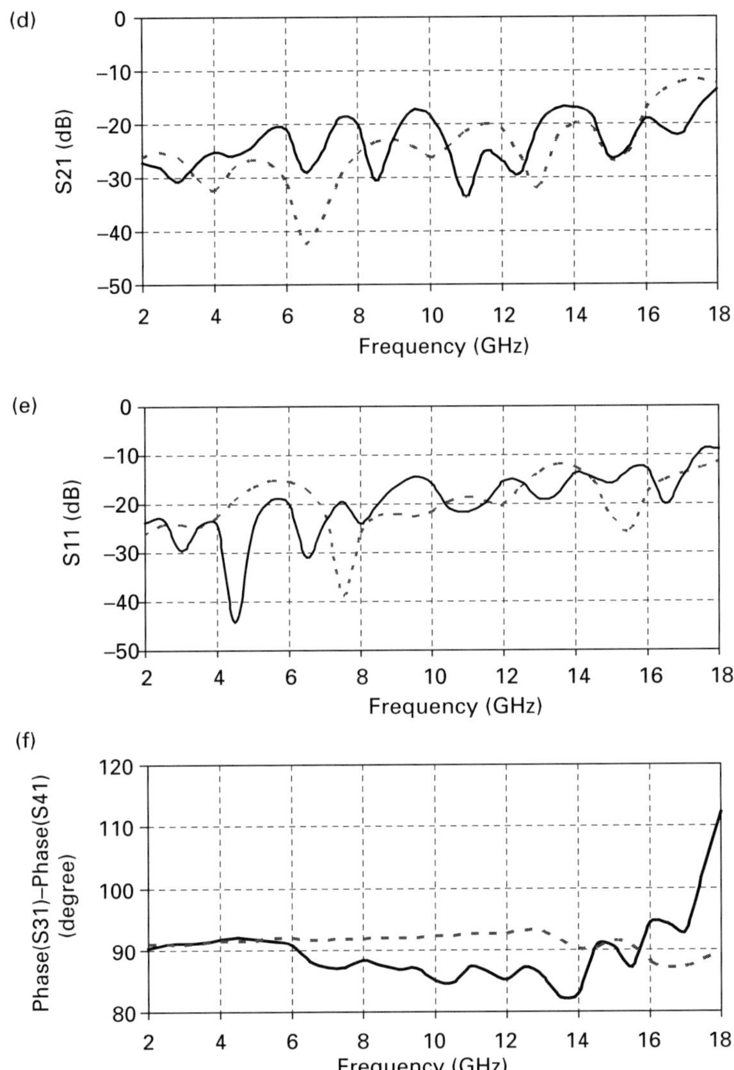

Fig. 6.33 *(cont.)*

6.3.3 Hybrid coupler summary

A wide-bandwidth folded coupler in multilayer LCP has been designed and characterized. The experimental results show that the coupler achieves a bandwidth ratio 9 : 1. The isolation is better than 15 dB and the phase difference is within ±7° from 2 to 17 GHz. A 42% size reduction is achieved by folding the coupler in a multilayer substrate. To the best of our knowledge, this coupler achieves the highest bandwidth ratio in proportion to its area reported in the literature to date.

6.4 Chapter summary

This chapter has provided examples of LCP broadband passives. LCP provides a convenient, readily available, thin low-dielectric integration platform that accommodates broadband design. A multilayer balun was demonstrated that provides excellent operation over 4–20 GHz. The use of LCP is critical for this design in providing the highly controlled thin low-loss dielectric that is crucial to its performance. Also demonstrated was a broadband power combiner–divider. Clearly, this topology is able to provide, to a high standard, resistive layers directly in an LCP package platform. Excellent performance is shown, with a greater than 10 dB return loss maintained from 2–18 GHz. Lastly, a wideband hybrid coupler was demonstrated employing novel three-dimensional meandering in a thin-film multilayer LCP stack. A 10 dB return loss is maintained with $90^{\circ}\pm7^{\circ}$ coupling over 2 to 17 GHz in a design that is reduced by 42% in area by novel LCP implementation.

References

[1] A. C. Chen, A.-V. Pham, R. E. Leoni, "Development of a low-loss multilayered broadband balun using twin-thickness thin-film," in *Proc. IEEE MTT-S Int. Microwave Symp. Dig.*, June 2005.

[2] N. Marchand, "Transmission line conversion transformers," *Electronics*, vol. **17**, no. 12, pp. 142–145, December 1944. (This design was originally implemented with coaxial cables.)

[3] J. W. McLaughlin, C. A. Dunn, R. W. Grow, "A wide-band balun," *IRE Transactions on Microwave Theory and Techniques*, vol. **6**, pp. 314–316, July 1958.

[4] J. W. Lee, K. J. Webb, "A low-loss planar microwave balun with an integrated bias scheme for push–pull amplifiers," in *Proc. IEEE MTT-S Int. Microwave Symp. Dig.*, 2001, pp. 197–200.

[5] J. Schellenberg, and H. Do-ky, "Low-loss, planar monolithic baluns for K/Ka-band applications," in *Proc. IEEE MTT-S Int. Microwave Symp. Dig.*, vol **4**, June 1999, pp. 1733–1736.

[6] K. S. Ang, I. D. Robertson, K. Elgaid, I. G. Thayne, "40 to 90 GHz impedance transforming CPW Marchand balun," in *Proc IEEE MTT-S Int. Microwave Symp. Dig.*, vol. **2**, June 2000, pp. 1141–1144.

[7] M. N. Tutt, H. Q. Tserng, A. Ketterson, "A low loss, 5.5 GHz – 20 GHz monolithic balun," in *Proc. IEEE MTT-S Int. Microwave Symp. Dig.*, 1997, pp. 933–936.

[8] A. M. Pavio, A. Kikel, "A monolithic or hybrid broadband compensated balun," in *Proc. IEEE MTT-S Int. Microwave Symp. Dig.*, 1990, pp. 483–486.

[9] R. Bawer, J. J. Wolfe, "A printed circuit balun for use with spiral antennas," *IRE Transactions on Microwave Theory and Techniques.*, pp. 319–325, May 1960.

[10] R. Schwindt, C. Nguyen, "Computer-aided analysis and design of a planar multilayer Marchand balun," *IEEE Transactions on Microwave Theory and Techniques*, vol. **MTT-42**, pp. 1429–1434, July 1994.

[11] K. Nishikawa, I. Toyoda, T. Tokumitsu, "Compact and broad-band three-dimensional MMIC balun," *IEEE Transactions on Microwave Theory and Techniques*, vol. **47**, no. 1, pp. 96–99, January 1999.
[12] W. K. Roberts, "A new wide-band balun," *Proc. IRE*, vol. **45**, pp. 1628–1631, December 1957.
[13] C.-W. Tang, M.-G. Chen, Y.-S. Lin, J.-W. Wu, "Broadband microstrip branch-line coupler with defected ground structure," *Electronic Letters*, vol. **42,** *no. 25*, December 2006.
[14] E. J. Wilkinson, "An *N*-way hybrid power divider," *IEEE Transactions on Microwave Theory and Techniques*, vol. **MTT-8**, no. 1, pp. 116–118, January 1960.
[15] S. B. Cohn, "A class of broadband three-port TEM-mode hybrids," *IEEE Transactions MTT*, vol. **MTT-16**, pp. 110–116, February 1968.
[16] S. Horst, R. Bairavasubramanian, M. M. Tentzeris, J. Papapolymerou, "Modified Wilkinson power dividers for millimeter-wave integrated circuits," *IEEE Transactions on Microwave Theory and Techniques*, vol. **55**, no. 11, pp. 2439–2446, November 2007.
[17] S. Yi, A. P. Freundorfer, "Broadband folded Wilkinson power combiner/splitter," *IEEE Microwave and Wireless Components Letters*, vol. **14**, no. 6, pp. 295–297, June 2004.
[18] C. Q. Li, S. H. Li, R. G. Bosisio, "CAD/CAE design of an improved, wideband Wilkinson power divider," *Microwave Journal*, pp. 125–130, November 1984.
[19] D. M. Pozar, *Microwave Engineering*, 3rd edition, Wiley, 2005
[20] A. C. Chen, M. J. Chen, A.-V. Pham, "Design and fabrication of ultra-wideband baluns embedded in multilayer liquid crystal polymer flex," *IEEE Transactions on Advanced Packaging*, vol. **30**, no. 3, August 2007.
[21] A. C. Chen, M. Chen, A.-V. Pham, "Development of microwave ultra-wide band balun using liquid crystal polymer flex," in *Proc. Electronic Components and Technology Conf.*, 2005, pp. 783–787.
[22] A. C. Chen, A.-V. Pham, R. E. Leoni III, "Development of low-loss broad-band planar baluns using multilayered organic thin Films," *IEEE Transactions on Microwave Theory and Techniques*, vol. **53**, no. 11, pp. 3648–3655, November 2005.
[23] A. C. Chen, A.-V. Pham, R. E. Leoni, "A novel broadband even-mode matching network for Marchand baluns," *IEEE Transactions on Microwave Theory and Techniques*, vol. **57**, no. 12, pp. 2973–2980, December 2009.
[24] H. H. Ta, A.-V. Pham, "Development of a defected ground structure wide bandwidth balun on multilayer organic substrate," in *Proc. Microwave Conf. APMC 2009*, December 2010, pp. 1641–1644.
[25] J.-C. S. Chieh, A.-V. Pham, "Development of a wide bandwidth wilkinson power divider on multilayer organic substrates," *Wiley Interscience Microwave and Optical Technology Letters*, vol. **52**, no. 7, pp. 1606–1609, July 2010.
[26] J.-C. S. Chieh, A.-V. Pham, "Development of a broadband Wilkinson power combiner on liquid crystal polymer," in *Proc. IEEE MTT-S Int. Microwave Symp. Dig.*, 2009, pp. 2068–2071.
[27] S. Uysal, A. H. Aghvami, "Synthesis and design of wideband symmetrical nonuniform directional coupler for MIC applications" in *Proc. IEEE MTT-S Symp. Dig.*, 1988, pp. 587–590.
[28] P. Salem, Chen Wu, M. C. E. Yahoub, "Non-uniform tapered ultra-wideband directional coupler design and modern ultra-wideband balun integration" in *Proc. Microwave Conf. APMC 2006*, December 2006, pp. 803–806.

[29] E. G. Cristal, and L. Young, "Theory and tables of optimum symmetrical TEM-mode coupled-transmission line directional couplers," *IEEE Transactions on Microwave Theory and Techniques*, vol. **13**, pp. 544–558, September 1965.

[30] http://www.sonnetsoftware.com/

[31] http://www.ggb.com/calsel.html

[32] H. H. Ta, A.-V. Pham, "Development of a compact broadband folded hybrid coupler on multilayer organic substrate," *IEEE Microwave and Wireless Components Letters*, vol. **20**, no. 2, pp. 76–78, 2010.

7 LCP for system design

Morgan J. Chen, Kunia Aihara, Andy C. Chen, Jia-Chi Samuel Chieh, Anh-Vu H. Pham

This chapter presents a number of subsystem-level modules that benefit from an LCP implementation. The first module, in section 7.1, is a long time delay (LTD) circuit with amplitude compensation. This module demonstrates the advantages of homogenous dielectric core and ply layers in a multilayer build. Known analytic solutions for transmission lines may be readily applied for first-pass success. In addition, the homogenous multilayer build achieves amplitude compensation through novel LCP transmission line implementations. Lastly, this module demonstrates LCP's surface mount (SMT) component compatibility with commercially available MEMS switches.

The second module, in section 7.2, is a push–pull amplifier. This module demonstrates how LCP's multilayer construction easily allows minimally short bondwires for high-performance chip interconnect. Further, this module integrates high-performance LCP baluns to achieve excellent even-mode distortion cancellation. This module also demonstrates how LCP lends itself naturally to the higher-level integration of LCP-enhanced passives.

Lastly, a receiver module with a built-in phased-array antenna is described in section 7.3. In this receiver module, LCP is demonstrated to provide a convenient platform for mechanically flexible electronics. Passive antenna structures are designed directly into the LCP build. Active semiconductor chips are packaged into this platform to show how LCP is ideally suited for building up large systems.

Each module represents advanced research based on an LCP platform that extends the electrical performance and mechanical functionality of today's subsystem modules.

7.1 Wideband thin-film amplitude-compensated LTD circuits implementing MEMS switches

Long time delay (LTD) circuits allow variable signal timing for arrays and are a key component in large broadband-array antenna systems. As the name implies, LTDs allow signals to undergo time delays that can be tuned to be electrically long on the order of hundreds of picoseconds. Unlike a phase shifter, LTDs are inherently wideband and thus are ideal for pulse transmissions on phased-array antenna (PAA) systems. In general, a PAA requires an LTD circuit for each antenna element within an array. As a result, the volume and weight scale with array size; both specifications are critical in space-flight applications.

Long time delay circuits are easily implemented using relay switches and true-time delay (TTD) elements. Multiple transmission lines with different lengths provide discrete time delays. For these circuits, the loss variations across different line lengths result in amplitude imbalance for the entire LTD. To solve this problem, high-frequency transmission lines with different loss characteristics can be easily constructed in LCP technology to achieve a highly integrated, amplitude-balanced, LTD. In developing such an LTD, the design challenges include thin-film LCP fabrication, low-loss design, and broadband amplitude balance.

Measurements on an amplitude-compensated two-bit LTD realized on LCP show a 5.5 dB insertion loss, with ±0.3 amplitude imbalance over DC to 10 GHz. This LTD is capable of a relative delay of 0, 200. 400, or 600 ps over DC to 10 GHz with return losses in excess of 7 dB. Amplitude compensation is achieved with a multilayer LCP stack-up allowing different signal and ground configurations while maintaining a high-frequency match. In section 7.1.1 we discuss the design of an LTD module, and in section 7.1.2 its fabrication and measurement; then we consider the results. In section 7.1.3 we will give our conclusions.

7.1.1 LTD design

Long time delay circuits, in their simplest form, comprise long transmission lines. Ideally, they are built on a single substrate for convenience and designed to achieve a specific time delay. Unfortunately, without further design considerations this simple build would result in an undesired amplitude mismatch. Amplitude-mismatch effects are especially severe for long time delays. Assuming that each line exhibits an equal loss per unit length, the amplitude imbalance scales linearly with the maximum time delay. In other words, long lines experience high losses, while short lines experience low losses. Amplitude compensation may be achieved by providing additional attenuation and/or gain elements, dielectric tuning, or negative-dielectric circuits to obtain negative group delays. However, the drawbacks of each method include poor broadband matching and/or complex fabrication.

In this chapter, we consider how amplitude compensation may be achieved by designing each delay line with different attenuation characteristics. Figure 7.1 illustrates a module stack-up that has been employed to enable four microstrip lines to be built with different loss characteristics while maintaining a 50 Ω match. Figure 7.2 illustrates the drill layers required for DC routing.

Closed-form line calculations

Amplitude compensation is achieved by using low-loss geometries for long delay lines and high-loss geometries for short delay lines. The amplitude imbalance is minimized when

$$\alpha_1 l_1 = \alpha_2 l_2 = \cdots = \alpha_n l_n \text{ (Np)}, \tag{7.1}$$

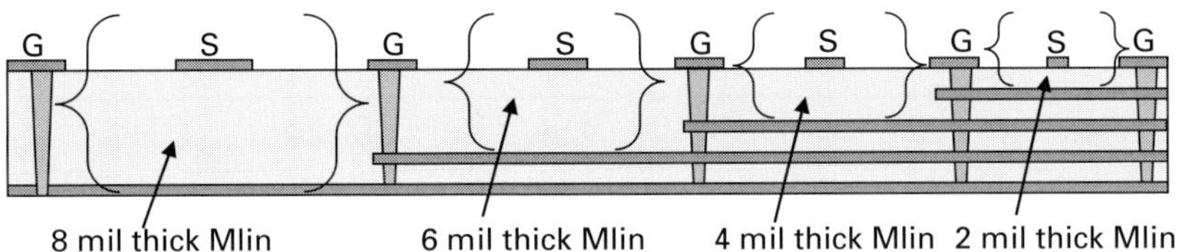

Fig. 7.1 Long time delay module stack, showing the various thickness of microstrip line (MLin) involved.

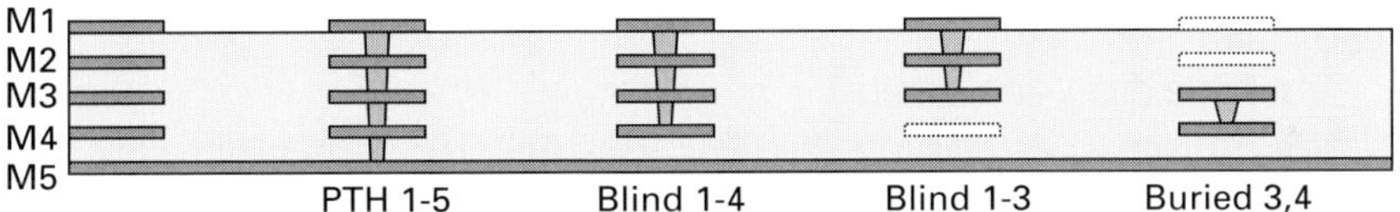

Fig. 7.2 Via types implemented on an LTD LCP stack; M1–M5 label the metal layers.

for n delay lines, where α is the attenuation constant in Np/m and l is the physical length in m. The attenuation constant α consists of the conductor-loss constant α_c and the dielectric-loss constant α_d. The total attenuation may be calculated by adding the conductor and dielectric losses:

$$\alpha = \alpha_c + \alpha_d \tag{7.2}$$

where α_c is the conductor loss and α_d is the dielectric loss (both in Np/m). A factor 20/ln 10 or 8.686 converts Np into dB. The dielectric loss is known to exhibit a dependence on the width and height geometries. In analytical form, it is calculated as [1]

$$\alpha_d = \pi \frac{\varepsilon_{ef} - 1}{\varepsilon_r - 1} \frac{\varepsilon_r}{\varepsilon_{ef}} \frac{\tan\delta}{\lambda_g} \text{ (Np/m)}, \tag{7.3}$$

where ε_{ef} is the frequency-dependent effective relative permittivity (dimensionless), ε_r is the relative permittivity (dimensionless), $\tan\delta$ is the loss tangent (dimensionless), and λ_g is the guide wavelength (in m).

The electrical permittivity as a function of frequency, ε_{ef}, may be calculated as described in [1]. The conductor loss may be calculated as in [2]:

$$\alpha_c = \frac{R_1 + R_2}{2Z_c}. \tag{7.4}$$

The normalized series distributed resistance, R_1, in the signal strip and normalized series ground resistance, R_2, may be calculated as described in [2].

In this LCP module, a 3 dB/m dielectric loss is calculated for each of four different microstrip lines. The conductive loss is calculated to vary from 8 to 35 dB/m. The dielectric loss contributes only a small portion of the total loss, owing to LCP's low loss tangent value, 0.0025. Therefore, this design relies on exploiting the conductive loss α_c to achieve amplitude compensation.

Table 7.1. Microstrip geometries and losses from ADS at 10 GHz

	Line 1	Line 2	Line 3	Line 4
Substrate height (μm)	50	100	150	200
Signal width (μm)	125	255	385	515
Signal length (cm)	5.0	9.0	13.0	17.0
α (dB/m)	38	21.1	14.2	11.4
Total line loss (dB)	1.9	1.899	1.846	1.938

True-time delay line length

For transmission lines, the physical delay length determines the transmission delay. The time delay may be stated as

$$T_d = \frac{l}{v_p} \tag{7.5}$$

where l is the physical delay-line length. The phase velocity v_p on a transmission line is given by

$$v_p = \frac{\omega}{k} = \frac{2\pi f}{2\pi/\lambda} = \frac{C}{\sqrt{\varepsilon_{reff}}}, \tag{7.6}$$

where ω is the angular frequency (in s), k is the wavenumber (in 1/m), λ is the wavelength (in m), c is the speed of light in vacuum (in m/s), and ε_{reff} is the relative effective dielectric constant (dimensionless).

Combining (7.5) and (7.6), the delay may be expressed as

$$T_d = \frac{l\sqrt{\varepsilon_{reff}}}{C}. \tag{7.7}$$

The relative time delay is calculated as the difference between delay times

$$T_{r,n} = T_{d,n} - T_{d,1}, \tag{7.8}$$

where $T_{r,n}$ is the relative time delay of line n (in s), $T_{d,n}$ is the absolute time delay of line n (in s), and $T_{d,1}$ is the absolute time delay of the reference line (in s).

Analysis and design are easily performed in the Agilent ADS Linecalc. The widths are tuned later, in the Sonnet simulation. Physical line lengths of 50, 90, 130, and 170 mm provide 0, 200, 400, and 600 ps relative delays on the proposed LCP stack-up, respectively. The design dimensions are shown in Table 7.1.

MEMS switches

Single-pole four-throw (SP4T) MEMS switches, consisting of four cantilever beams tied together at a common input, are employed. The devices are built on alumina and then packaged within resin. The pad connections consist of one RF input, four RF outputs, four controls, and one ground. The switches actuate electrostatically

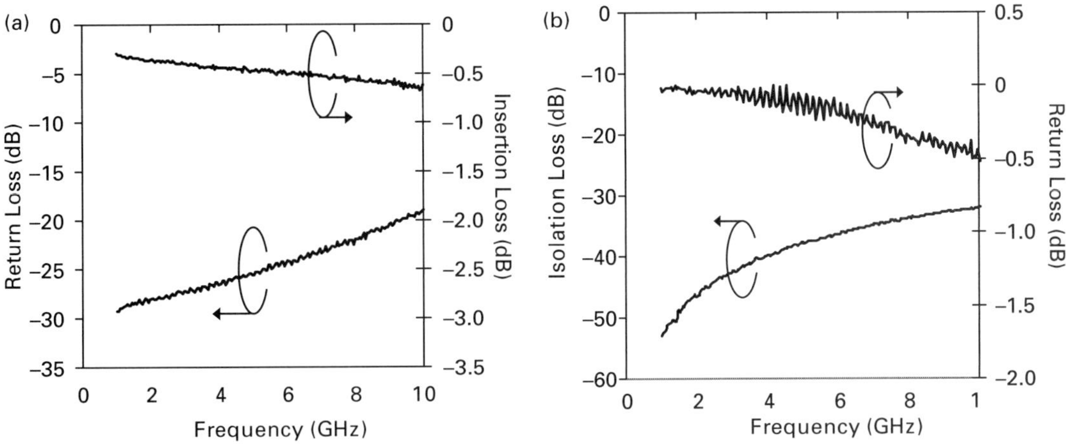

Fig. 7.3 The MEMS SP4T (a) on-state insertion loss and return loss [25] and (b) isolation and off-state return loss.

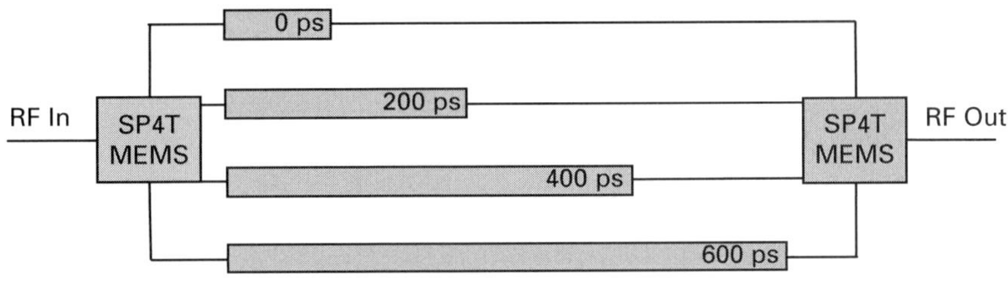

Fig. 7.4 Long time delay circuit block diagram [25].

under 65 V and 0 A. Typical measured switch characteristics are plotted in Fig. 7.3. The graphs show less than 0.6 dB on-state insertion loss across DC to 10 GHz. The return loss is seen to be greater than 18 dB over the same band. The off-state isolation is measured as better than 30 dB up to 10 GHz.

Long time delay design

A complete LTD module is formed when SP4T switches are added to connect different delay lines. Figure 7.4 illustrates an LTD circuit schematic. An ADS simulation is performed using empirical S-parameter models for the MEMS switches and built-in analytical models for the delay elements. To demonstrate the amplitude-balance improvement, Fig. 7.5 shows a 600 ps line with and without amplitude compensation. Simulated results on a complete LTD module are shown in Fig. 7.6, demonstrating less than 3 dB insertion loss, ±0.3 dB amplitude compensation, and better than 15 dB return loss at 10 GHz. The simulated linear insertion phase and the group delay are shown in Fig. 7.7 and Fig. 7.8, respectively, to show the intended phase and group delay diversity. The relative group delay is readily ascertained to be 200 ps by comparing the difference between intermediate states.

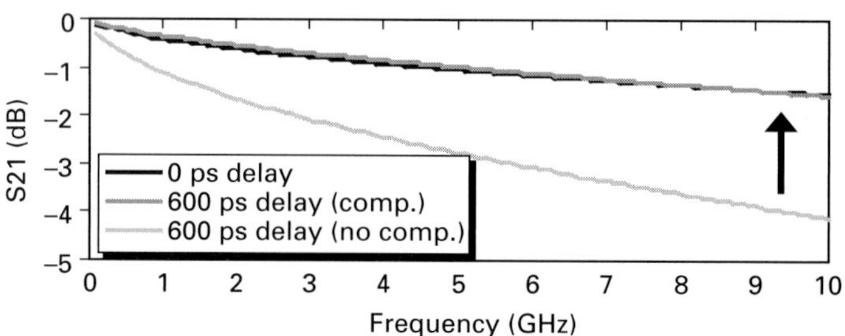

Fig. 7.5 Loss versus frequency for a 600 ps line with and without amplitude compensation [25].

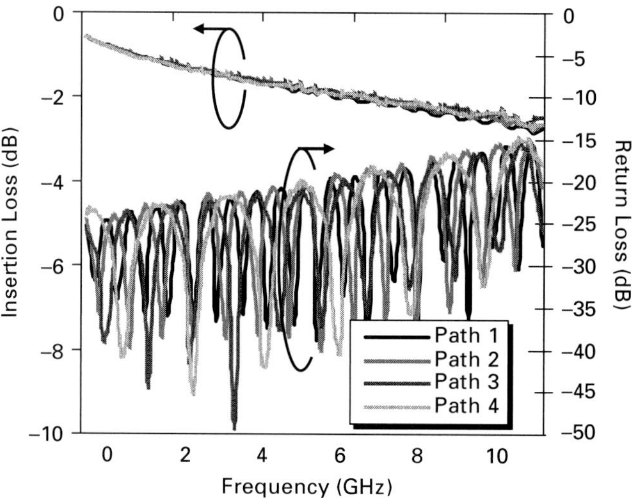

Fig. 7.6 Simulated LTD unit insertion loss and return loss [25], [26] (© 2008 IEEE).

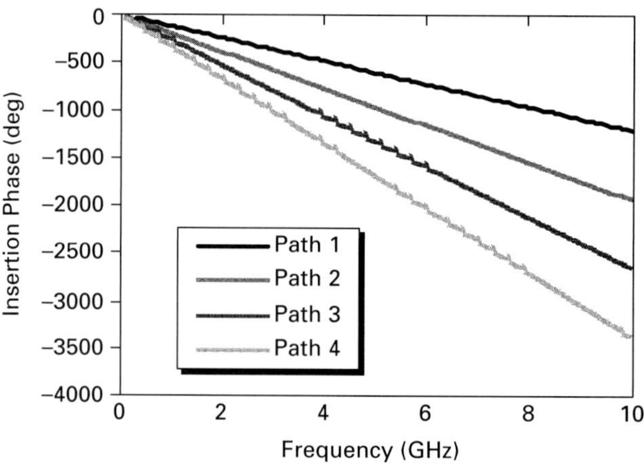

Fig. 7.7 Simulated LTD unit insertion phase [25], [26] (© 2008 IEEE).

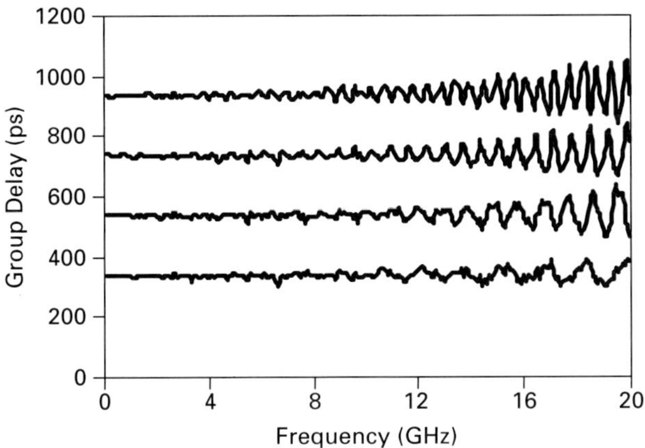

Fig. 7.8 Simulated LTD unit group delay [25].

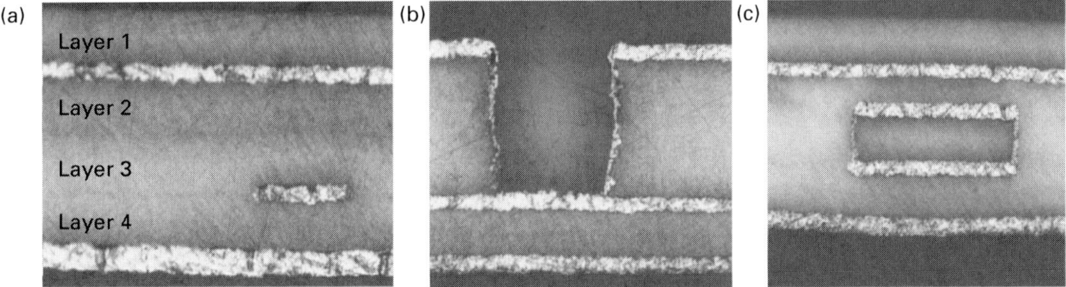

Fig. 7.9 LCP board cross-sections showing (a) a four-LCP-layer board with three metals [25], (b) a blind via through three LCP layers, (c) a buried via between two metal planes.

7.1.2 Fabrication and measurement results

A multilayer LCP board is formed with stacks of 2 mil thick Nippon Steel Chemical Corporation LCP laminates. The internal metal layers are made by 9 μm thick copper deposition. The outer copper layers are made 12 μm thick. The metal layers are then etched to define the circuitry. A multilayer structure is laminated with alternating low- and high-melting-temperature LCP. Heat is applied to melt the low-melting-temperature LCP, which acts as an adhesive, and the structure is then cooled to form a multilayer stack. Vias are formed by laser drilling and plating. Cross-section images of the module are provided in Fig. 7.9.

In order to screen for known-good-die (KGD) switches, each MEMS component is DC tested at each control with an ohmmeter. Each SP4T device requires four tests for each of the four outputs. Burn-in (also known as burn-off in this particular application) may be required to eliminate surface residue, which could cause a high series resistance between the RF ports, or to unglue a stuck-on switch. In the case where there is a high series resistance, a burn-in consists of 200 mA applied through

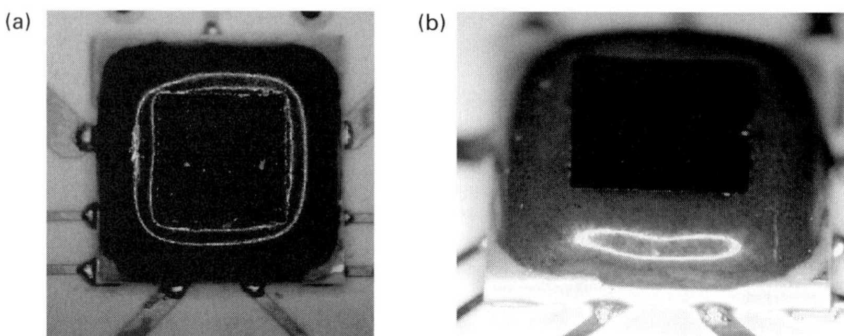

Fig. 7.10 Surface mounted MEMS: (a) top-down view [26] (© 2008 IEEE); (b) perspective view [26] (© 2008 IEEE).

a closed-state RF switch at the operation voltage. In the present design, the operation voltage is defined as 5 V above the actuation voltage when a measured through path is below 1 MΩ. Typically, burn-in is able to convert a contaminated through path with 1 kΩ resistance to less than 6 Ω after 5–10 seconds application. In a stuck-on switch, the cantilever tips may have become stuck to the contact pads owing to contamination. A gentle 10 mA burn-in may be applied across the RF contacts for several seconds to unstick a switch. Typically, this failure is not seen again after burn-in has been applied. Assembly is completed by surface-mounting the KGD MEMS switches on LCP board, as shown in Fig. 7.10.

Measurements on the complete LTD module were performed on a Cascade Microtech RF-1 probe station with 150 μm GGB picoprobes and Gore RF test cables. The S-parameter values were recorded using an Agilent 8361A (PNA). Load–reflect–match (LRM) calibration established a probe tip reference plane. J-micro CPW-to-microstrip substrate adapters were used to probe a two-bit section on the right-hand side of the fabricated LTD circuit module shown in Fig. 7.11. Isolation vias are located between the transmission lines to minimize cross-talk. The switches are electrostatically actuated with a 70 V/0 A bias and a 0 V/0 A bias to close and open the circuit, respectively.

Long time delay insertion loss and return loss plots are provided in Fig. 7.12. The phase shifts and group delays are plotted in Fig. 7.13. Insertion loss is measured to be 3.9 dB with ±0.4 dB variation at 10 GHz. The return loss is greater than 14 dB at 10 GHz. The results vary from the simulation by an additional 1 dB insertion loss at 10 GHz. The return loss is greater than 7.5 dB over DC to 10 GHz. The differences between the measurement and the simulation are likely to be due to the wire bond fixtures implemented for measurement purposes. The amplitude imbalance also increases, from ±0.1 to ±0.5 dB, owing to RF fixture mismatch.

7.1.3 LTD module summary

We have presented an LTD module having a unique amplitude compensation scheme integrated on a multilayer PCB composed of thin-film LCP. The amplitude

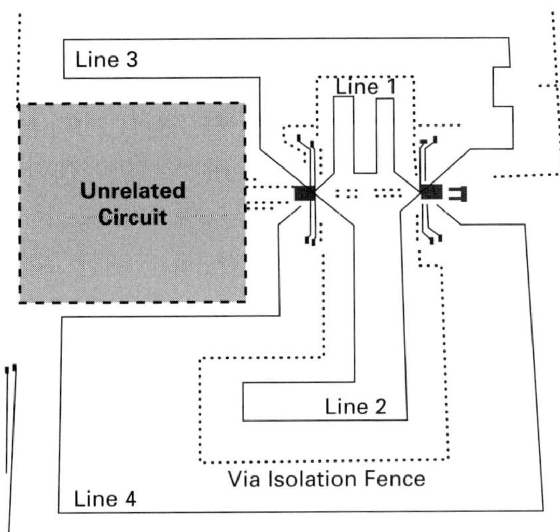

Fig. 7.11 LCP board with surface mounted components [25, 26] (© 2008 IEEE).

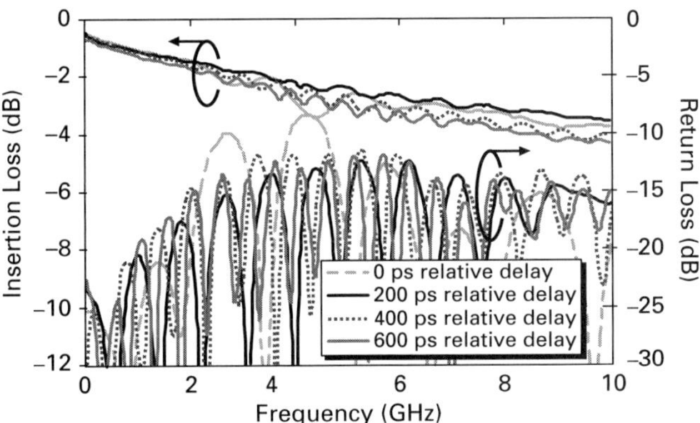

Fig. 7.12 Measured LTD losses showing amplitude compensation [25, 26] (© 2008 IEEE).

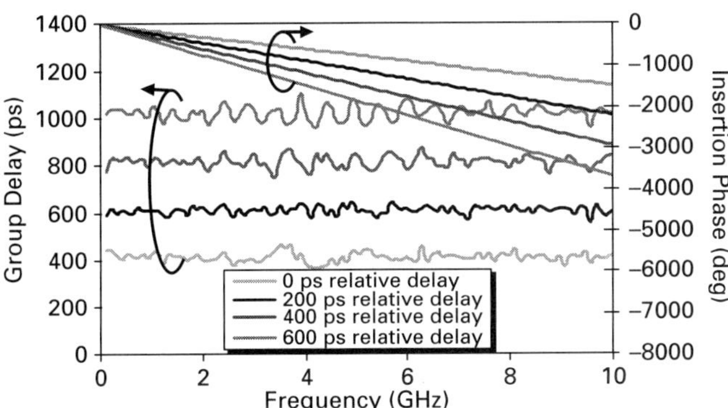

Fig. 7.13 Two bits of the measured LTD insertion phase and group delay [25, 26] (© 2008 IEEE).

compensation shows a less than ±0.4 dB amplitude imbalance at 10 GHz. The return loss is measured to be greater than 14 dB at 10 GHz and better than 8.5 dB over DC to 10 GHz. The return loss may be improved with better matching at the input and output fixtures. This LTD circuit achieves linear phases shifts at 10 GHz of 1480°, 2198°, 2930°, and 3680° for 0, 200, 400, and 600 ps relative time delays, respectively.

7.2 Broadband push–pull PA

A broadband push–pull power amplifier (PA) is another electrical module that benefits from LCP integration. A push–pull PA comprises a pair of PA devices fed with signals 180° apart. In this case, the LCP allows this design to be realized by using high-performance broadband baluns to create the 180° split, as discussed in chapter 6. Push–pull PA module design and fabrication is discussed below in section 7.2.1. Measurements are given for push–pull PAs without and with balun even-mode matching networks (EMMNs) in sections 7.2.2 and 7.2.3, respectively. The quantities measured include the small-signal S-parameters, output harmonics, and intermodulation distortions. Nonlinear measurements were performed by sweeping the input and output frequencies across a wide frequency range.

7.2.1 Push–pull PA design and fabrication

Using baluns to provide a 180° split allows inherent even-mode distortion cancellation. LCP enhances the design by allowing the ablation of precise chip cavities. Placing PA devices into a cavity minimizes parasitic wire bond effects. Two builds are demonstrated to compare the design with and without an EMMN. A diagram of the push–pull PA's operation and even-order distortion cancellation is given in Fig. 7.14.

Figure 7.15 shows a photograph of a push–pull PA module implemented in LCP without an EMMN. The two amplifiers used in this module are identical Fairchild RMPA61810 pHEMT GaAs PA MMICs. Each MMIC provides 31 dBm typical output power at 1 dB gain compression (P_{1dB}), a 21 dB small signal gain, and 22% power added efficiency (PAE) at P_{1dB}; all are specified for 6–18 GHz bandwidth. A copper–molybdenum alloy (Cu–Mo) is used as a carrier base to provide a coefficient of thermal expansion (CTE) match to the GaAs and to avoid heat-induced stress. The cutouts in the LCP were formed by laser with 2 mil over-milling to accommodate the die attach. The MMIC Z-axis locations were designed to minimize the gold ribbon length.

Two 1 mil diameter gold wire bonds are used to connect the PA's gate voltage V_G bond pads to the adjacent 100 pF bypass capacitors. Three gold wire bonds are used to connect the drain voltage V_D to the bypass capacitors, which supply a stable drain voltage and prevent PA oscillation. The RMPA61810 devices have a high reverse isolation, which promotes the overall push–pull stability of the PA.

Figure 7.16 presents a cross-section of the PA module. The baluns are implemented across five LCP layers and three copper layers. Broadside-coupling

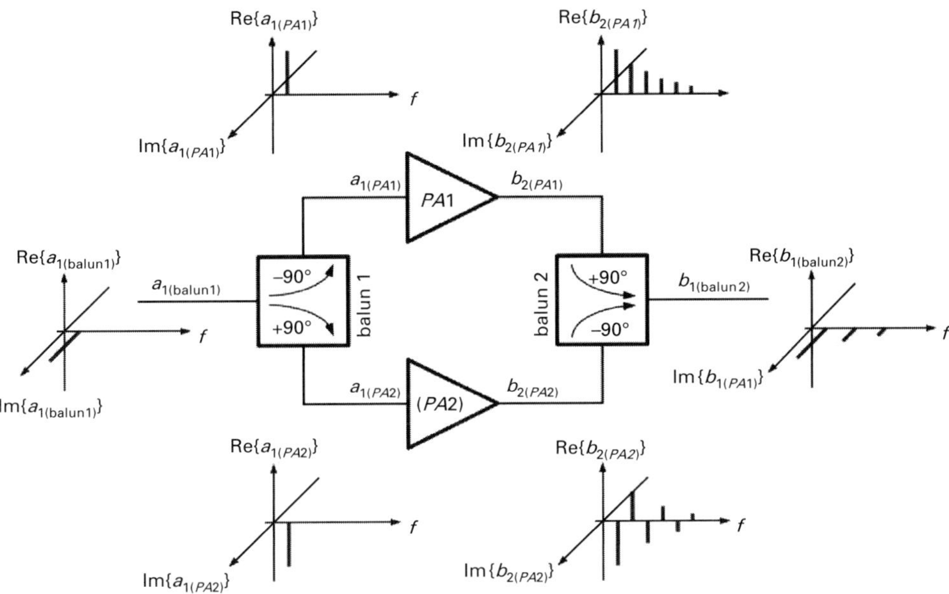

Fig. 7.14 Push–pull PA concept showing its operation and even-order distortion cancellation [8] (© 2007 IEEE).

Fig. 7.15 Photograph of a broadband push-pull PA module without EMMN [8] (© 2007 IEEE).

transmission lines are implemented on the top two metal layers. Power supply DC connections are also implemented in the same LCP substrate. A realized broadband push–pull PA module with EMMN is shown in Fig. 7.17.

7.2.2 Push–pull PA module measurement without EMMN

Measurements of the S-parameters for a push–pull PA and a single-ended PA are discussed in this subsection. A single-ended PA is a single RMPA61810 device. Both configurations are biased with gate voltages $V_G = -0.58$ V and drain voltages

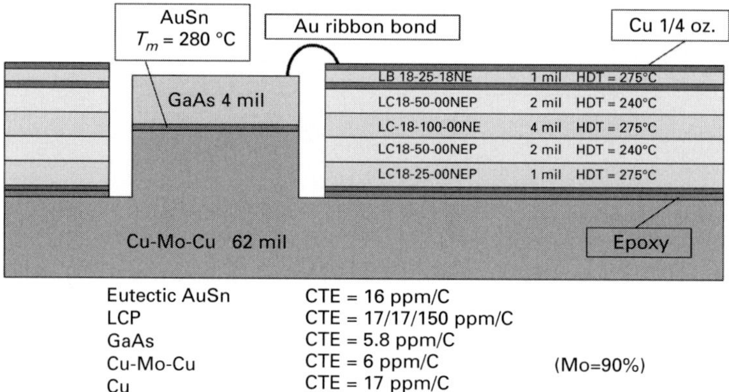

Fig. 7.16 Cross-section of a push–pull PA module [8] (© 2007 IEEE).

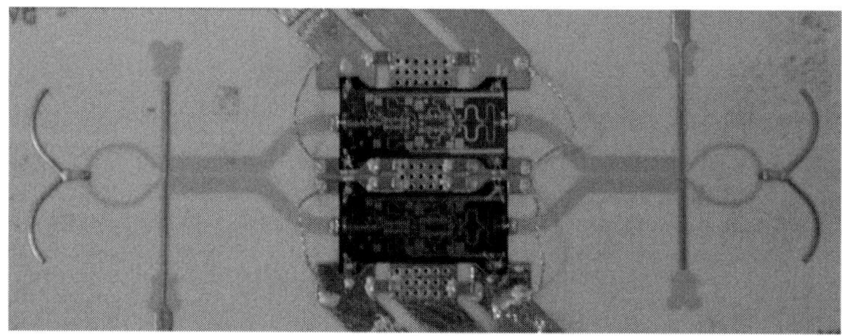

Fig. 7.17 Photograph of a wideband push–pull PA module with an EMMN.

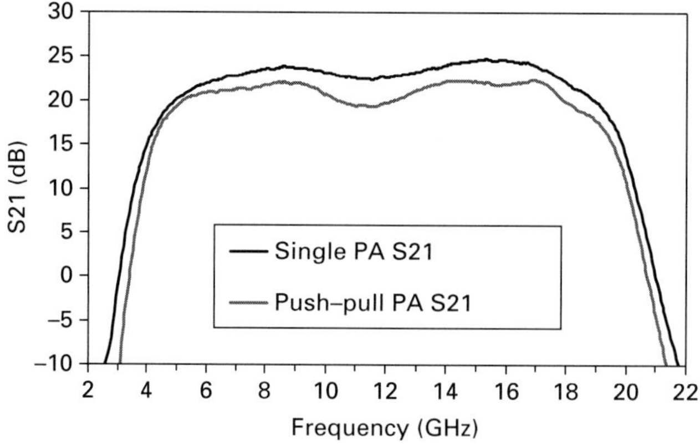

Fig. 7.18 Gain versus frequency for a push–pull and for a single-ended PA [8] (© 2007 IEEE).

V_D = 8 V. The corresponding DC drain currents are I_D = 580 mA for each RMPA61810 device at low-input-power levels and I_D = 800 mA when they are each pushed to the P_{1dB} output power level. Figures 7.18 through 7.20 show the measured S-parameters for a push–pull PA and those for a single PA. The

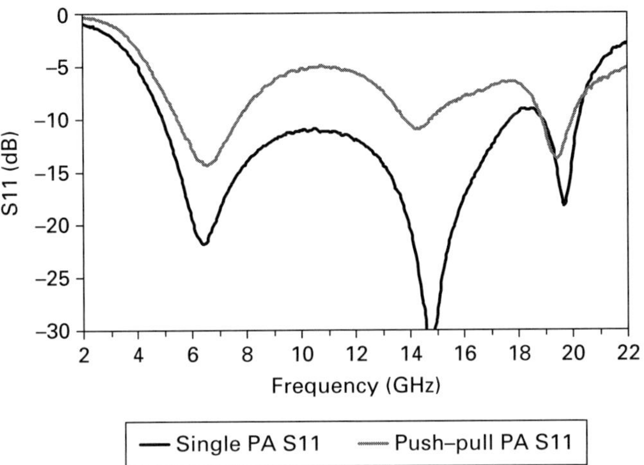

Fig. 7.19 Input return loss versus frequency for a push–pull and for a single-ended PA [8] (© 2007 IEEE).

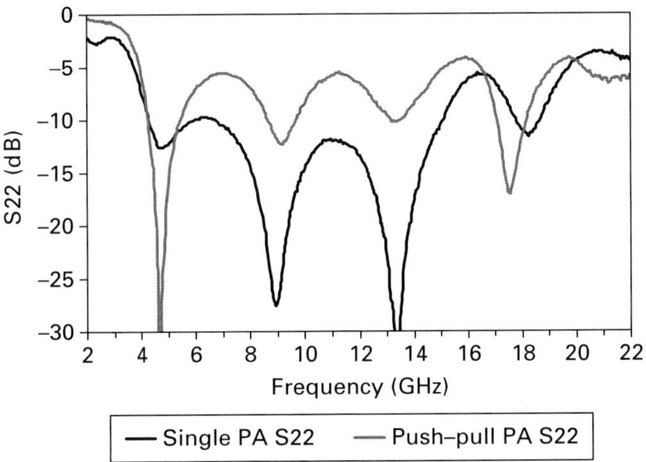

Fig. 7.20 Output return loss versus frequency for a push–pull and for a single-ended PA [8] (© 2007 IEEE).

push–pull gain is roughly 1 to 3 dB lower than that for a single PA, owing to conductor losses within the input and output baluns and power reflection between the baluns and amplifier MMICs.

Harmonic distortion measurement

We now turn our attention to large-signal and nonlinear measurements. Figure 7.21 shows the measured output P_{1dB} and the second-order harmonic distortion (HD2) level versus frequency for a push–pull PA together with that for a single PA. Sets of 1 dB gain compression output power (P_{1dB}) values are plotted along the

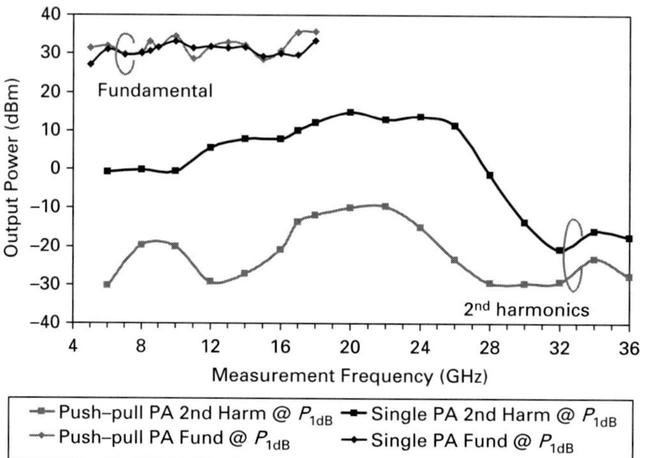

Fig. 7.21 Output P_{1dB} and $HD2$ levels versus measurement frequency for a push-pull PA compared with a single-ended PA [8] (© 2007 IEEE).

fundamental frequency, f_0, and the corresponding output, HD2, is plotted along $2f_0$. Plots are provided for both a push–pull and a single-ended PA. This figure illustrates that HD2 has been reduced by more than 20 dB with a push–pull PA implementation from 6 to 28 GHz. Above 28 GHz, HD2 remains well suppressed. This plot also shows that $HD2$ is below −40 dBc at the P_{1dB} output-power level.

A spectrum analyzer (SA) was used to measure the full output spectrum from the push-pull PA to check for possible instabilities. Neither oscillations nor any spurious spectrum other than harmonics were revealed. Figure 7.22 gives plots of P_{out}, the second-order harmonic distortion HD2, and the third-order harmonic distortion HD3 versus P_{in} for both the push–pull and the single PA. The PAs were measured with an input signal tone at $f_0 = 8.5$ GHz. From Fig. 7.22, the push–pull PA has a 15 dB larger spurious free dynamic range (SFDR), corresponding to the 30 dB lower output HD2 seen when the PAs are compared at the same input-power level. The output power and harmonics were measured with an SA at 1 kHz resolution and synchronized to a signal generator (SG) source with a 10 MHz reference frequency. The HD3 slope for a single PA did not exactly follow a 3 dB/dB slope at this particular frequency, as it experiences some expansion [3].

Table 7.2 gives for comparison push–pull PA module results together with results from earlier publications.

Intermodulation distortion measurement

Two source generators are used to generate the signals required to measure the intermodulation distortion (IMD). A Wilkinson power combiner serves to combine the input test signals. The individual input tones are set to less than 1 dB difference. The signal generator and spectrum analyzer are all synchronized through 10 MHz reference signals, and the SA resolution is set to 2 kHz. All the equipment

Table 7.2. Performance comparison of published wide-band push–pull amplifiers.

Author	Bandwidth	Gain	Output power	HD2/IMD2 reduction	Reference
Hagensen	47–862 MHz	50 dBΩ (transimpedance)	Optical (receiver) front-end	Average 29 dB	[4]
Meharry *et al.*	4–16 GHz	6–10 dB	5 dBm @ P_{1dB}	~ 20–40 dB*	[5]
Lee *et al.*	4–8.5 GHz	8–10 dB	36 dBm @ P_{3dB}	0–20 dB (not wideband)	[6]
Hsu *et al.*	5.4–10 GHz	3.5–5 dB	19 dBm @ P_{1dB}	—	[7]
Hsu *et al.*	2–4 GHz	10–11 dB	17 dBm @ P_{1dB}	—	[7]
Chen *et al.*	6–18 GHz	19–23 dB	31 dBm @ P_{1dB}	> 20 dB up to P_{1dB}	[8]

*Estimated from the OIP2 value for a 6–18 GHz conventional amplifier.

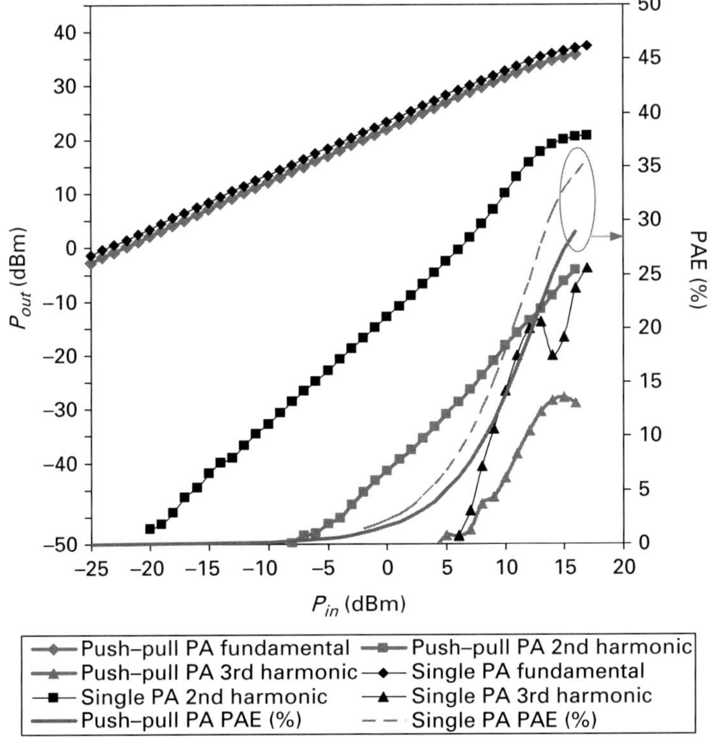

Fig. 7.22 Plots of P_{in} versus P_{out}, output HD2 and HD3 levels, and power added efficiency (PAE) curves for the push–pull PA, together with those for a single-ended PA. Measurements were made with an input single tone at $f_0 = 8.5$ GHz [8] (© 2007 IEEE).

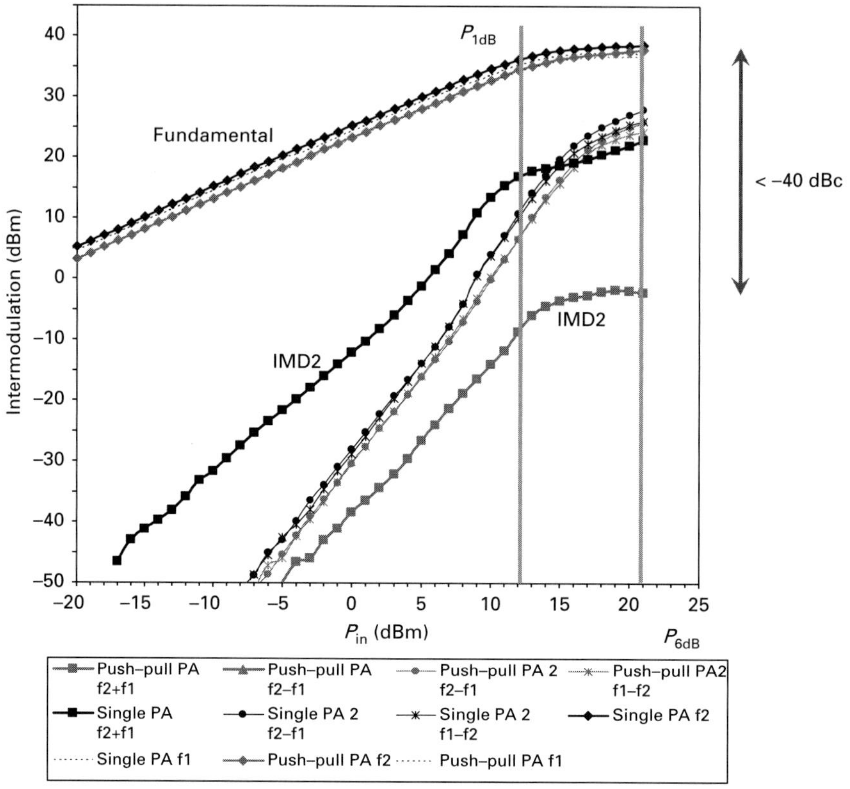

Fig. 7.23 Plots of P_{in} versus P_{out}, output, IMD2 and IMD3 levels for a push–pull PA together with those a single-ended PA. Measurements were taken with two input tones, at $f_1 = 8.5$ GHz and $f_2 = 8.520$ GHz.

is controlled by automated PC software through general purpose interface bus (GPIB) interfaces.

The fundamental output, the second-order intermodulation distortion (IMD2), and the third-order intermodulation distortion (IMD3) powers were plotted versus the input power (one of the two tones) for both a push–pull and a single PA. Figure 7.23 presents power sweeps at $f_1 = 8.5$ GHz and $f_2 = 8.520$ GHz with the output power swept to the 6 dB compression point (P_{6dB}), as indicated in the plot.

Each input two-tone signal is swept up to 15 dBm for a single-ended PA (equivalent to 18 dBm if a single tone were used) and to 18 dBm for a push–pull PA. These measurements also show greater than 20 dB IMD2 suppression.

7.2.3 Measurement results for a push–pull PA with EMMN

Adding an EMMN to the baluns improves the input and output return losses over a broad bandwidth for a push–pull PA, as shown in Figs. 7.24 and 7.25. However, the gain only improves near 11 GHz and needs to be further improved (Fig. 7.26).

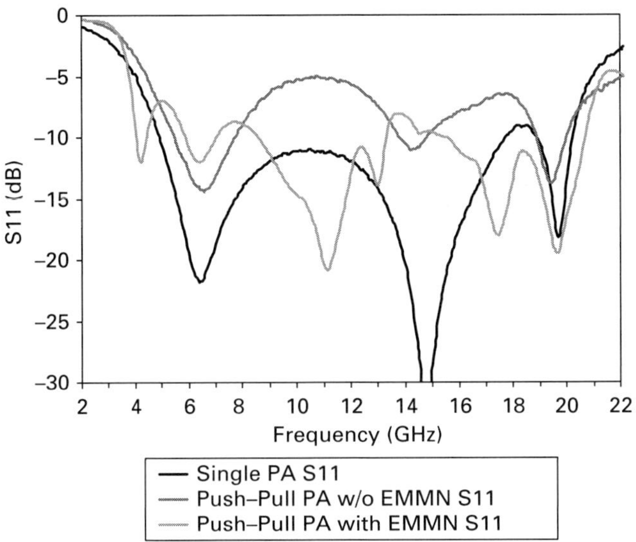

Fig. 7.24 Input return loss versus frequency for a push–pull PA with EMMN and without EMMN and for a single-ended PA.

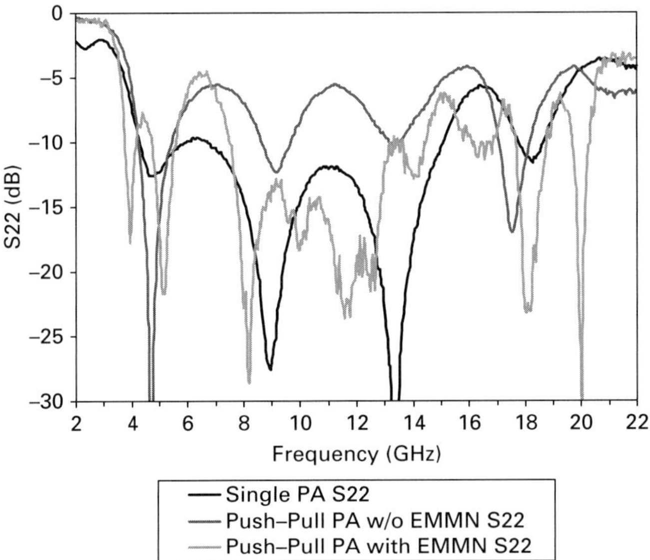

Fig. 7.25 Output return loss versus frequency for a push–pull PA with EMMN and without EMMN and for a single-ended PA.

7.2.4 Push–pull PA enabled by LCP summary

This section has demonstrated a 6 to 18 GHz push–pull PA module achieving excellent even-order distortion cancellation on LCP. Broadband baluns were integrated into an LCP substrate and incorporated into this push–pull PA module. The typical

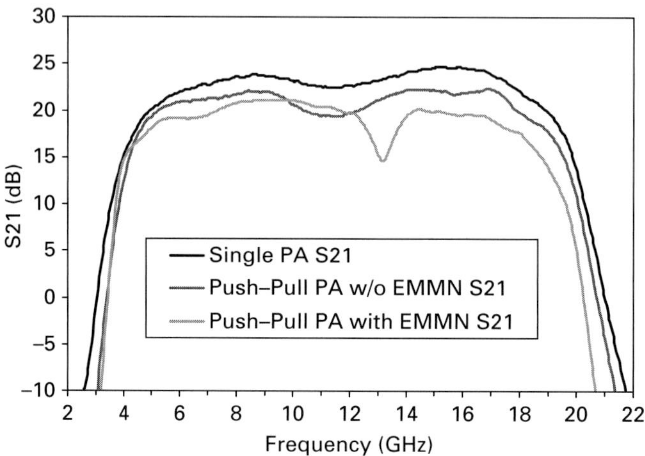

Fig. 7.26 Gain versus frequency for a push–pull PA with EMMN and without EMMN and for a single-ended PA.

P_{1dB} output power is 31 dBm. The output HD2 and IMD2 levels were measured to be –40 dBc at P_{1dB} and below. Incorporating an EMMN into this PA module shows a measured broadband-match improvement.

7.3 Receiver module with phased-array antenna

A phased-array antenna system offers the ability to scan an antenna beam electronically by providing rapid beam steering, and reliability against element failure, to many high-performance transmit and receive (T/R) systems. Applications of the phased array are numerous and include radar [9], SATCOM [10], microwave imaging [11], and, more recently, commercial W-PAN 60GHz links [12]. Phased-array antennas integrated on an organic substrate provide a way to make a system lightweight and compact [13, 14]. LCP enables advanced phased-array antennas with the ability to support multiple-layer boards on a homogeneous substrate media. Moreover, LCP is an organic substrate that is mechanically strong, light, and flexible; these are very important qualities for space-based applications.

Moreover, it has been widely reported in the literature that the phased array can improve sensitivity by a factor $10\log n$ and reduce interference from external signals by increasing spatial directivity [15]. The directivity of a synthesized beam increases when the antenna elements are sufficiently well spaced. In fact, the array gain or synthesized aperture beamwidth scales proportionally with the number of array elements. Moreover, the reception of signals from an undesired direction can be mitigated with increased spatial directivity. Figure 7.27 demonstrates this concept. For the reasons described above, many modern phased-array systems use hundreds to thousands of elements, making a scalable design highly desirable.

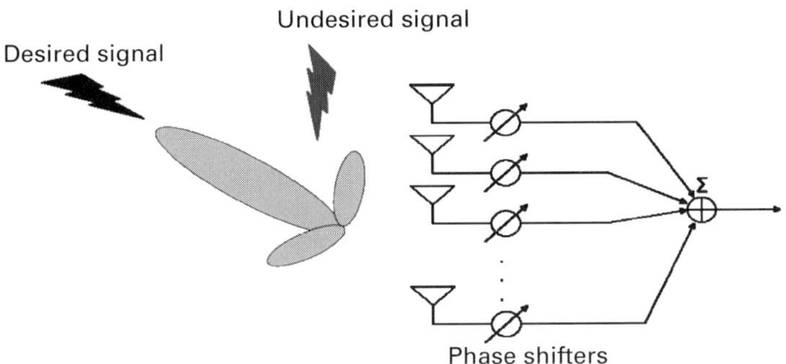

Fig. 7.27 A phased-array system with antenna directed towards a desired signal.

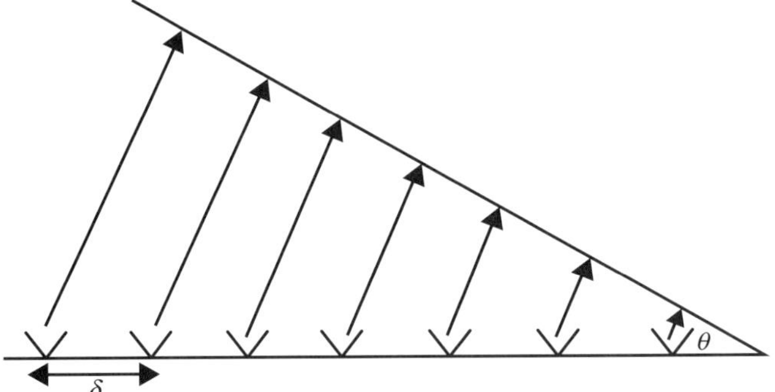

Fig. 7.28 Beam-steering through coherent phase addition.

The active electronically scanned array (AESA) has seen renewed interest in recent years in part because of improvements in very large scale integration (VLSI). The latter allows multiple T/R elements to be realized on a single chip, thus providing a compact and cost-effective solution [16–18]. However, although VLSI has greatly reduced the number of receiver system components, the principal antenna array component has a fundamental size limitation that trades performance, complexity, and cost with aperture size. This sets a primary restriction on the level of integration that can be achieved. Brick module architectures traditionally suffer from complicated interconnects and are typically extremely heavy and bulky. Below, a lightweight and highly integrated brick-type phased-array module is presented. LCP enables this phased-array module to be lightweight, highly dense, and highly scalable.

7.3.1 Overview of phased-array technology

A beam is electronically steered through the control of individual radiating elements in such a way that the total signal is coherently in phase in the desired beam direction. Figure 7.28 demonstrates this concept.

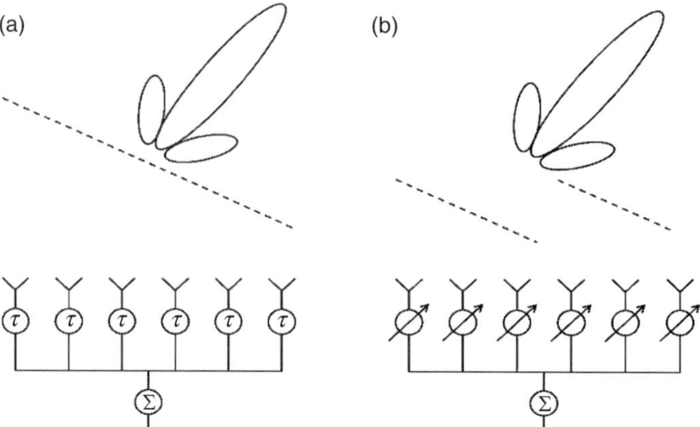

Fig. 7.29 Wavefronts for an AESA using (a) time delay elements, (b) a phase shifter.

The path length difference ΔX can be defined as

$$\Delta X = \delta \sin \theta, \tag{7.9}$$

where δ is the interelement spacing and θ is the beam angle relative to the normal. If the radiated signal frequency is known then one can convert this path length to a phase difference. This can be defined as

$$\Delta\phi = 2\pi(\Delta X/\lambda) \tag{7.10}$$

where $\Delta\phi$ is defined as the phase difference. Clearly, the beam direction can be adjusted by simply maintaining the phase difference between radiating elements.

The phased array is named aptly because of the predominantly important phase shifter used to steer the beam. Ideally, a frequency-independent time delay would be used in summing the signals from each antenna, creating a linear wavefront as shown in Fig. 7.29a. For sinusoidal signals it is well known that a phase shift is sufficient. However, because phase shifters operate with a 360° modulus, a discontinuous wavefront is formed, as shown in Fig. 7.29b. Although the above approximation is good at narrow bandwidths the phase shift is independent of frequency, which results in spatial distortion in wideband signals. The distinction between the need for true-time delay elements rather than a phase shifter has been documented in [19] and can be explained in terms of intersymbol interference (ISI).

In an ideal situation that uses time delay elements, the relationship between the phase and frequency is given by

$$f(t-\tau) \leftrightarrow F(\omega)e^{-j\omega\tau}. \tag{7.11}$$

Thus

$$\phi(\omega) = -\omega\tau. \tag{7.12}$$

As can be seen in the above ideal relationship, the time delay is independent of frequency (although the phase is not). Unfortunately, this is not the case for designs

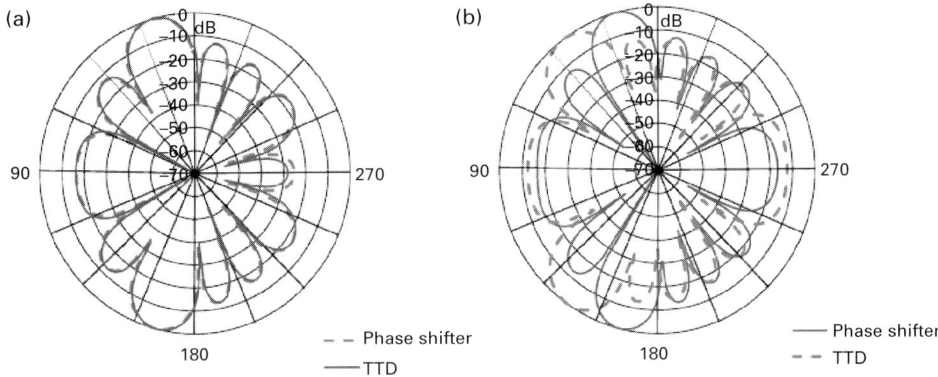

Fig. 7.30 Array factor for (a) high fractional bandwidth, (b) low fractional bandwidth.

implemented with phase shifters. To see the effects of the phase shift approximation, one has only to look at the array bandwidth described in [20].

It is well known that the array factor (Fig. 7.30) for isotropic sources is defined as

$$F(\theta, f) = \frac{\sin\left[N\pi\frac{d}{c}(f\sin\theta - f_0\sin\theta_0)\right]}{N\sin\left[\pi\frac{d}{c}(f\sin\theta - f_0\sin\theta_0)\right]}, \tag{7.13}$$

where the maxima occur at $f\sin\theta - f_0\sin\theta_0$. When $f = f_0$ and the Taylor expansion approximation

$$\sin\theta - \sin\theta_0 \approx (\theta - \theta_0)\cos\theta_0 \tag{7.14}$$

is applied then the following relationship appears:

$$f\Delta\theta\cos\theta_0 = 0. \tag{7.15}$$

Now, setting $\theta = \theta_0$ gives the relationship

$$\Delta f\sin\theta_0 = f\Delta\theta\cos\theta. \tag{7.16}$$

Using (7.13) and (7.14), the fundamental relationship between frequency, bandwidth, and beam shift can be determined:

$$\Delta\theta = \left(\frac{\Delta f}{f}\right)\tan\theta \tag{7.17}$$

This relationship dictates the bandwidth for a phased array. For a given bandwidth the scan angle of the array will depend on the frequency. This angle is also known as the beam squint. Since it is dependent on the fractional bandwidth, it can be seen that, for a given bandwidth, the beam squint will always be more severe at lower frequencies.

7.3.2 Antenna design

An antipodal tapered slot antenna that has been extensively studied [21] was chosen for use in this phased-array design because of its multi-octave bandwidth capabilities and its convenient printed circuit board implementation. One challenge that is often faced when designing a tapered slot antenna is determining the substrate thickness. It has been found experimentally that the effective thickness should be kept between $0.005\lambda_0$ and $0.03\lambda_0$ [22]. Above the upper bound, unwanted substrate modes can degrade the radiation pattern and lead to reduced efficiency. When the required effective thickness becomes less than 5 mil, materials such as LTCC or Duriod lose mechanical strength and become brittle, which makes those designs unrealistic. For example, using a material such as Rogers 4003 ($\varepsilon_r = 3.5$) at 45 GHz center frequency would yield a 3.0 mil upper limit for the effective substrate thickness, which is not readily available for rigid materials. Furthermore, for rigid materials, thicknesses of less than 5 mil result in serious mechanical strength reduction. One method to overcome this problem is by selectively drilling holes in a thick substrate to reduce the effective dielectric constant and thus to emulate a thin substrate. LCP, however, has the benefit of being flexible while maintaining mechanical strength even on very thin substrates.

In a paper by Schaubent *et al.*, various RF substrates with various thicknesses were investigated [22]. It was determined through parametric analysis that "well behaved" antennas, which closely resembled an optimum traveling-wave antenna with low sidelobes, were those on a very thin substrate. Intuitively, the reason is that in the millimeter-wave spectrum a thin substrate with low dielectric constant is necessary to prevent unwanted substrate modes. The equation governing the effective substrate thickness may be expressed as

$$t_{eff} = t\left(\sqrt{\varepsilon_r} - 1\right), \tag{7.18}$$

where t_{eff} should be kept between $0.005\lambda_0$ and $0.03\lambda_0$ to minimize substrate modes. In the present design a 12 mil thickness was used, which meets these criteria. A three-sectioned Chebyshev transformer was used at the antenna input so that the coupled-line balun could be as streamlined as possible. Figure 7.31 shows the designed antenna element, with dimensions $L_1 = 19$ mm, $L_2 = 6.82$ mm, and $W = 4.57$ mm.

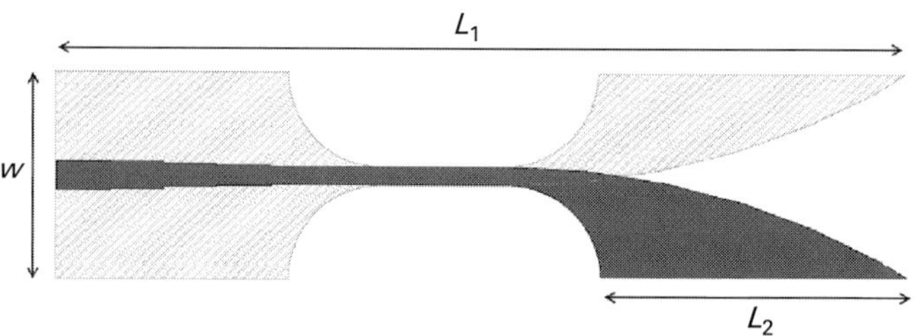

Fig. 7.31 Designed tapered slotted antenna element [13] (© 2010 IEEE).

7.3.3 Passive antenna array design

The design of the feed network is critical for beam-forming arrays. A Wilkinson power combiner is especially well suited for this purpose because it has good channel-to-channel isolation, which prevents re-radiation from the receiver. In our design, a multiple sectioned Wilkinson power combiner with integrated thin film resistors was used. Integrated thin-film resistors can also be beneficial for microwave and millimeter-wave designs. Surface mount components can add unexpected parasitics from both the packaging and the solder necessary to mount the devices. Resistor size can be kept to a minimum if a high sheet resistance is chosen for the thin-film resistors. In our design a 100 Ω/□ NiCr resistor was used. Simulations were completed in Sonnet and measurements made using an Agilent E8361A network analyzer in conjunction with a Cascade microwave probe station. The back-to-back structure is useful for verifying excess insertion loss in the feed network. Figure 7.32 shows the designed back-to-back power combiner, and Fig. 7.33 shows the measured excess insertion loss.

Using a broadband Wilkinson corporate-feed network, a passive tapered slot antenna array was designed, manufactured, and measured. It is well known [23] that to avoid grating lobes the optimum antenna-to-antenna element spacing *d*,

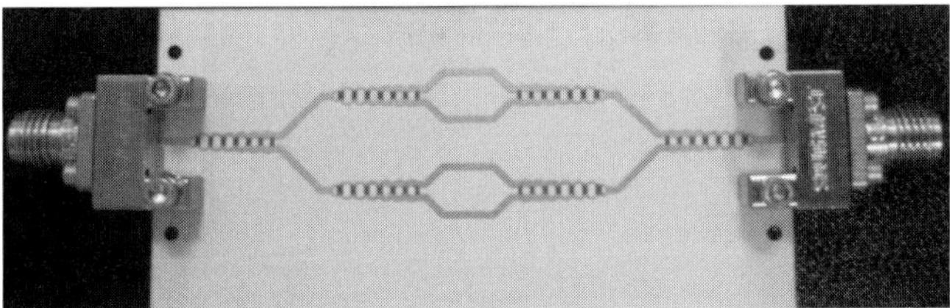

Fig. 7.32 Back-to-back Wilkinson power combiner.

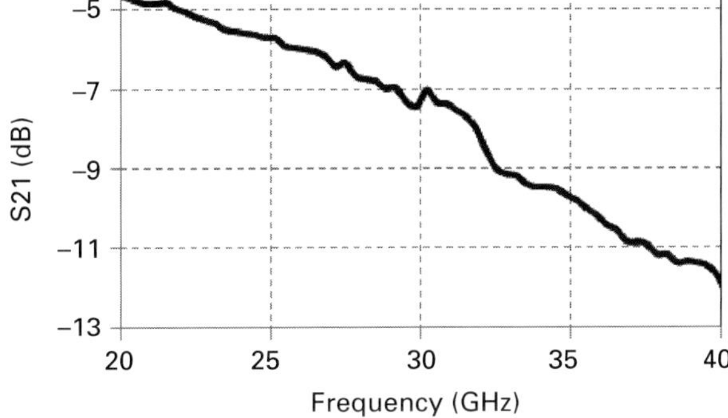

Fig. 7.33 Measured insertion loss of 1-4-1 power divider, i.e., a four-way power divider with a four-way power combiner.

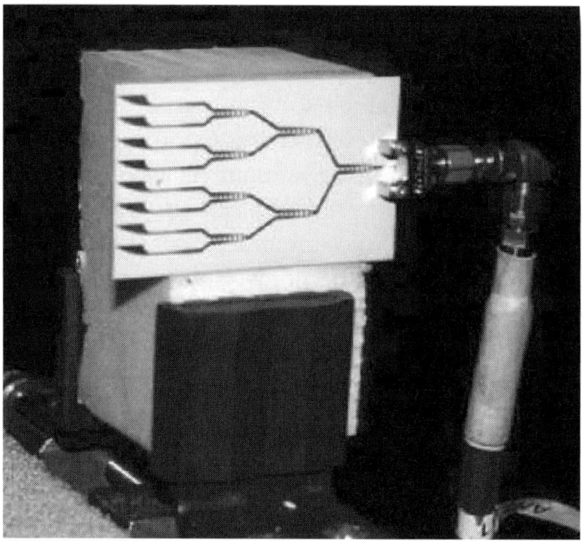

Fig. 7.34 Passive antenna array on LCP.

assuming isotropic sources, is given by

$$\frac{d}{\lambda} < \frac{1}{1+|\sin\theta_0|} \tag{7.19}$$

To satisfy the criterion for no grating lobes, a 185 mil antenna element spacing is used, which corresponds to 0.5λ at 32 GHz and 0.58λ at 37 GHz. This corresponds to a maximum scan angle, with isotropic sources, of 80° at 32 GHz and 46° at 37 GHz [24].

The designed passive antenna array and test setup is shown in Fig. 7.34 and the measured radiation patterns in Fig. 7.35. The measured return loss is given in Fig. 7.36 and exhibits a VSWR of better than 1.9:1 from 30 to 50 GHz. The radiation pattern measurements were made within a compact range and the gain measurements were made with a standard-gain horn antenna. The graphs were plotted with simulations that were completed in HFSS. As can be seen, the measured sidelobes are asymmetric, which indicates a phase- or amplitude-imbalance gradient across the aperture. The antipodal tapered slot antenna is especially sensitive to whether the input transition is balanced. The measured half-power beamwidth (HPBW) is about 28°, and the maximum measured gain is 11.5 dBi at 38 GHz.

7.3.4 Active phased-array module

Many modern phased-array modules use a tile-based array architecture. Integrating all control distribution networks onto the subarray itself is essential to reduce the complexity of tile-based arrays. For this, a multilayer board is of paramount necessity, and the advantages of using a cost-effective multilayer organic substrate are

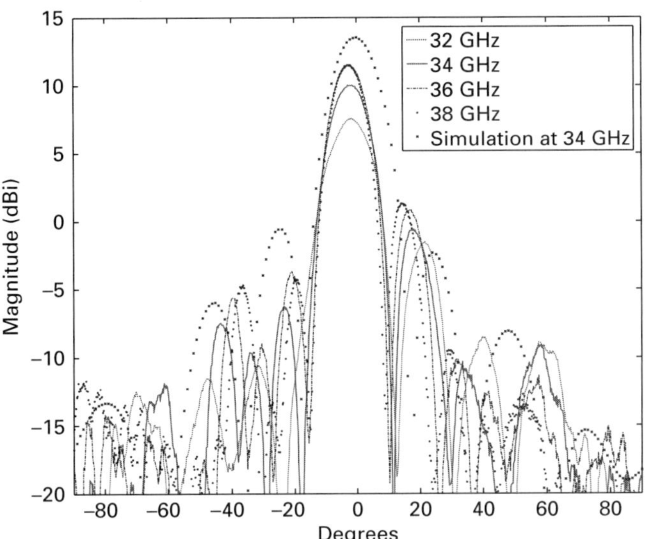

Fig. 7.35 Measured radiation pattern and gain.

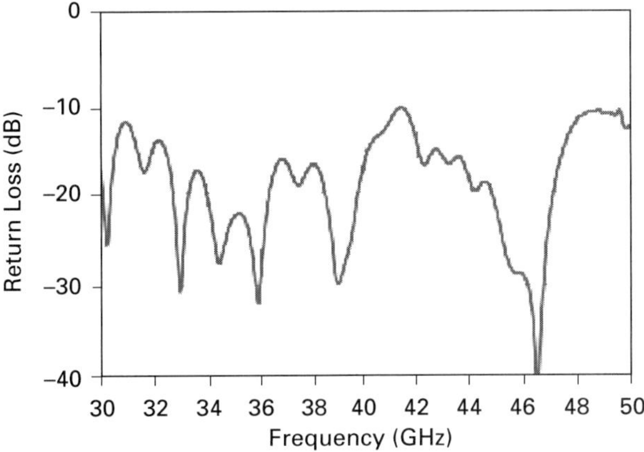

Fig. 7.36 Measured return loss of passive array [13] (© 2010 IEEE).

clear. In this design, the top layer is utilized for passive circuit fabrication and MMIC chip-on-board assemblies. Digital control chips used to interface with a five-bit phase shifter are surface mounted to the bottom layer. An intermediate layer is employed for control-line routing purposes. Finally, a ground plane layer is used to shield the RF from noisy digital circuits. A multilayer stackup is shown in Fig. 7.37.

The completed wideband phased-array receiver module and the chip assembly are shown in Fig. 7.38. In this system, a Hittite ALH369 LNA chip with a nominal gain of 22 dB covering the bandwidth 24–40 GHz is used. The phase shifter

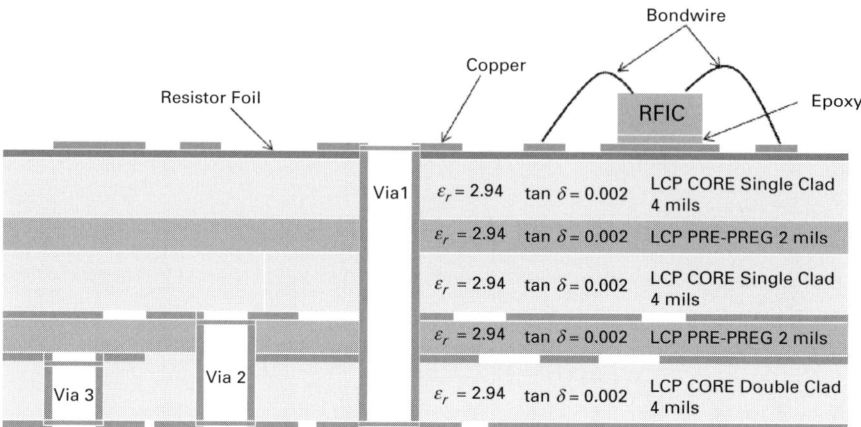

Fig. 7.37 Multilayer LCP substrate [13] (© 2010 IEEE).

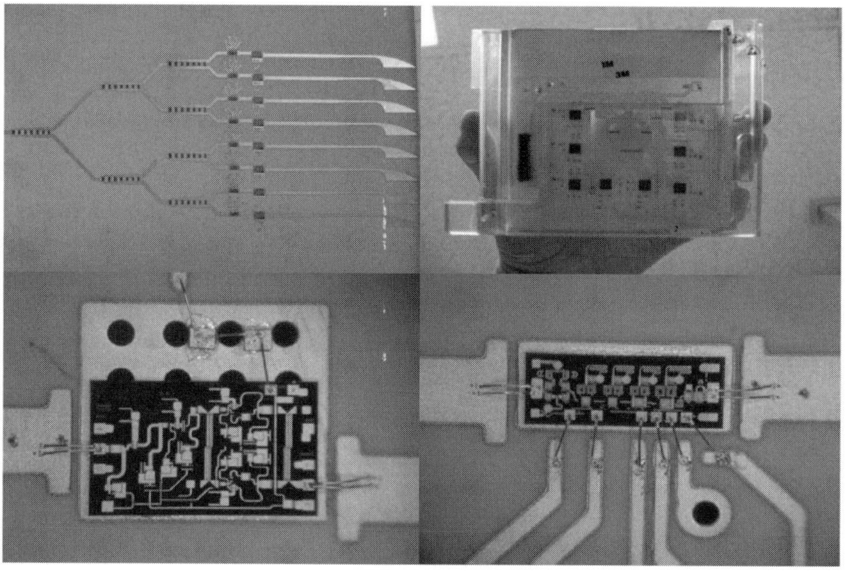

Fig. 7.38 Phased array-module top side (top left), phased-array module bottom side (top right), LNA assembly (bottom left), phase shifter assembly (bottom right) [13] (© 2010 IEEE).

MMIC is Triquint's TGP-2102EPU with five bits of control, 7 dB nominal insertion loss, and 3.5° RMS phase error at center frequency. It covers the frequency span 32–37 GHz. Finally, a Hittite HMC677LP5 six-bit serial-to-parallel controller with a built-in level shift is used to interface with the GaAs components to simplify the control routing. A three-wire serial peripheral interface (SPI) is possible when multiple chips are cascaded. A field programmable gate array (FPGA) with custom Verilog is used to interface with the controller chip to control the phase setting.

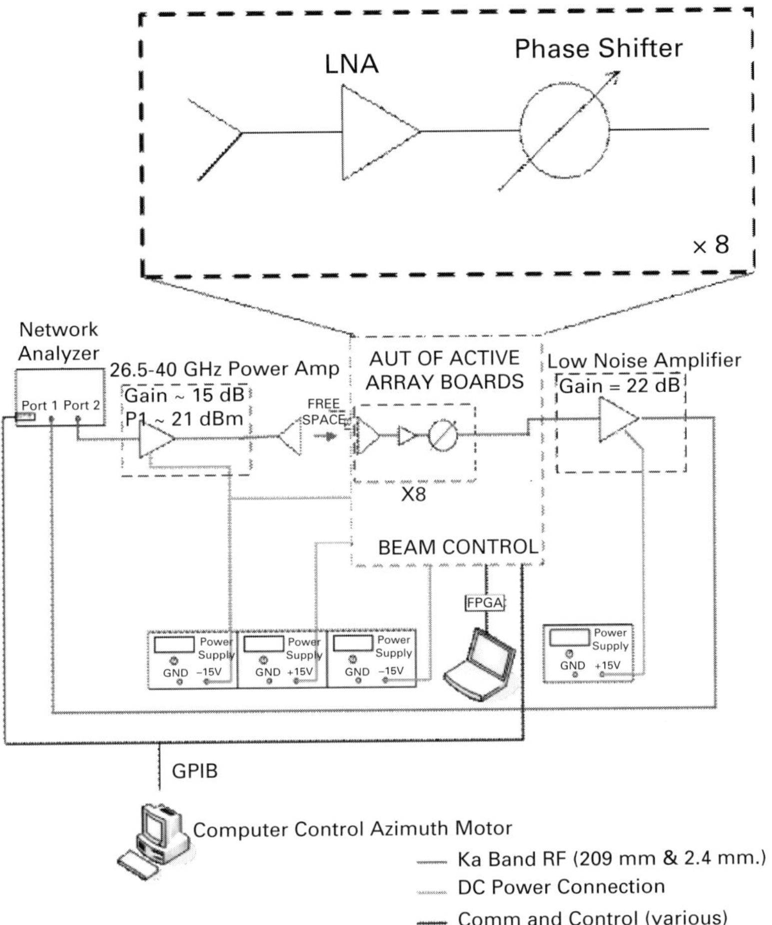

Fig. 7.39 Receiver module test setup [13] (© 2010 IEEE).

For radiation pattern measurements, the phased-array antenna receiver module was tested, using the setup shown in Fig. 7.39, with an HP8510C network analyzer in conjunction with computer-controlled azimuth rotators. Calibration was performed using a WR-28 standard-gain horn antenna. The system's dynamic range is increased by using a power amplifier on the transmit side as well as a low-noise amplifier (LNA) on the backend of the receive side. Finally, a computer controls the phase states for proper beam steering by interfacing with an FPGA.

Measurements were recorded at 1 GHz frequency and 1° azimuth angle steps. Figure 7.40 shows the measured radiation patterns at 32, 34, and 37 GHz with the beam steered from ±30° in 15° steps. From the measured data it may be observed that the amplitudes for each beam steering angle are dissimilar. This is due to the tapered slot radiator's asymmetric beam pattern [23] and can be corrected with gain-control circuitry.

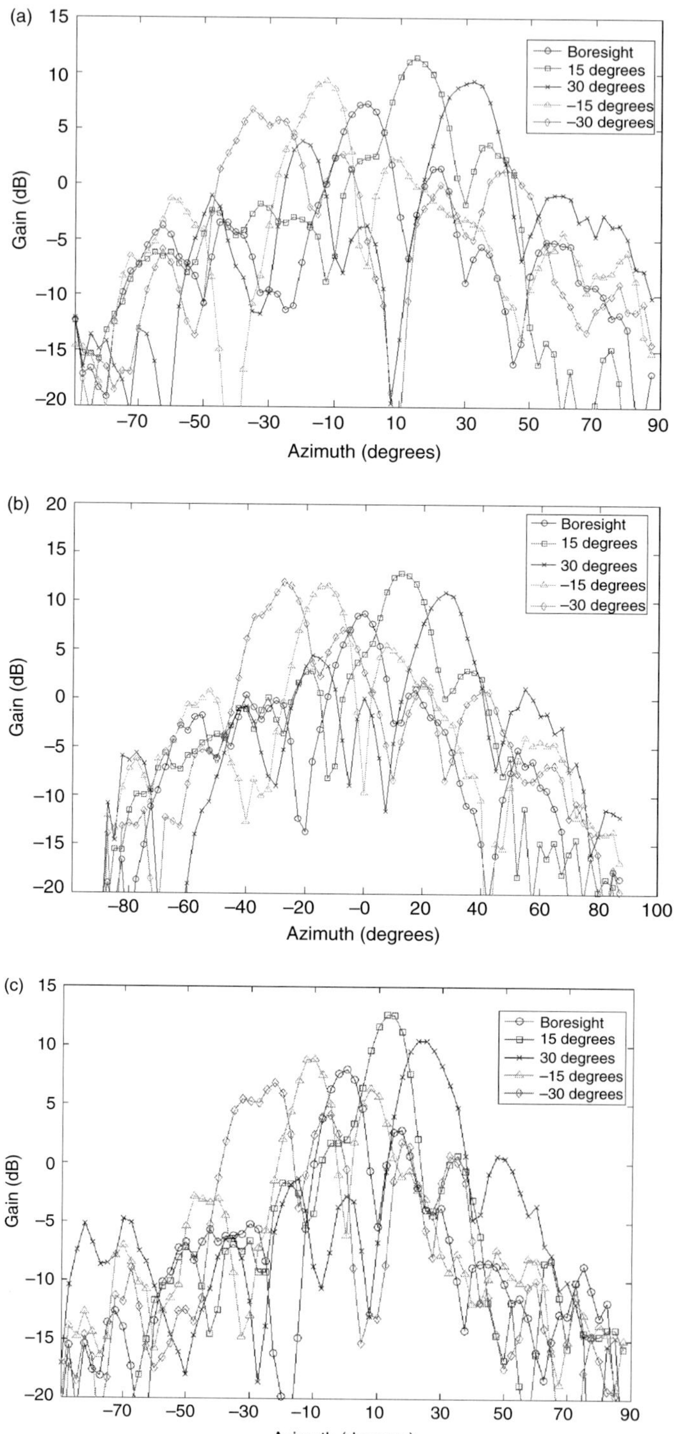

Fig. 7.40 Measured radiation patterns at (a) 32 GHz [13] (© 2010 IEEE), (b) 34 GHz [13] (© 2010 IEEE), (c) 37 GHz [13] (© 2010 IEEE).

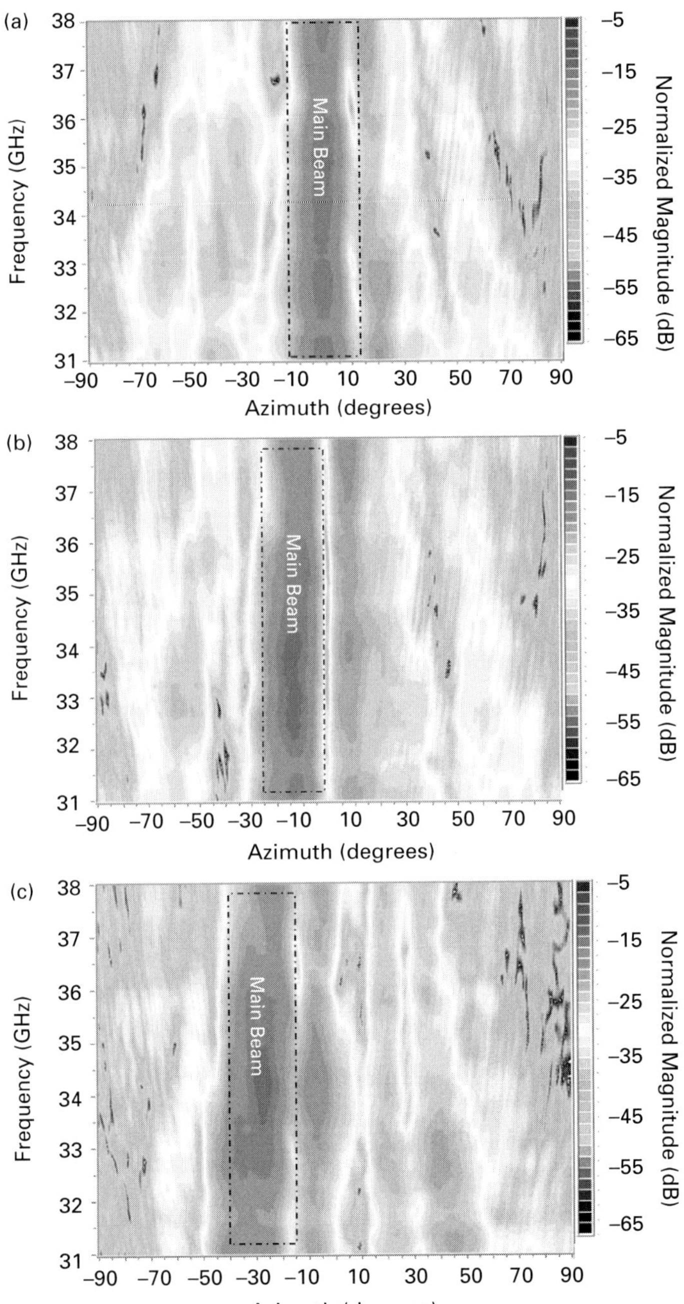

Fig. 7.41 Measured three-dimensional antenna pattern for scan angles at (a) boresight [13] (© 2010 IEEE), (b) −15°, (c) −30°.

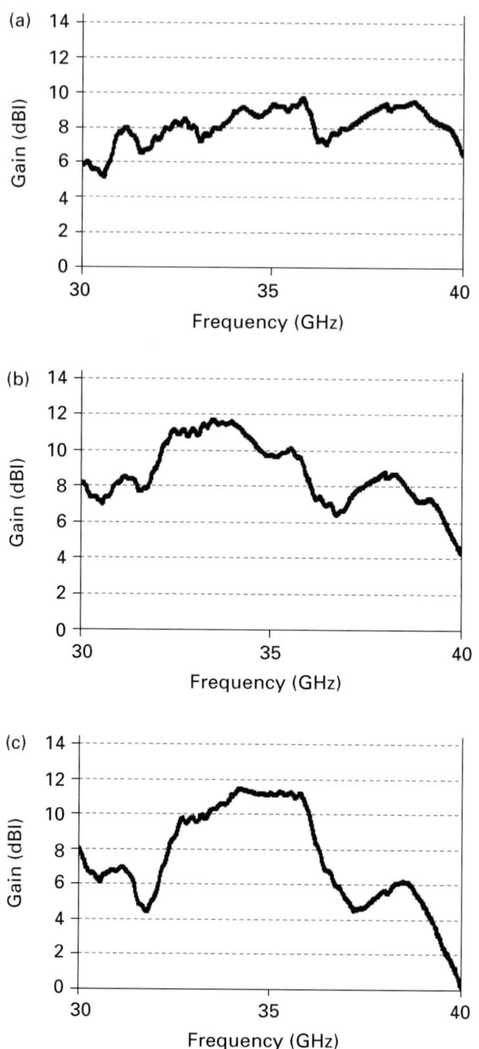

Fig. 7.42 Calibrated gain measurements at (a) boresight, (b) –15° (c) –30°.

Three-dimensional representations of the radiation pattern can be useful for observing how the beam pattern varies over frequency, as shown in Fig. 7.41. As can be seen, at 34 GHz the pattern is well defined and there is a less than 2° beam shift from the intended beam angle. The reason is that at the center frequency the phase is proportional to frequency and the beam squint is minimal. When the active array is at boresight, the radiation pattern is fairly constant and does not shift over frequency. However, when the beam is steered towards –15°, there is a slight beam squint at 32 and 37 GHz, which correspond to the operating frequency edges. When the beam is steered towards –30°, the beam squint becomes a little more obvious; however, the observed beam squint is less than 5° at 32 GHz. At the high-frequency end, at 37 GHz, the observed beam squint is less than 6.75°. Also observed from the measurements is the presence of high sidelobes when the beam is steered to –15° and –30°, regardless of frequency, as shown in Figs. 7.41b and 7.41c. This suggests a phase error in this particular phase state. Imperfections in array measurements may be attributed to numerous reasons, including the limitations in dynamic range caused by frequency limitations in test instruments, mutual coupling between antenna elements, power reflection due to mismatches, nonlinear phase error from the phase shifter, and antenna-array phase and amplitude imbalance. Pattern degradation due to phase errors can be remedied by proper channel calibration; an additional attenuator for each channel would allow for this.

Calibrated gain measurements of the active array plotted versus frequency are shown in Fig. 7.42. The measured gain flatness at boresight is within 3 dBi from 32–37 GHz. When the array is steered towards –15°, the gain flatness is within 5 dBi. Finally, when the array is steered towards –30°, a 6.7 dBi maximum gain flatness is measured. The gain flatness can also be improved by proper channel calibration through the use of weighted attenuators and variable gain blocks.

7.3.5 Phased-array antenna receiver module conclusions

In this section, a flexible, highly scalable, phased-array antenna module has been demonstrated on LCP. The multilayer LCP build allows for a compact design. Main beam formation has been successfully shown at boresight, ±15°, and ±30° over 32 to 37 GHz. The beam squint and gain flatness are measured to be less than 6.75° and less than 6.7 dBi, respectively, over the above tuning range.

7.4 Chapter summary

This chapter has demonstrated examples of system modules that have benefited from an LCP platform. Firstly, an amplitude-compensated long time delay was demonstrated. This system directly utilizes LCP's convenient multilayer thin-film structure to accomplish a multi-transmission-line amplitude-compensation approach. This technique demonstrated measured imbalances under ±0.5 dB over the DC to 10 GHz band and for a 600 ps tuning range.

Secondly, a broadband push–pull PA was demonstrated. This structure achieves broadband performance through the use of high-performance integrated LCP baluns. Further, the use of multilayer cavity-drilled LCP provides a convenient platform to minimize the bond length between chip and package. The performance measured over 6 to 18 GHz shows P_{1dB} at 31 dBm with greater than 20 dB IMD2 suppression for single-amplifier implementation.

Thirdly, a phased-array receiver module was demonstrated on LCP. Thin-film LCP enables low-cost tapered-slot antennas to be used as receiving elements at high frequencies. Further, thin-film LCP provides mechanical flexibility. In combination with broadband LCP Wilkinson power combiners, a phased-array receiver was demonstrated having a measured beam formed over ±30° from 32 to 37 GHz.

References

[1] F. Gardiol, *Microstrip Circuits*, John Wiley and Sons, 1994.

[2] R. E. Collin, *Foundations for Microwave Engineering*, McGraw-Hill, 1992.

[3] N. B. Carvalho, J. C. Pedro, "Large- and small-signal IMD behavior of microwave power amplifiers", *IEEE Transactions on Microwave Theory Techniques*, vol. **45**, no. 12, pp. 2150–2152, December 1997.

[4] M. Hagensen, "Influence of imbalance on distortion in optical push–pull frontends," *J. Lightwave Technology*, vol. **13**, no. 4, pp. 650–657, April 1995.

[5] D. E. Meharry, J. E. Sanctuary, B. A. Golja, "Broad bandwidth transformer coupled differential amplifiers for high dynamic range," *IEEE Journal of Solid-State Circuits*, vol. **34**, no. 9, pp. 1233–1238, September 1999.

[6] J.-W. Lee, L. F. Eastman, K. J. Webb, "A gallium-nitride push–pull microwave power amplifier," *IEEE Transactions on Microwave Theory and Techniques*, vol. **51**, no. 11, pp. 2243–2249, November 2003.

[7] P. S. Hsu, C. Nguyen, M. Kintis, "Uniplanar broad-band push–pull FET amplifiers," in *Proc. IEEE Transactions on Microwave Theory and Techniques*, vol. **47**, no. 12, pp. 2364–2374, December 1999.

[8] A. C. Chen, A.-V. Pham, R. E. Leoni III, "A 6–18 GHz push–pull power amplifier with wideband even-order distortion cancellation in LCP module," in *Proc IEEE MTT-S Int. Microwave Symposium Dig.*, Honolulu, June 2007, pp. 1079–1082.

[9] E. Brookner, *Practical Phased-Array Antenna Systems*, Artech House, 1991.

[10] R. R. Kunath, R. Q. Lee, K. S. Martzaklis, K. A. Shalkhauser, A. N. Downey, R. Simons, "A vertically integrated Ka-band phased array antenna," in *Proc. 14th Int. Communication Satellite Systems Conf.*, March 1992.

[11] B. D. Steinberg, *Microwave Imaging with Large Antenna Arrays*, Wiley-Interscience, 1983.

[12] L. Caetano, 60 GHz architecture for wireless video display, SiBEAM [online].

[13] J. S. Chieh, A.-V. Pham, T. W. Dalrymple, D. G. Kuhl, B. B. Garber, K. Aihara, "A light weight 8-element broadband phased array receiver on liquid crystal polymer," in *Proc. IEEE Int. Microwave Symp.*, Anaheim, May 2010.

[14] N. Kingsley, G. E. Ponchak, J. Papapolymerou, "Reconfigurable RF MEMS phased array antenna integrated within a liquid crystal polymer (LCP) system-on-package," *IEEE Transactions on Antennas and Propagation*, vol. **56**, no. 1, pp. 108–118, January 2008.

[15] A. Hajimiri, H. Hashemi, A. Natarajan, X. Guan, A. Komijani, "Integrated phased array systems in silicon," *Proceedings of the IEEE*, vol. **93**, no. 9, pp. 1637–1655, September 2005. D. Parker, D. C. Zimmermann, "Phased arrays – part 1: theory and architectures," *IEEE Transactions on Microwave Theory and Techniques*, vol. 50, no. 3, pp. 678–687, March 2002.

[16] G. Xiang, H. Hashemi, A. Hajimiri, "A fully integrated 24-GHz eight-element phased-array receiver in silicon," *IEEE Journal of Solid-State Circuits*, vol. **39**, no. 12, pp. 2311–2320, December 2004.

[17] H. Hashemi, G. Xiang, A. Komijani, A. Hajimiri, "A 24-GHz SiGe phased-array receiver-LO phase-shifting approach," *IEEE Transactions on Microwave Theory and Techniques*, vol. **53**, no. 2, pp. 614–626, February 2005.

[18] K. Kwang-Jin, G. M. Rebeiz, "An X- and Ku-band 8-element phased-array receiver in 0.18-μm SiGe BiCMOS technology," *IEEE Journal of Solid-State Circuits*, vol. **43**, no. 6, pp. 1360–1371, June 2008.

[19] D. Parker, D. C. Zimmermann, "Phased arrays–part II: implementations, applications, and future trends,"*IEEE Transactions on Microwave Theory and Techniques*, vol. **50**, no. 3, pp. 688–698, March 2002.

[20] H. Hashemi, C. Ta-shun, J. Roderick, "Integrated true-time-delay-based ultra-wide-band array processing,"*IEEE Communications Magazine*, vol. **46**, no. 9, pp. 162–172, September 2008.

[21] K. F. Lee, W. Chen, *Advances in Microstrip and Printed Antennas*, John Wiley and Sons, 1997.

[22] D. Schaubert, E. Kolberg, T. Korzeniowski, T. Thungren, J. Johansson, K. Yngvesson, "Endfire tapered slot antennas on dielectric substrates," *IEEE Transactions on Antennas and Propagation*, vol. **33**, no. 12, pp. 1392–1400, December 1985.

[23] J. D. S. Langley, P. S. Hall, P. Newham, "Balanced antipodal Vivaldi antenna for wide bandwidth phased arrays," *IEE Proceedings – Microwaves, Antennas and Propagation*, vol. **143**, no. 2, pp. 97–102, April 1996.

[24] J. L. Allen, "The theory of array antennas (with emphasis on radar applications)," Massachusetts Institute of Technology, Lincoln Laboratory, Technical Report No. 323, 25 July 1963.

[25] M. J. Chen, Z. Zhang, A.-V. Pham, D. Hyman, "Design and development of a broad-band amplitude compensated long time delay circuit on thin-film liquid crystal polymer," *Wiley InterScience Microwave and Optical Technology Letters*, vol. **51**, no. 4, pp. 1060–1063, April 2009.

[26] M. J. Chen, Z. Zhang, A.-V. Pham, D. Hyman, "Thin-film LCP amplitude compensated long time delay circuit," in *Proc. IEEE MTT-S Int. Microwave Symp. Dig.*, Atlanta, June, 2008.

8 LCP reliability

In this chapter we evaluate the long-term functionality of LCP packages and the protection they offer against external environments. This emerging flex material has ultra-low moisture absorption and permeation close to that of glass. It is an attractive material for making hermetic packages that can provide reliability in a low-cost and lightweight platform. Prototype LCP packages for RF to millimeter-wave frequencies have been reported recently [1–13]; the emphasis in these publications has been primarily on electrical characteristics, with less focus on the reliability aspects. Among the limited list of publications [7–13], some authors claim that LCP can be used for hermetic packages with long-term reliability. Other groups, including the authors of [13], have performed environmental tests such as measuring the water absorption of LCP-cavity packages by submerging them in water. In this chapter, a variety of reliability tests and results on an LCP package will be reported using standard tests recognized as being required for military and commercial products.

A primary hurdle for LCP packaging, or any hermetic packaging in general, is achieving a high-quality lid-seal process. This hurdle can be exacerbated by LCP's inert chemical properties, which require a careful approach to processing. In a commercially available LCP-molded lead-frame package, a lid may be attached to a base using epoxy. In an epoxy-sealed package, a typical lid attachment process includes a cure cycle, i.e. 5 psi at 165 °C for one hour [6]. Moisture can readily and detrimentally pass through this epoxy layer. In ultrasonic-welded packages, the width of the lid interface to the base must be narrow enough for it to accumulate sufficient ultrasonic energy and melt the LCP. In addition, molding small features in LCP is challenging, as the features may be too thin to form a reliable seal. For these various reasons, and given LCP's short history, it is not surprising that the leak rate and reliability of LCP packages have not been reported extensively in the literature.

Using the sealing technique introduced in Chapter 3, we were able to make a first-pass leak-rate assessment and evaluation of reliability through the exposure of all-LCP encapsulated air cavity packages for MMIC packaging to different environmental conditions. Good results were obtained. The present chapter expands on those package techniques in an attempt to quantify LCP for reliable packaging. Table 8.1 outlines environmental tests that are of interest for quantifying the reliability of this sealing technique. Many of these tests simulate

Table 8.1. Reliability tests for amplifier packages

Test no.	Test type	Method
1	Operating humidity exposure	Standard 85/85 testing
2	Non-operating temperature step stressing	Mil-Std-883F method 2010.8 test condition C
3	Non-operating rapid thermal transition	
4	Non-operating high-temperature storage	Mil-Std-883F method 1008.2 test condition F
5	Non-operating moisture resistance	Mil-Std-883F method 1004.7
6	Non-operating thermal shock	
7	Freeze expansion stressing	
8	Non-operating mechanical shock	Mil-Std-883F method 2002.4 test condition C
9	Non-operating vibration	Mil-Std-883F method 2007.3 test condition A

real-life environments, including combinations of temperature cycles, temperature extremes, humidity, mechanical shock, and vibration in accelerated form. One other major challenge faced when one is using LCP packaging is a *Z*-direction coefficient of thermal expansion (CTE) of 150 ppm. One may question whether vias can survive the *Z*-direction expansion and contraction of LCP packages during temperature cycles. One of this chapter's goals is to investigate via reliability in LCP substrates by exposing 15 LCP packages to the environmental conditions outlined below.

Section 8.1 introduces the LCP air-cavity QFN-type SMT package used to evaluate long-term reliability. Section 8.2 reports temperature-related environmental tests and results. Section 8.3 reports environmental tests that involve both moisture and temperature stressing, including operating humidity exposure (at 85 °C/85%), non-operating moisture resistance, and freeze-expansion stressing. Section 8.4 reports mechanical-related stress tests including non-operating mechanical shock and non-operating vibration tests. Lastly, section 8.5 provides concluding remarks.

8.1 Package vehicle for environmental testing

An LCP multi-chip module (MCM) is presented. The package houses an amplifier and two bypass capacitors. Figure 8.1 illustrates a cross-section of this multilayer LCP package type. Sets of such packages were taken through a series of environmental tests, as described in the subsequent sections of this chapter. Five layers of LCP film with respective thicknesses 1, 2, 4, 2, and 1 mil are laminated to form the package base, with a laser-drilled cavity inside which an amplifier can be mounted. Then, after chip assembly, the package lid and base are laminated using a low-melting-temperature LCP film at the interface, to form a seal. Plated through vias 10 mil in diameter with 1 μm Au finish are drilled through the package base at ~1.5 mm from the edges. Once fully sealed they become blind vias, since one end is covered with an LCP lid.

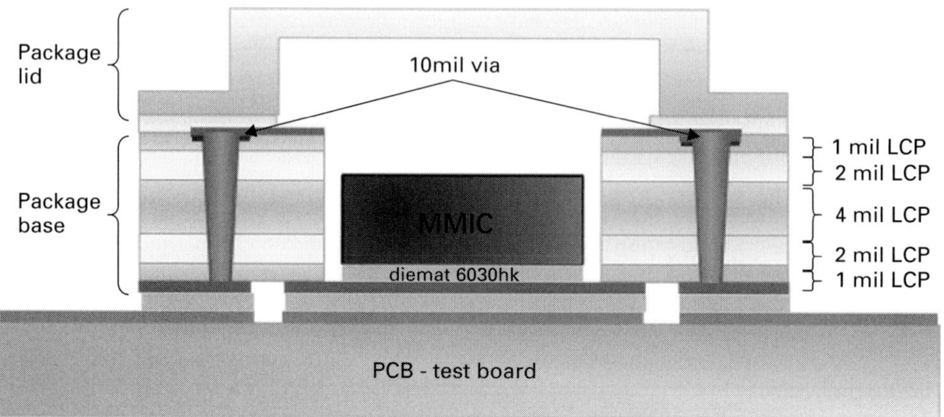

Fig. 8.1 Cross-section of the reliability test board for a single chip package.

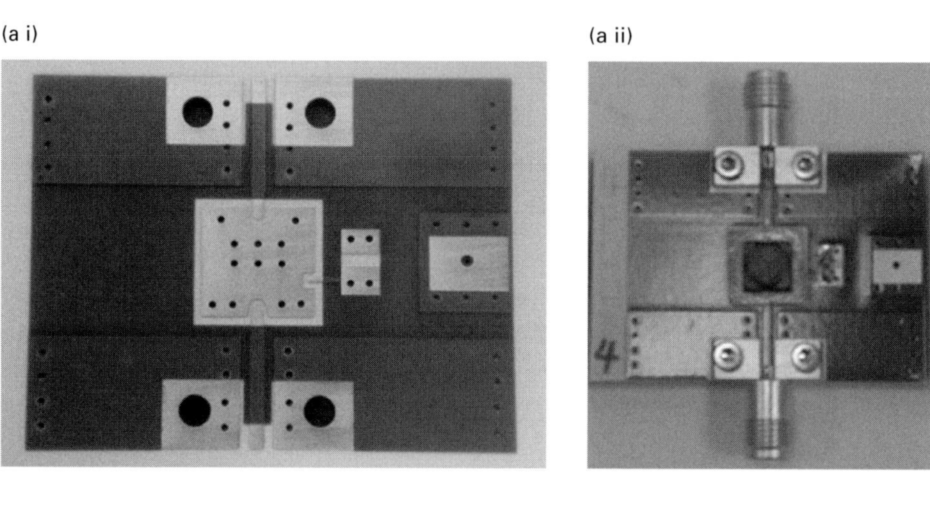

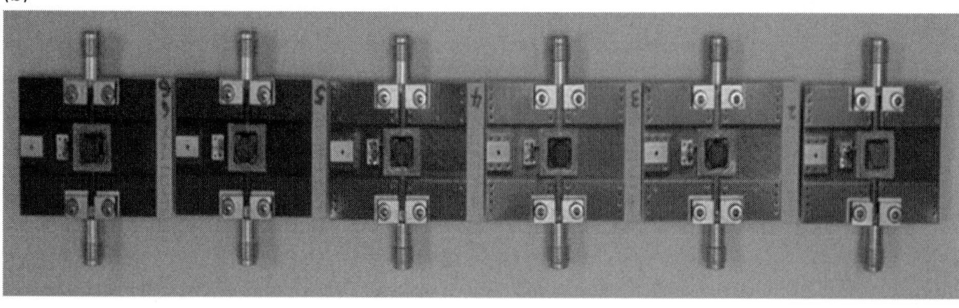

Fig. 8.2 Reliability test board for (a) a single-chip package [17] (with permission of Springer Science+Business Media), and (b) Six of the 15 single-chip packages used in the tests. [13] (© 2011 IEEE).

Three sets of five packages were assembled and mounted on a set of motherboards with coaxial connectors for measurements. The packaged amplifier modules were then subjected to a series of environmental tests. Each set of packages went through one-third of the tests. Fig. 8.2ai shows an unassembled RO4003 test board and Fig. 8.2aii shows LCP-cavity packages mounted on the board with coaxial connectors. Each package houses a Hittite HMC564 amplifier with operating frequency from 6 to 10 GHz. The amplifier bias circuit requires two 100 pF capacitors to be installed inside the package cavity and a 0.1 μF capacitor to be mounted on a test board. End-launch SMA connectors were mounted on single-chip-package test boards.

Each environmental test requires pre- and post-test data measurements for comparison. Thus, the amplifier gain was measured before and after each test from 6–10 GHz and the gain reduction evaluated. The amplifiers were biased at 2.5 volts and drew ~50 mA current.Through-reflect-line (TRL) calibration [15] was used to calibrate the RF cables that connect the PNA ports to the end-launch connectors on the test board. Any degradation more than 1 dB in gain (S21) is considered to indicate that the part has failed. The plots in the following sections will show post-test minus initial-test gain data. In this method we represent the gain drop as a negative value.

8.2 Temperature testing

In this section, the thermal effects on LCP packaging are considered. The effects of extreme and transient conditions are reported. Aspects are examined and reported for an LCP SMT multi-chip module. Section 8.2.1 presents thermal cycling. Section 8.2.2 presents non-operating rapid thermal transition. Section 8.2.3 presents non-operating high-temperature storage. Section 8.2.4 presents non-operating thermal shock.

8.2.1 Thermal cycling

A temperature cycle test was carried out to determine the modules' ability to withstand cyclical exposures to temperature extremes. The testing consisted of subjecting packages to alternating low and high temperatures for 100 cycles according to the US military standard 883F method 2010.8 test condition C [16]. The testing first involved initial electrical measurements at room temperature. The test units are placed within an environmental chamber, and their temperature was lowered to –65 °C. The temperature was held at that value for 10 minutes before and then raised to 150 °C within approximately 54 minutes (a rate of 4 °C/min). The chamber was held for 10 minutes at 150 °C and then subsequently cooled to –65 °C again. This alternation between temperature extremes occurred for 99 additional cycles (in total, 100). Lastly, electrical measurements were performed at room temperature. Five Hittite HMC564 packages were tested. All packages were measured as passing the test. Figure 8.3 shows the measured gain differences. Even though the LCP

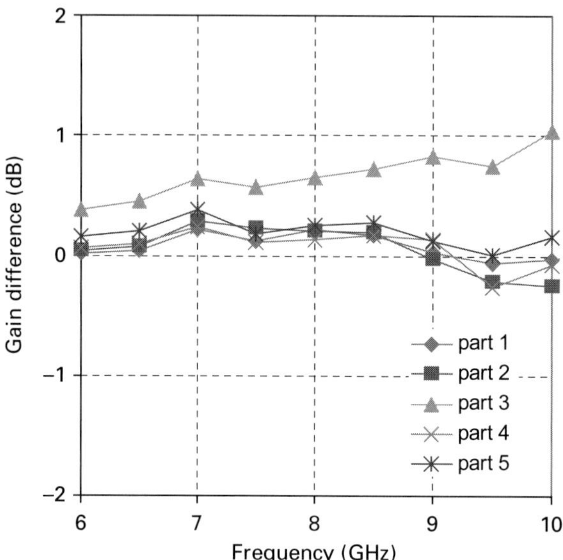

Fig. 8.3 Non-operating temperature-cycle test result for five amplifier packages [17] (with permission of Springer Science+Business Media).

laminates have a Z-axis CTE of 150 ppm, the test result indicates that reliable vias can be fabricated in thin-film LCP packages.

8.2.2 Non-operating rapid thermal transition

The non-operating rapid thermal transition test between extremely low and high temperatures in cyclical exposure is just like the temperature cycle test described above but with a much faster temperature-ramping rate per minute (25 °C/min) than the non-operating temperature cycle test (4 °C/min) and with less extreme temperature limits (the lower limit was –50 °C rather than –65 °C and the higher limit was 90 °C rather than 150 °C). The non-operating rapid thermal transition test involves the following steps.

1. Electrical measurements are taken at room temperature.
2. The units are placed inside the chamber, and the temperature is lowered to –50 °C.
3. The temperature is held at –50 °C for 10 minutes.
4. The temperature is then raised to 90 °C within approximately 6 minutes (25 °C/min).
5. The temperature is at 90 °C for 10 minutes, and then the chamber is cooled to –50 °C in 6 minutes (25 °C/min).
6. Steps 3 to 5 are repeated for 47 cycles (in total, 48).
7. The unit is rested at room temperature, and electrical measurements are performed.

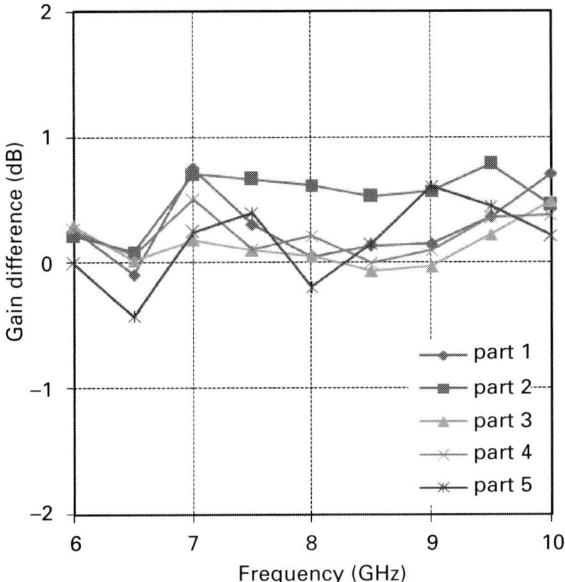

Fig. 8.4 Non-operating rapid thermal transition test result for five amplifier packages.

Figure 8.4 shows drop in gain for the amplifier packages after non-operating rapid thermal transition. Five packages were tested, and all five showed a less than 1 dB gain drop after testing, compared with the initial data, and therefore passed, the test.

8.2.3 Non-operating high-temperature storage

When a system is used in an environment where it sees a constant hot temperature, for example, it is in the sun all day, package reliability within the system can be questionable. A system may not be in use constantly but nevertheless it may have to withstand heat for many hours every day. A non-operating high-temperature storage (HTS) test determines the effect of long-term storage at elevated temperatures for packaged devices without any applied electrical stress. The purpose of the HTS test is to assess the long-term reliability of packaged devices under a high-temperature environment. This test is carried out following the US military standard 883F method 1008.2 test condition F [16] and involves the following steps.

1. For 24 hours before the test the units should be rested at ambient room temperature for electrical measurements.
2. The units are placed in the chamber and the temperature is raised to 300 °C.
3. The units remain in the chamber for a minimum of 1 hour at 300 °C.
4. The units are taken out of the chamber and rested for 72 hours.
5. The final electrical measurement are taken within 96 hours of removal from the chamber.

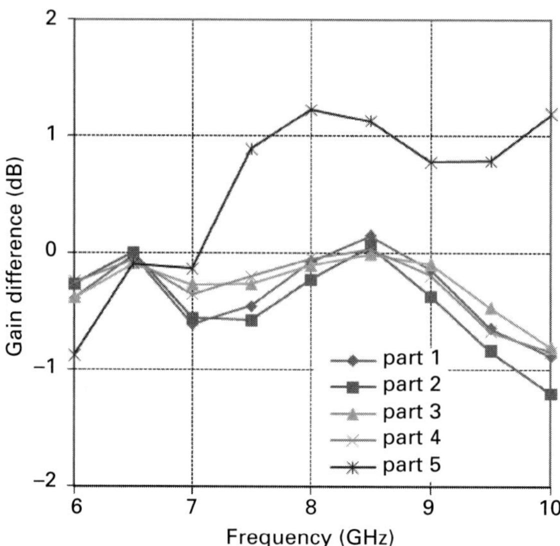

Fig. 8.5 Non-operating high-temperature-storage test result for five amplifier packages.

Figure 8.5 shows the difference in gain before and after the non-operating high-temperature-storage test for the amplifier multi-chip module packages. As can be seen, all five tested parts have a measured gain drop less than 1 dB after the test. One test unit shows a gain difference greater than 1 dB but, since it is in the positive region, this means that the gain measured after testing is greater.

8.2.4 Non-operating thermal shock

A non-operating thermal shock test was performed to evaluate a package's resistance to sudden changes in temperature (greater than 400 °C/min). The parts underwent a specified number of thermal cycles, starting at the ambient temperature. The parts were submerged in liquid nitrogen, then placed on a hot plate, and finally re-submerged in liquid nitrogen. The detailed steps for conducting non-operating thermal shock first involve holding the test substrate (with die attached) by a corner with tweezers at room temperature. The substrate is then immersed in liquid nitrogen for 1 minute (+30/–0 seconds) while held with tweezers. The substrate is then removed from the liquid nitrogen and released onto a 300 °C (+20 °C/–0 °C) aluminum hot plate for 1 minute (+30/–0 seconds); the transfer from the liquid nitrogen to the hot plate must be completed in less than 15 seconds. The substrate is removed from the hot plate with tweezers and again immersed in liquid nitrogen for 1 minute (+30/–0 seconds) while held with tweezers; the transfer from hot plate to liquid nitrogen must be completed in under 15 seconds. This is repeated at least eight times. Final liquid-nitrogen-to-hot-plate and hot-plate-to-room-temperature transitions are carried out and measurements are taken when ambient temperature has been reached. Five LCP-packaged Hittite HMC564s were tested, and all parts

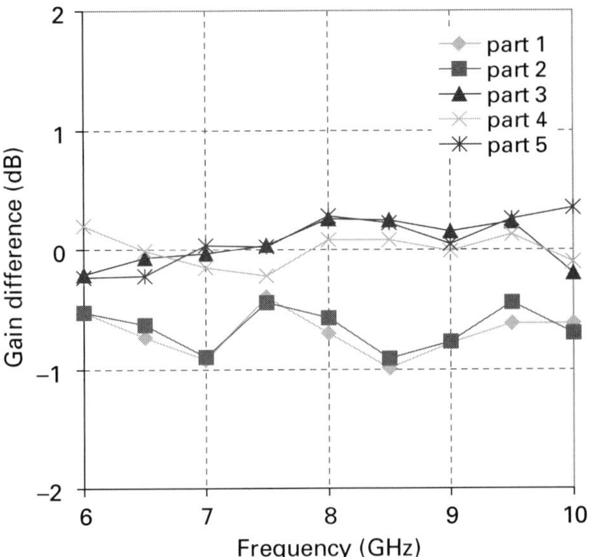

Fig. 8.6 Non-operating thermal shock test result for five amplifier packages [17] (with permission of Springer Science+Business Media).

passed the test. Figure 8.6 shows the gain differences before and after the thermal shock test.

8.3 Moisture and temperature stressing

Electronics are especially known for their sensitivity to moisture. Section 8.3.1 presents a test for operating-humidity exposure under 85 °C/85% relative humidity. Section 8.3.2 presents a test for non-operating moisture resistance. Section 8.3.3 presents a test for freeze-expansion stressing.

8.3.1 Operating humidity exposure (85 °C/85%)

In this section, test samples go through an environment at 85 °C and with 85% relative humidity for more than 1000 hours while electrically biased. The applied bias accelerates metal corrosion and dendrite formation on chip metal when moisture enters a package cavity. The electrical performance degrades when the chip metal corrodes or dendrites form, causing a packaged chip to short/open circuit. Operating a humidity exposure test involves first placing the test components inside an environmental chamber and applying a bias to all test components. The chamber is sealed and set to 85 °C and 85% relative humidity. Current readings for all packages are checked every 24 hours to make sure that the proper bias is being applied to each chip. If the current readings drop by more than 10% of the initial value, the

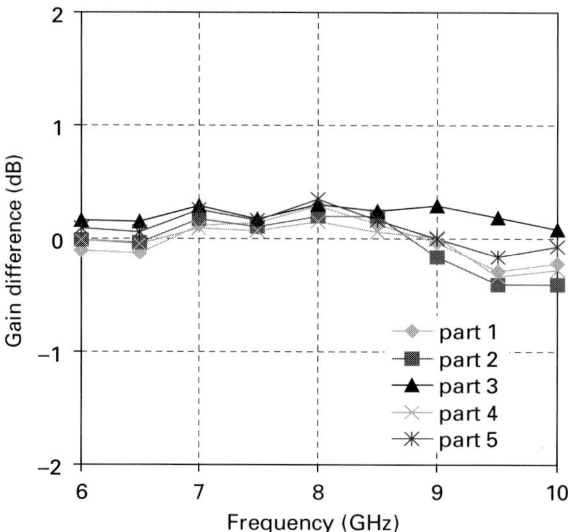

Fig. 8.7 Operating humidity exposure test result for five amplifier packages [17] (with permission of Springer Science+Business Media).

package is considered to have failed. Intermediate measurements are taken at 24, 96, and 500 hours to ensure that the parts remain functional. Lastly, post-test data are taken after 1000 hours.

Figure 8.7 shows the test result of operating humidity exposure. Five LCP packaged HMC 564 amplifiers were tested. All five tested parts measured less than 1 dB gain drop after the test. This result indicates that the LCP amplifier packages have good moisture resistance.

8.3.2 Non-operating moisture resistance

Non-operating moisture-resistance testing evaluates a part's resistance to high humidity and cyclical heat; most tropical degradation is a result of moisture absorption and associated corrosion due to surface wetting. The test also includes a low-temperature subcycle that allows the detection of cracks and fissures caused by small cracks during freeze expansion. The test conditions follow military standard 883F method 1004.7 [16]. The initial electrical measurements are taken at ambient room temperature. The test samples are placed in an environmental chamber and ramped to 65 °C while relative humidity is kept at 90%–100%. The parts remain at 65 °C for 3 hours minimum while relative humidity is kept at 90%–100%. The chamber is then ramped down to 25 °C while the relative humidity is kept at 80%–100%. This cycle is repeated nine times. In at least five out of the 10 cycles it is required that a subcycle is run that consists of the following requirements: (a) the temperature is ramped down to −10 °C, (b) the temperature is held at −10 °C for minimum of 3 hours, (c) any humidity is eliminated, (d) the temperature is

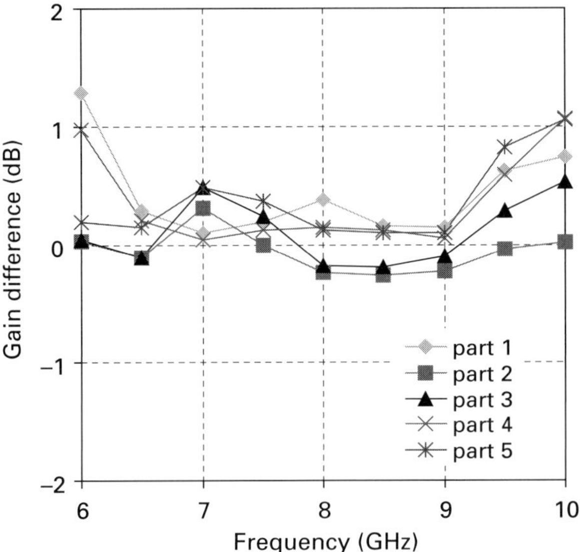

Fig. 8.8 Non-operating moisture resistance test result on LCP packaged HMC564 amplifier modules [13] (© 2011 IEEE)

returned to 25 °C and the relative humidity to 80%–100%, (e) the units are rested at ambient for 24 hours upon their removal from the chamber. The electrical measurements are performed within 48 hours of removing the units from the chamber. Figure 8.8 shows the test results for non-operating moisture resistance. Again, five LCP-packaged HMC564s were tested. All five tested parts measured a less than 1 dB gain drop after the test.

8.3.3 Freeze–expansion stressing

Freeze–expansion stressing detects the weak sealing of a package. In this test, a cyclical mechanical stress is applied to a package by means of the contraction and expansion of water between the solid and liquid states. The test involves first placing samples at the bottom of a small aluminum cup. Between 0.5 to 5.0 cm^3 of water is poured into the cup to cover the samples. The cup is placed onto a hot plate set to 165 °C and covered with a petri dish. Next, the water is brought to boiling point. Then the cup is placed onto a cold plate pre-cooled to −50 °C. The unit is allowed to cool until the water is frozen solid. The expansion pressure may cause weak package seals to break, so that, on melting, water will be forced into the package, which would cause the packaged parts to fail. The cup is then returned to the hot plate, and the freeze–expansion cycle is repeated at least four more times. Lastly, the cup is emptied of water and returned to the hot plate until the samples are dry. The samples are then allowed to cool to room temperature before being tested.

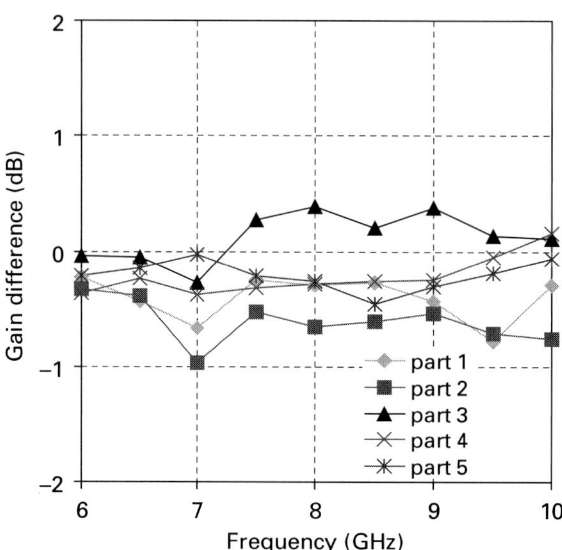

Fig. 8.9 Non-operating freeze–expansion stressing test results for five HMC564 packages [13] (© 2011 IEEE).

Figure 8.9 shows the gain difference before and after the non-operating freeze–expansion test. Six LCP packaged HMC564 amplifiers were tested. Five of the six packaged amplifiers were shown to pass the test. One package failed due to a broken bond wire in the RF connection, which was found after delaminating the lid. This was due either to a lack of pressure during wire bonding or to an improperly cleaned bond pad, which caused the wire to fall off during the rapid temperature change. Also, wire bonds sometimes do not stick well to smooth metal surfaces. In short, the open circuit was due not to a malfunctioning LCP package but, rather, to imperfect prototype assembly.

8.4 Mechanical tests

In this section, mechanical tests are described and the results reported. Section 8.4.1 presents a non-operating mechanical shock test. Section 8.4.2 presents a non-operating vibration test.

8.4.1 Non-operating mechanical shock

Non-operating mechanical-shock testing determines the ability of semiconductor devices to withstand shock from suddenly applied forces or abrupt changes in motion encountered during mishandling, improper transportation, or field operation. Shocks of this type can cause devices to degrade in performance or to become permanently damaged. Package lid-seal strength is also evaluated. The

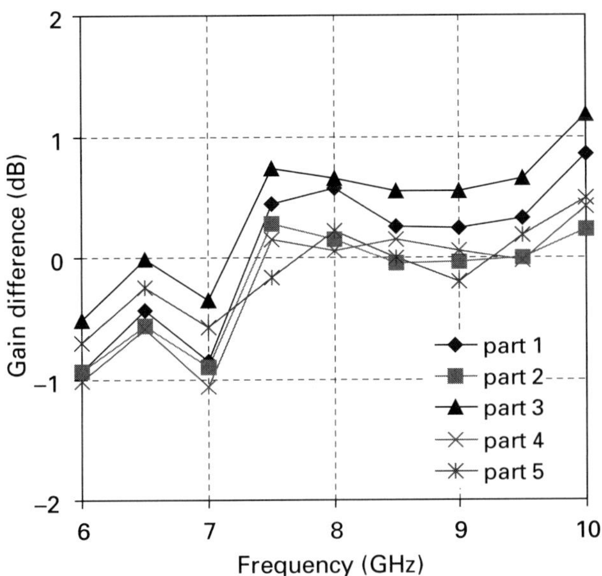

Fig. 8.10 Non-operating mechanical shock test results for five HMC564 packages [13] (© 2011 IEEE).

test follows military standard 883F method 2002.4 test condition C [16]. The test involves exposing the packages to five shock pulses with a 0.3 ms half-sine waveform at 3000*g* acceleration in six directions, two in the *X* direction, two in the *Y* direction, and two in the *Z* direction. Five LCP-packaged HMC564s were tested, and all parts were found to pass the test. Figure 8.10 shows the test results for non-operating mechanical shocks.

8.4.2 Non-operating vibration

Vibration occurs in airplanes, automobiles, trains, and all other situations where motors and engines are used. The rotation of an engine or motor causes a system to shake constantly at various frequencies depending on the rotational speed. A non-operating vibration test may be performed to determine the effects of mechanical vibration within a specified frequency range on packaged semiconductor devices. Military standard 883F method 2007.3 test condition A [16] was followed and includes the following steps.

1. The device is rigidly fastened to the vibration platform.
2. The device is vibrated with harmonic motion having peak acceleration of 25*g*.
3. The vibration frequency is varied logarithmically from 20 to 2000 Hz and returned to 20 Hz in a four-minute minimum sweep time up and down. (The total up-and-down time is eight minutes minimum).
4. The cycle is performed four times in each of the *X*, *Y*, and *Z* orientations.
5. Data is recorded before and after testing.

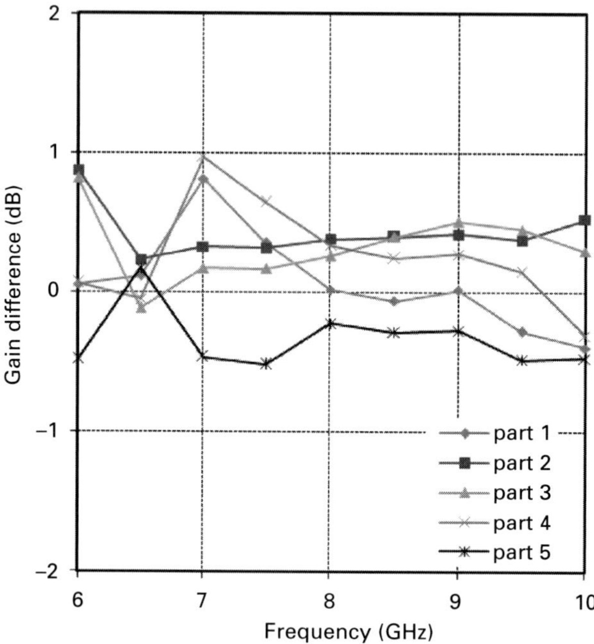

Fig. 8.11 Non-operating vibration test rest results for five amplifier packages.

Figure 8.11 shows the difference in gain measured before and after the non-operating vibration test on the amplifier packages. Five parts were tested. As can be seen, all five parts show a gain drop of less than 1 dB after testing and hence pass the test.

8.5 Conclusions

Reliable performance over time and varying conditions is a significant hurdle for electronics operating at high frequencies. This chapter has reported results from extensive environmental tests on LCP-packaged MMICs. Table 8.2 shows a summary of the test results for amplifier MMIC packages, and Table 8.3 summarizes all the environmental tests that have been investigated in this chapter. Single-chip packages passed almost all the tests except for one unit in the freeze–expansion stressing test, where the bond wire connection between the chip and the package RF trace was lost after the test. After this series of environmental tests, it can be predicted that the package-sealing technique developed in Chapter 3 shows promising results for the long-term reliability of MMIC packages. In addition, these LCP packages contained 10 mil diameter, 10 mil long, 1/2 oz Cu plated through vias. Even with a 150 ppm Z-direction CTE for LCP, none of the vias broke after being exposed to the temperature cycling conditions shown in the tables. Hence, we conclude that vias can be reliably used to span a 10 mil thick LCP-package base.

Table 8.2. Summary of reliability test results on the 15 amplifier packages

Test type	Test results
Operating humidity exposure	5/5 passed
Non-operating temperature cycle	5/5 passed
Non-operating rapid thermal transition	5/5 passed
Non-operating high-temperature storage	5/5 passed
Non-operating moisture resistance	5/5 passed
Non-operating thermal shock	5/5 passed
Freeze–expansion stressing	5/6 passed
Non-operating mechanical shock	5/5 passed
Non-operating vibration	5/5 passed

Table 8.3. Summary of reliability tests on amplifier packages. The data recorded were the gain and DC current draw. The failure criterion was a > 1 dB drop in gain or 10% change in DC current

Test no.	Test type	Qty	Method	Test condition
1	Operating humidity exposure	Five	Standard 85 °C /85% testing	Temperature 87±2 °C Humidity 87%±2% Duration 1000 hours
2	Non-operating temperature step stressing	Five	Mil-Std-883F method 2010.8 test condition C	Start at 25 °C Typical rate of change 4 °C/min Dwell 10 min or until temperature is stable Limits −65 °C, +150 °C 10 cycles
3	Non-operating rapid thermal transition	Five		Start at 25 °C Typical rate of change 25 °C/min Dwell 10 min or until temperature is stable Limits 90 °C, −50 °C 48 cycles
4	Non-operating high-temperature storage	Five	Mil-Std-883F method 1008.2 test condition F	Temperature 200 °C Atmosphere room air Time at high temperature: 1 hour
5	Non-operating moisture resistance	Five	Mil-Std-883F method 1004.7	10 cycles
6	Non-operating thermal shock	Five		Start at 20 °C Typical rate of change > 400 °C/min Dwell 1 min or until stable Limits: −200 °C, +200 °C 10 cycles

Table 8.3. (*cont.*)

Test no.	Test type	Qty	Method	Test condition
7	Freeze–expansion stressing	Six		Submerge in boiling water Freeze while submerged Bring to boil before removing Five cycles
8	Non-operating mechanical shock	Five	Mil-Std-883F method 2002.4 test condition C	Acceleration 3000g Duration: 0.3 ms half-sine pulse Number of pulses: five Directions: ±Y, ±X, ±Z
9	Non-operating vibration	Five	Mil-Std-883F method 2007.3 test condition A	Peak acceleration 20g Freq. tested: 20–2000 Hz swept logarithmically Sweep time: 4 minutes min. up and 4 minutes min. down Number of orthogonal directions: 3 Number of sweep cycles/direction: 4

References

[1] Z. Aboush, J. Benedikt, J. Priday, R. J. Tasker, "DC-50 GHz low loss thermally enhanced low cost LCP package process utilizing micro via technology," in *Proc. IEEE MTT-S Int. Microwave Symp. Dig.*, San Francisco, June 2006, pp. 961–964.

[2] H. Kanno, H. Ogura, K. Takahashi, "Surface-mountable liquid crystal polymer package with vertical via transition compensating wire inductance up to V-band," in *Proc. IEEE MTT-S Int. Microwave Symp. Dig.*, Philadelphia, June 2003, pp. 1159–1162.

[3] K. Kitazawa, S. Koriyama, H. Minamiue, M. Fujii, "77-GHz-band surface mountable ceramic packages," *IEEE Transactions on Microwave Theory and Techniques*, vol. **48**, no. 9, pp. 1488–1491, September 2000.

[4] V. Palazzari, D. Thompson, N. Papageorgiou *et al.*, "Multi-band RF and mm-wave design solutions for integrated RF functions in liquid crystal polymer system-on-package technology," in *Proc. 54th Electronic Components and Technology Conf.* Las Vegas, June 2004, pp. 1658–1663.

[5] D. Thompson, N. Kingsley, G. Wang, J. Papapolymerou, M.M. Tentzeris, "RF characteristics of thin film liquid crystal polymer (LCP) packages for RF MEMS and MMIC integration," in *Proc. IEEE MTT-S Int. Microwave Symp. Dig.*, Long Beach CA, June 2005, pp. 857–860.

[6] K. Aihara, A.-V. Pham, J.W. Roman, "Development of molded liquid crystal polymer QFN packages for Ku-band applications," in *Proc. Conf. on IMAPS Device Packaging*, Arizona, March 2006.

[7] K. Aihara, A.-V. Pham, "Development of thin-film liquid crystal polymer surface mount packages for Ka-band applications," in *Proc. IEEE MTT-S Int. Microwave Symp. Dig.*, San Francisco, June 2006, pp. 956 – 959.
[8] K. Aihara, A.-V. Pham, D. Zeeb, T. Flack, E. Stoneham, "Development of multi-layer liquid crystal polymer Ka-band receiver modules," *Microwave and Optical Technology Letters*, vol. **51**, no. 2, pp. 364–367, February 2009.
[9] M. P. Mcgrath, K. Aihara, A.-V. Pham, S. R. Nelson, "Development of LCP surface mount package with a bandpass feedthrough at K-band," in *Proc. IEEE MTT-S Int. Microwave Symp. Dig.*, Atlanta, June 2008, pp. 93–96.
[10] J-C. S. Chieh, A-V. Pham, T. W. Dalrymple, D. G. Kuh, B. B. Garber, K. Aihara, "A light weight 8-element broadband phased array receiver on liquid crystal polymer," in *Proc. IEEE MTT-S Int. Microwave Symp. Dig.*, Anaheim CA, May 2010, pp. 1024–1027.
[11] K. Aihara, M. J. Chen, A-V. Pham, "Development of thin-film liquid crystal polymer surface mount packages for Ka-band applications" *IEEE Transactions on Microwave Theory and Techniques*, vol. **56**, no. 9, pp. 2111–2117, September 2008.
[12] K. Aihara, M. J. Chen, A-V. Pham, "Reliability of liquid crystal polymer air cavity packaging," *IEEE Transactions on Advanced Packaging*, accepted for publication.
[13] D. Thompson, M. Tentzeris, J. Papapolymerou, "Experimental analysis of the water absorption effects on RF/mm-wave active/passive circuits packaged in multilayer organic substrates," *IEEE Transactions on Advanced Packaging*, vol. **30**, no. 3, pp. 551–557, August 2007.
[14] A.-V. Pham, J. Laskar, J. Schappacher, "Development of on-wafer microstrip characterization techniques," in *Proc. 47th IEEE ARFTG Conf. Dig.*, June 1996, pp. 85–94.
[15] Test Method Standard Microcircuits: Mil-Std-883, February 28, 2006.
[16] http://www.decatur.de/javascript/dew/index.html
[17] K. Kuang, F. Kim, S. S. Cahill, *RF and Microwave Microelectronics Packaging*, Springer, 2010.

Abbreviations, acronyms, and symbols

This list provides definitions for abbreviations, acronyms, and symbols used in the book. Some abbreviations may be listed several times, where multiple meanings are possible depending on the context.

%	percent
β	propagation phase constant (in Np/m)
ω	angular frequency (in rad/s)
□	length/width
μ	magnetic permeability
μm	micrometer (10^{-6} meters)
α	attenuation constant
α_c	conductor loss (in dB/unit length, i.e. Np/m)
α_d	dielectric loss (in dB/unit length, i.e. Np/m)
Γ	reflection coefficient
δ	inter-element spacing
δ	skin depth
ΔX	path length difference
$\Delta\Phi$	phase difference
ε_{r}	relative dielectric constant (dimensionless)
ε_{re}	effective dielectric constant
ε_{reff}	relative effective dielectric constant (dimensionless)
θ	phase shift (in degrees)
θ	beam angle relative to normal
θ_{JC}	thermal resistance "theta-JC"
λ	wavelength (in m)
μ_0	series inductance per unit length (in H/m)
σ	shunt conductance per unit length (in mho/m)
Ω	ohm
ω	angular frequency (in rad/s)
A	area
A	ampere
Å	angstroms (10^{-10} m)
ADC	analog-to-digital converter
ADS	advanced design system

AESA	active electronically scanned array
Al	Aluminum
AlN	Aluminum Nitride
ASTM	American Society for Testing and Materials
atm	atmospheres
Au	gold
BeO	beryllium oxide
BGA	ball grid array
BVH	blind via hole
BWR	bandwidth ratio
c	speed of light in a vacuum (in m/s)
C	shunt capacitance per unit length (in F/m)
C4	controlled collapse chip connection
CAD	computer-aided design
CAE	computer-aided engineering
CHE	coefficient of hydroscopic expansion
cm	centimeter (10^{-2} meters)
CO_2	carbon dioxide
CoB	chip-on-board
CoF	chip-on-flex
CoG	chip-on-glass
CPW	coplanar waveguide
CPWG	coplanar waveguide with ground backing
Cr	chromium
CSP	chip-scale package
CST	computer simulation technology
CTE	coefficient of thermal expansion
Cu	copper
CVCM	collected volatile condensable materials
dB	decibel
dBc	power level in dB with respect to carrier
dBi	dB relative to an isotropic source
dBm	absolute power normalized to 1 mW
DC	direct current
DCA	direct chip attach
deg	degrees
°C	degrees Celsius
DF	dissipation factor/loss tangent
DGS	defected ground structure
DIP	dual in-line package
DK	relative dielectric constant
DoD	US Department of Defense
DUT	device under test

e-beam	electron beam
EBPVD	electron beam physical vapor deposition
EM	electromagnetic
EMMN	even-mode matching network
ENIG	electroless nickel, immersion gold
ENIPIG	electroless nickel, immersion palladium, immersion gold
F	fail
F	farads
f	femto prefix (10^{-15})
f	frequency (in Hz)
f_0	fundamental frequency
FCCL	foil-copper-clad laminates
FEM	finite element method
fF	femto farad (10^{-15} farrad)
f_H	high-side passband cutoff frequency
f_L	low-side passband cutoff frequency
FPGA	field programmable gate array
FR-4	flame-resistant type-4 self-extinguishing material commonly used to refer to low-frequency PCBs
ft	foot
g	gram
G	shunt conductance per unit length (in mho/m)
GaAs	gallium arsenide
GE	General Electric
GHz	gigahertz (10^9 hertz)
GND	ground
GSG	ground–signal–ground
H	henry
h	thickness of dielectric substrate
H_2O	water
HAST	highly accelerated stress test
HBA	p-hydroxybenzoic acid
HC	hydrocarbon
HD2	second-order harmonic distortion
HD3	third-order harmonic distortion
HDI	high-density interconnect
HDT	heat distortion temperature
He	helium
HEPA	high-efficiency particulate arresting
HFSS	software formerly known as a High Frequency Structure Simulator
HNA	6-hydroxy-2-napthoic acid
HPBW	half-power beamwidth

HTS	high-temperature storage
Hz	hertz
IEEE	Institute of Electrical and Electronics Engineers
IF	intermediate frequency
IL	insertion loss
IMC	intermetallic compound
IMD2	second-order intermodulation distortion
IMD3	third-order intermodulation distortion
IMS	International Microwave Symposia
in	inch
IP3	third-order intercept
IPC	industrial organization formerly called the Institute for Interconnecting and Packaging Electronic Circuits
ISI	intersymbol interference
ISS	impedance standard substrate
j	imaginary number (dimensionless)
JEDEC	industrial organization formerly called the Joint Electron Devices Engineering Council
JESD	JEDEC standard
K	thermal conductivity
k	Wavenumber (in m^{-1})
Ka, Ka-band	26.5 to 40 GHz frequency band
K-connector	2.92 mm coaxial connector
KGD	known good die
L	inductance value (in H)
L	series inductance per unit length (in H/m)
l	line length (in m)
lb	pounds
LCP	liquid crystal polymer
LGA	land grid array
LNA	low-noise amplifier
LO	local oscillator
LPI	liquid photoimageable
LRM	load–reflect–match calibration
LTCC	low-temperature co-fired ceramic
LTD	long time delay
m	meter
m	milli prefix (10^{-3})
MCM	multi-chip module
MEMS	microelectromechanical system
microns, μm	micrometer (10^{-6} meters)
mil	25.4 μm, 1/1000th of 1 inch
Mil-Std	military standard

MIM	metal–insulator–metal
min	minute
ML	mismatch loss
mm	millimeter (10^{-3} meters)
MMIC	monolithic microwave integrated circuit
Mo	molybdenum
MoM	method of moments
MPa	10^6 pascal pressure
MSDS	material safety data sheet
MTT	microwave theory and techniques
MTTF	mean time to failure
MTT-S	Microwave Theory and Techniques Society
N	newton
N	number of cascaded sections
NF	noise figure
Ni	nickel
Np	nepers
org.	organics
oz	ounce
P	pass
p	pico prefix (10^{-12})
P_{1dB}	1 dB compression point
PA	power amplifier
PAA	phased-array antenna
PAE	power-added efficiency
Pb	lead
PCB	printed circuit board
Pd	palladium
PEN	polyethylene naphthalate
PGA	pin grid array
pH	picohenry (10^{-12} henry)
pHEMT	pseudomorphic high-electron-mobility transistor
PNA	performance network analyzer
ppm	parts per million (10^{-6})
PSI	pounds per square inch
PTH	plated through hole
PVD	physical vapor deposition
QFN	quad flat no-lead
QFP	quad flat package
R	series resistance per unit length (in ohm/m)
RF	radio frequency
RFI	RF input
RFO	RF output

RGA	residual gas analysis
RH	relative humidity
RIE	reactive ion etching
RL	return loss
RMS	root-mean-square
RPM	rotations per minute
R_s	series resistance per unit length (in ohm/m)
s	second (of time)
S	siemen
S	signal
S	signal width
SA	spectrum analyzer
SATCOM	satellite communications
SEM	scanning electron microscope
SFDR	spurious free dynamic range
SG	signal generator
Si	silicon
SIP	single in-line package
SiP	system-in-package
SMT	surface mount technology
Sn	tin
SOIC	small-outline integrated circuit
SOLT	short–open–load–through calibration
SP4T	single-pole 4-throw
S-parameters	scattering parameters
SPI	serial peripheral interface
SPST	single-pole single-throw
SWR	standing wave ratio
T-parameters	transmission parameters
t	thickness of conductor
T/R	transmit and receive
tan d, tan δ	dissipation factor/loss tangent
TDR	time-domain reflectometry
TEC	thermal electric cooler
TEM	transverse electromagnetic
TH	through hole
Ti	titanium
TL	transmission line
T_m	melting temperature
TML	total mass loss
TOI	third-order intercept
TRL	through–reflect–line
TTD	true-time delay

UWB	ultra-wideband
V	voltage
V-band	50 to 75 GHz frequency band
V_d	drain voltage
VDC	DC voltage
V_g	gate voltage
VLSI	very large scale integration
v_p	propagation velocity (in m/s)
VSWR	voltage standing wave ratio
W	watt
W	width of microstrip
W-band	75 to 110 GHz frequency band
W_g	coplanar waveguide signal-to-ground gap
WL-CSP	wafer-level chip-scale package
W-PAN	wireless personal area network
WVR	water vapor regain
X-band	8 to 12 GHz frequency band
YAG	yttrium aluminum garnet
Y-parameters	admittance parameters
Z	vertical position
Z_0	characteristic impedance
Z_{0cp}	characteristic impedance of CPW
Z-parameters	impedance parameters

Index